D0024572

Algebra for College Students

Third Edition

RICHARD S. PAUL
ERNEST F. HAEUSSLER, Jr.
The Pennsylvania State University

A RESTON BOOK
PRENTICE-HALL, INC., Englewood Cliffs, New Jersey 07632

Library of Congress Cataloging in Publication Data

Paul, Richard S.
Algebra for college students.

Includes index.
1. Algebra. I. Haeussler, Ernest F. II. Title.
QA152.2.P37 1985 512.9 84-26231
ISBN 0-8359-9179-2

Interior production and design by Karen Winget
Cover design by Debbie Balboni
Cover art: "The Evil Genius of a King" by Giorgio de Chirico

10 9 8 7 6 5 4

Printed in the United States of America

Contents

Preface

This third edition continues to be designed for students who need to develop manipulative skills in algebra. It can be used either by students taking a terminal course in mathematics or by those who are preparing for further work in mathematics. Included are functions and their applications, as well as other topics that we feel are most basic to students' needs. Every attempt has been made to write in "down-to-earth" language with short paragraphs, simple vocabulary, and conversational style.

In this edition, real numbers and several other topics have been rewritten for greater clarity and student understanding. Zero and negative exponents now appear earlier. All the basic operations with algebraic expressions are now covered prior to solving equations. Linear inequalities now occur immediately following the solution of linear equations. Factoring has been divided into two sections. The hyperbola is now discussed in a separate section. A section on graphing quadratic functions has been added. We express our thanks to The Mathematical Association of America and to The National Council of Teachers of Mathematics for permission to use fourteen problems from *A Sourcebook of Applications of School Mathematics* (1980).*

At the end of each chapter is a Review Section that contains a list of important terms and symbols, together with the numbers of the sections where these items first appear. Following the list are numerous review problems. At the back of the book are answers to all odd-numbered problems.

---

* This book contains the following acknowledgement and disclaimer statement: "The material was prepared with the support of National Science Foundation Grant No. SED72–01123 A05. However, any opinions, findings, conclusions, or recommendations expressed herein are those of the authors and do not necessarily reflect the views of NSF."

A skull and crossbones symbol is employed to highlight errors commonly made by students. It is intended to alert them and cause them to beware. We feel that it is just as important for students to know what they cannot do as to know what they can do.

Available from the publisher is an extensive instructor's manual. In addition to many worked-out solutions, the manual offers examination questions and their answers.

Finally, we express our thanks to Karen Winget, our production editor, for her efficiency, competent assistance, and enthusiastic cooperation.

Richard S. Paul
Ernest F. Haeussler, Jr.

1 Real Numbers

1.1 THE REAL NUMBER LINE

In mathematics there are different types of numbers. To describe them we'll use the idea of a *set*.

> A **set** is a collection of objects or numbers, called **elements** or **members** of the set.

We can specify a set by listing its elements within braces, { }. For example, the set of numbers 1, 3, and 12 may be written

$$\{1, 3, 12\}.$$

Thus 1, 3, and 12 are elements of this set. The order in which we list the elements is not important. So we can also write this set as {3, 12, 1}. Can you think of other ways?

Probably the numbers most familiar to you are 1, 2, 3, 4, 5, and so on, which are used in counting. They are called **positive integers**.

> **SET OF POSITIVE INTEGERS**
>
> $\{1, 2, 3, 4, 5, \ldots\}$

The three dots above mean "and so on." This set has infinitely many elements.

The numbers $-1, -2, -3, \ldots$ are called **negative integers**. Putting together the positive integers, the negative integers, and zero, we get the set of **integers**.

SET OF INTEGERS

$$\{\ldots, -2, -1, 0, 1, 2, \ldots\}$$

A number that can be written as an integer divided by an integer is called a **rational number**.* Example 1 shows some rational numbers.

EXAMPLE 1

Some rational numbers.

a. $\frac{2}{3}$.

b. $\frac{9}{5}$.

c. 6, since $6 = \frac{6}{1}$.

d. -6, since $-6 = \frac{-6}{1}$.

e. 0, since $0 = \frac{0}{1}$.

f. $3\frac{1}{8}$, since $3\frac{1}{8} = \frac{25}{8}$.

In parts c, d, and e, you can see that an integer is a rational number because it can be written with a denominator of 1.

All rational numbers can be represented by decimal numbers that *terminate*, such as $\frac{3}{4} = .75$ and $\frac{3}{2} = 1.5$, or by *nonterminating repeating* decimals (a group of digits repeats without end), such as $\frac{2}{3} = .666\ldots$ and $\frac{-4}{11} = -.3636\ldots$. Numbers represented by *nonterminating*, *nonrepeating* decimals are called **irrational numbers**. An irrational number cannot be written as an integer divided by an integer. For example, the numbers π (pi) and $\sqrt{2}$ are irrational. Sometimes π is replaced by $\frac{22}{7}$ or 3.14 in a calculation, but these numbers are just approximations of π.

Putting together the rational numbers and irrational numbers, we get the set of **real numbers**. That is, the real numbers consist of all decimal numbers. See Fig. 1-1. Some examples of real numbers are $2, -5, 0, \frac{2}{3}, 16\frac{1}{2}, \pi$, and $\sqrt{2}$.

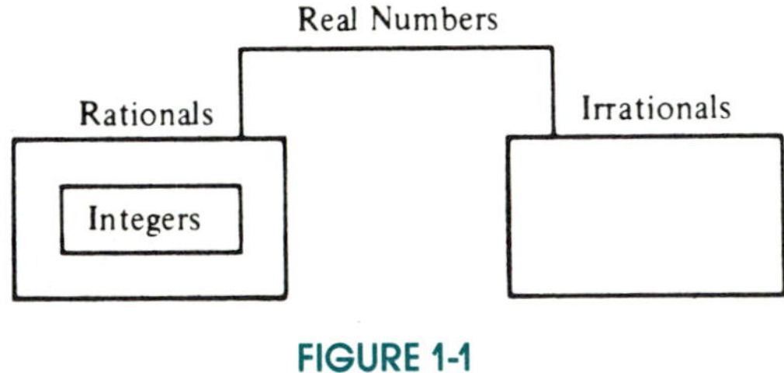

FIGURE 1-1

---

*$\frac{1}{0}$ appears to be an integer divided by an integer, but division by 0 is not defined. So $\frac{1}{0}$ isn't a number at all.

Real numbers can be represented by points on a straight line, like the markings on a thermometer. On a line we choose a point, called the **origin**, to represent the number 0. See Fig. 1-2(a). Then we mark off equal distances to the right and left of the origin. These points represent the positive and negative integers. See Fig. 1-2(b). With any point on this line we can associate a real number that depends on the position of the point with respect to the origin.

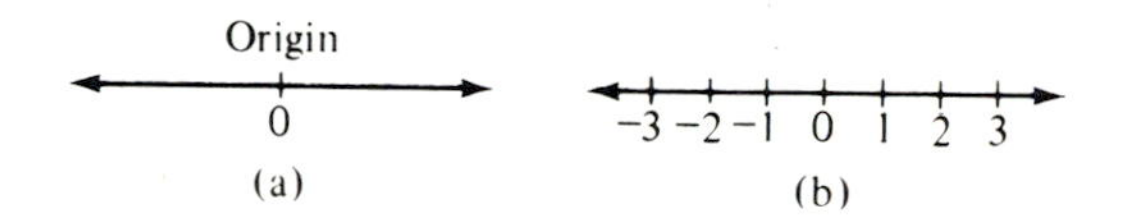

FIGURE 1-2

Positions to the right of the origin are considered to be *positive*, and positions to the left are considered to be *negative*. For example, with the point $\frac{1}{2}$ unit to the *right* of the origin, we associate the real number $\frac{1}{2}$. See Fig. 1-3. In fact, we

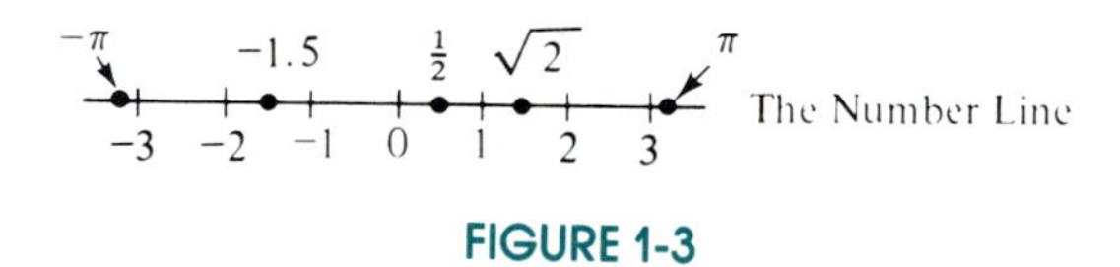

FIGURE 1-3

sometimes write $\frac{1}{2}$ as $+\frac{1}{2}$ to stress that it is to the *right* of the origin. We associate the real number -1.5 with the point that is 1.5 units to the *left* of the origin. The rest of the points represent all other rational and irrational numbers. Thus for each point there corresponds a real number, and clearly for each real number there corresponds a point on the line. Because of these correspondences we shall feel free to speak interchangeably of the *point* 3 and the *number* 3. For obvious reasons we call this line the (real) **number line**. You may think of real numbers as *signed numbers*: that is, either positive (+), negative (−), or 0.

In the future, whenever we use the word *number* we shall mean *real number*.

Exercise 1.1

In Problems **1–7**, *complete the statements.*

1. A collection of objects is called a ______________ .

2. The objects in a set are called its members or ______________ .

3. The numbers 1, 2, 3, . . . are called ______________ integers.

4. The number of elements in $\{1, 3, 0, -6\}$ is ______________ .

5. The set of integers consists of the positive integers, the negative integers, and

________________ .

6. Every real number is either positive or negative except for the number

________________ .

7. Real numbers that are not rational numbers must be ________________ numbers.

8. Is 5 a rational or an irrational number?

In Problems **9–12**, *you are given two numbers. Determine the number that is farther from* 0 *on the number line.*

9. 3, 4. **10.** 7, -8. **11.** -3, 1. **12.** -5, -8.

In Problems **13–16**, *you are given two numbers. Determine the number that is to the right of the other number on the number line.*

13. 6, -5. **14.** -3, -4. **15.** -13, -12. **16.** 0, -2.

In Problems **17–22**, *determine whether each statement is true or false.*

17. 56 is a positive number.

18. $\frac{2}{3}$ is an integer.

19. $\frac{9}{2}$ is a rational number.

20. -123 is an integer.

21. The sets $\{1, 2, 3\}$ and $\{2, 3, 1\}$ are not the same.

22. All real numbers are irrational.

1.2 INEQUALITIES AND ABSOLUTE VALUE

We often represent numbers by letters, such as a, b, x, or y. Letters used in this way are called **literal numbers**.

There is a way to say that one number is "smaller" than another. If a lies to the *left* of b on the number line, we say that a is **less than** b. In symbols we write $a < b$, where the symbol $<$ is read "is less than." See Fig. 1-4.

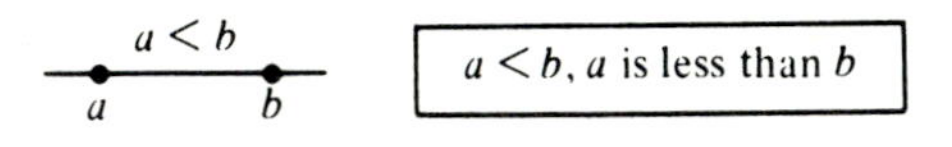

FIGURE 1-4

EXAMPLE 1

In Fig. 1-5 you can see that $4 < 7$, since 4 is to the left of 7 on the number line.

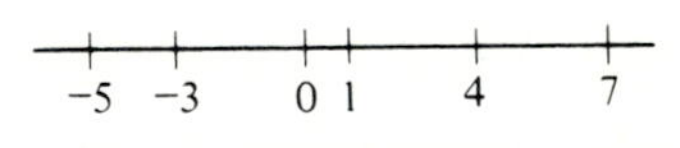

FIGURE 1-5

Similarly,

$$-3<1, \qquad 0<4, \qquad -5<-3.$$

If a is less than b, we also say that b is **greater than** a and write $b>a$. This means b is to the *right* of a on the number line. See Fig. 1-6. Thus from Fig. 1-5,

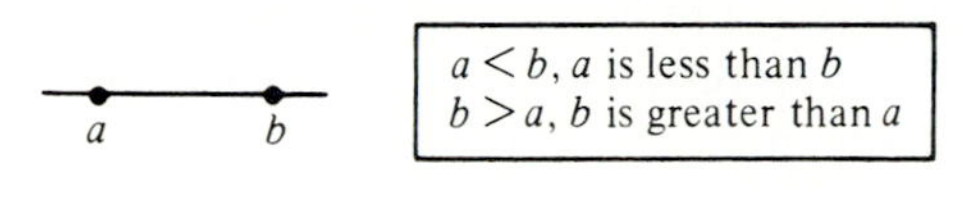

FIGURE 1-6

$4<7$ and $7>4$. The statements $a<b$ and $b>a$ are called **inequalities**. If you think of the inequality symbols $<$ and $>$ as pointers, then they should always point to the "smaller" number. Keep in mind that

> $a<b$ and $b>a$ have the same meaning.

EXAMPLE 2

Referring back to Fig. 1-5, we see that

$$7>0, \qquad 0>-3, \qquad 4>-5, \qquad -3>-5.$$

There is a way to describe positive and negative numbers with inequalities. On the number line all positive numbers are to the right of 0. Thus

> a is positive and $a>0$ mean the same.

All negative numbers are to the left of 0, so

> a is negative and $a<0$ mean the same.

There are two more inequality symbols, $\leq$ and $\geq$. Each plays a double role.

$a \leq b$ is read "a is less than or equal to b";
$a \geq b$ is read "a is greater than or equal to b."

Look at the double roles:

> $a \leq b$ means that ***either*** $a<b$ ***or*** $a=b$;
> $a \geq b$ means that ***either*** $a>b$ ***or*** $a=b$.

EXAMPLE 3

a. $7 \geq 6$, since $7 > 6$. Similarly, $5 \geq 2$.

b. $3 \leq 3$, since $3 = 3$. Similarly, $\frac{1}{2} \leq .5$.

c. $-4 \leq 3$, since $-4 < 3$. Similarly, $-3 \leq -2$.

Sometimes we're interested in the distance of a number from 0 on the number line. This distance is called the **absolute value** of the number. We use two vertical bars to indicate absolute value.

> $|a|$ denotes the absolute value of a.

For example, both 3 and -3 are three units from 0. See Fig. 1-7. Thus $|3| = 3$ and $|-3| = 3$.

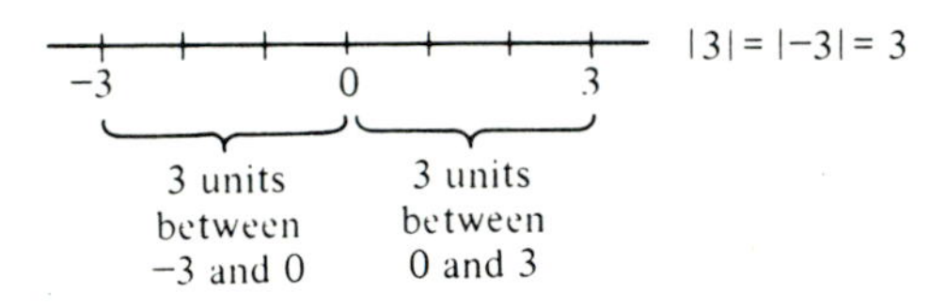

FIGURE 1-7

Since the absolute value of a number is a distance, it is always *positive* or 0. It is *never* negative.

> $|a| \geq 0.$

EXAMPLE 4

$$|5| = 5, \qquad |-8| = 8, \qquad |0| = 0.$$

$$\left|-\frac{2}{3}\right| = \frac{2}{3}, \qquad |7-4| = |3| = 3, \qquad |-\pi| = \pi.$$

In the next example we combine inequalities and absolute value.

EXAMPLE 5

a. $|-8| > -7$, because $|-8| = 8$ and $8 > -7$.

b. $|-5| < |8|$, because $|-5| = 5$, $|8| = 8$, and $5 < 8$.

c. $-3 < 2$ but $|-3| > |2|$, because $|-3| = 3$, $|2| = 2$, and $3 > 2$.

Since $|3| = 3$, you might think that $|x| = x$. However, it isn't always true that $|x| = x$, because we don't know whether x is positive or negative. For example, if $x = -2$, then

$$\begin{aligned} |x| &= |-2| = 2 \\ &= -(-2) \\ &= -x. \end{aligned}$$

Thus if x is negative, then $|x| = -x$, which is positive. But if $x = 3$, then

$$|x| = |3| = 3 = x.$$

Thus if x is positive, $|x| = x$. Of course, if $x = 0$, then $|x| = 0$. Now we can define $|x|$ without referring to a number line:

$$\begin{aligned} |x| &= x, && \text{if } x \geq 0; \\ |x| &= -x, && \text{if } x < 0. \end{aligned}$$

Exercise 1.2

In Problems **1–7**, *complete the statements.*

1. The statement $a > b$ is read "a is ______________ than b."

2. The statement $x < y$ means that x is to the ______________ of y on the number line.

3. Using the symbol $>$, we can write $3 < 4$ as ______________ .

4. If $a > 0$, then a is a ______________ number.

5. Since 4 and -4 are four units from 0, they have the same ______________ .

6. $a \geq 6$ means that either $a = 6$ or ______________ .

7. The absolute value of a number must be either positive or ______________ . It is never ______________ .

In Problems **8–13**, *put "<" or ">" in the blank so that the inequality is true.*

8. 7 ____ 16. **9.** 8 ____ 0. **10.** -5 ____ -7.

11. -3 ____ 2. **12.** 0 ____ -2. **13.** -4 ____ -1.

In Problems **14–19**, *determine the values.*

14. $|-6|$. **15.** $|270|$. **16.** $\left|-\frac{7}{2}\right|$.

17. $|0|$. **18.** $\left|3 \cdot \frac{1}{3}\right|$. **19.** $|3 - 1|$.

In Problems **20–22**, *you are given a pair of numbers. Determine the one that has the greater absolute value.*

20. 5, 3. **21.** $-5, -3$. **22.** $-4, 2$.

23. Arrange the numbers $-6, -3, 0, 4, 5$ in a list so that each number has a smaller absolute value than the numbers that follow it in the list.

24. Which of the following inequalities are *false*?

a. $|-2| > |-1|$. b. $|-1| \leq 0$. c. $|-9| < -8$. d. $3 \geq -1$.

1.3 BASIC LAWS OF REAL NUMBERS

The **sum** of the numbers a and b is written $a + b$. Whenever a mathematical expression involves a sum of numbers, the numbers are called **terms** of the expression. Thus in $a + b$, both a and b are *terms*. The **difference** of a and b involves subtraction and is written $a - b$. The **product** of a and b is written $a \cdot b$, and a and b are called **factors** of the product. Other ways to write the product $a \cdot b$ are ab, $a(b)$, $(a)b$, and $(a)(b)$. Thus we may write the product of 4 and x as $4x$. The **quotient** of a and b involves division and is written $a \div b$. We call a the **dividend** and b the **divisor**. The quotient $a \div b$ can also be written as the fraction $\frac{a}{b}$ (or a/b). Here we call a the **numerator** and b the **denominator**.

Real numbers obey three special laws: the *commutative*, *associative*, and *distributive laws*. The commutative laws are:

COMMUTATIVE LAWS

$a + b = b + a$	(commutative law of addition),
$ab = ba$	(commutative law of multiplication).

The commutative laws state that we can add (or multiply) numbers in any *order*. For example, $2 + 5 = 5 + 2$ because $2 + 5 = 7$ and $5 + 2 = 7$. Similarly, $6 \cdot 8 = 8 \cdot 6$.

The associative laws are:

ASSOCIATIVE LAWS

$a + (b + c) = (a + b) + c$	(associative law of addition),
$a(bc) = (ab)c$	(associative law of multiplication).

The associative laws state that when adding (or multiplying), we can *group* numbers in any way. For example, to find $7 + 2 + 3$, we can begin by adding either the last two numbers or the first two numbers. Thus $7 + 2 + 3 =$

$7 + (2 + 3) = 7 + 5 = 12$, or $7 + 2 + 3 = (7 + 2) + 3 = 9 + 3 = 12$. That is, $7 + (2 + 3) = (7 + 2) + 3$. Similarly, $7(2 \cdot 3) = (7 \cdot 2) \cdot 3$.

The distributive laws deal with a product in which one factor is a sum:

DISTRIBUTIVE LAWS

$$a(b + c) = ab + ac,$$
$$(a + b)c = ac + bc.$$

We say that multiplication distributes through addition. To apply the distributive law $a(b + c) = ab + ac$ to the product $2(3 + 5)$, we multiply the first factor, 2, by each term in the second factor, $3 + 5$, and add the results:

$$2(3 + 5) = 2 \cdot 3 + 2 \cdot 5.$$

first factor

Notice that the product on the left side is $2 \cdot 8$, or 16, which is equal to the sum on the right side, $6 + 10$, or 16. Similarly, by the distributive law $(a + b)c = ac + bc$, we have

$$(7 + 3)(5) = 7 \cdot 5 + 3 \cdot 5 = 35 + 15 = 50.$$

The commutative, associative, and distributive laws are useful when you work with *algebraic expressions*. **Algebraic expressions** are numbers, including literal numbers, or numbers combined by operations. Examples are 12, $4x$, $x + 7$, $5 - y$, and $\frac{x + 1}{3}$.

EXAMPLE 1

Using laws of real numbers.

a. By the commutative laws,

$$2x + y = y + 2x \quad \text{and} \quad y(2x) = (2x)y.$$

b. By the associative law (of multiplication),

$$2(6x) = (2 \cdot 6)x = 12x.$$

c. By the distributive laws,

$$2(x + 3) = 2 \cdot x + 2 \cdot 3 = 2x + 6$$

and

$$(7 + x)y = 7 \cdot y + x \cdot y = 7y + xy.$$

In Example 1c, *note that* $2(x+3) \neq 2x+3$. (*The symbol* $\neq$ *means "is not equal to."*) *You must also multiply* 3 *by* 2.

EXAMPLE 2

Simplifying.

a. $(2+x)+4 = (x+2)+4$ [commutative law]

$= x + (2+4)$ [associative law]

$= x + 6.$

b. $(3x)(4y) = 3 \cdot x \cdot 4 \cdot y$

$= 3 \cdot 4 \cdot x \cdot y$ [commutative; We usually write the numerical product $3 \cdot 4$ in front of the literal product $x \cdot y$.]

$= 12xy.$

c. $(4x)(3y)(2z) = 4 \cdot x \cdot 3 \cdot y \cdot 2 \cdot z = 4 \cdot 3 \cdot 2 \cdot x \cdot y \cdot z = 24xyz.$

The distributive laws can be generalized to include cases where one factor has more than two terms. For example,

$$a(b+c+d) = ab + ac + ad.$$

Notice that we multiply the first factor, a, by each term in the second factor, $b+c+d$. Thus

$$2(x+y+4) = 2 \cdot x + 2 \cdot y + 2 \cdot 4 = 2x + 2y + 8.$$

The following are useful properties related to the distributive laws.

$$a(b-c) = ab - ac,$$
$$(a-b)c = ac - bc.$$

We also call these distributive laws. By combining the distributive laws, we can write statements like

$$x(y+z-w) = xy + xz - xw.$$

EXAMPLE 3

Multiply.

a. $x(y-6) = xy - x(6)$

$= xy - 6x$ [commutative law].

b. $(2 - 3a)b = 2b - (3a)b = 2b - 3ab$.

c. $(2x)(y - 3z + 4w) = (2x)(y) - (2x)(3z) + (2x)(4w) = 2xy - 6xz + 8xw$.

Exercise 1.3

In Problems **1–8**, *name the law that is used in the given statement.*

1. $2(x + y) = 2x + 2y$.

2. $(x + y) + 5 = 5 + (x + y)$.

3. $2(3x) = (2 \cdot 3)x$.

4. $x + (x + y) = (x + x) + y$.

5. $2(x - y) = (x - y)(2)$.

6. $5(4 + 7) = 5(7 + 4)$.

7. $(w + 3)z = wz + 3z$.

8. $(-1)[-3 + 4] = (-1)(-3) + (-1)(4)$.

In Problems **9–26**, *perform the indicated operations.*

9. $6(3x)$.

10. $(5x)(4)$.

11. $(8 + x) + 2$.

12. $(x + 3) + 5$.

13. $4(2 + x)$.

14. $(x + 2)y$.

15. $(6x)(4y)$.

16. $x(2y)(2z)$.

17. $(2x)(3y)(4z)$.

18. $(3x + 4y)z$.

19. $(5a)(x + 3)$.

20. $(5x + 2y)(3z)$.

21. $2(3x - 2)$.

22. $(4x - 3)y$.

23. $(y - 6)(2x)$.

24. $3(4 + x - y)$.

25. $5x(2 - y - 3z)$.

26. $(x + 3y + 4)(2z)$.

1.4 OPERATIONS WITH REAL NUMBERS

Let's review how to add, subtract, multiply, and divide real (or signed) numbers. We begin with addition.

> **ADDITION WITH LIKE SIGNS**
>
> To add two positive numbers or two negative numbers, add their absolute values and place their common sign in front of the result.

EXAMPLE 1

a. $(+6) + (+7) = +(6 + 7) = +13$. Usually we omit the plus sign in front of 13. Thus $(+6) + (+7) = 13$.

b. $2 + 3 = +(2 + 3) = 5$.

c. $(-4) + (-3) = -(4 + 3) = -7$.

d. $(-2) + (-9) + (-3) = -(2 + 9 + 3) = -14$.

ADDITION WITH UNLIKE SIGNS

To add a positive number and a negative number, first find the absolute values of the numbers. Subtract the smaller absolute value from the larger absolute value. Write the sign of the number with the larger absolute value in front of this result. When two numbers with unlike signs have the same absolute value, their sum is zero.

EXAMPLE 2

Find each of the following sums.

a. $(+2) + (-9)$.

Since $|+2| = 2$ and $|-9| = 9$, -9 has the larger absolute value. Thus

$$(+2) + (-9) = -(9 - 2) = -7.$$

b. $(-7) + (10) = +(10 - 7) = 3$.

c. $-15 + 5 = -(15 - 5) = -10$.

d. $4 + (-4) = 0$, because 4 and -4 have unlike signs and the same absolute value.

Subtraction involves the fact that every real number has an *opposite*. The **opposite** (or *negative*) **of *a***, written $-a$, is that number which, when added to a, gives 0. Thus $a + (-a) = 0$. For example,

$$\text{the opposite of 4 is } -4,$$

because $4 + (-4) = 0$. Similarly, we can find the opposite of -4. Observe that $(-4) + 4 = 0$, so

$$\text{the opposite of } -4 \text{ is } 4.$$

Because the opposite of -4 is represented by $-(-4)$, we have just shown that $-(-4) = 4$. In general, the opposite of the opposite of a number is that number.

$$-(-a) = a.$$

Finally, the opposite of 0 is 0, because $0 + 0 = 0$.

EXAMPLE 3

a. $-(-3) = 3$.

b. To simplify $-(-xy)$, we let xy play the role of a in the rule $-(-a) = a$.

$$-(-xy) = xy.$$

We define subtraction using addition and an opposite.

> **SUBTRACTION**
>
> $$a - b = a + (-b).$$
>
> To subtract b from a, *add* the opposite of b to a.

For example,

opposite

$$7 - (9) = 7 + (-9) = -2.$$

change to addition

EXAMPLE 4

a. $5 - (-3) = 5 + (3) = 8$.

b. $-10 - 8 = -10 + (-8) = -18$.

Some problems involve a "string" of additions and subtractions, such as $7 - 2 + 6 - 4$. This represents the sum

$$7 + (-2) + 6 + (-4).$$

Using the commutative and associative laws, we may "shuffle" and group these numbers so that positive numbers are together and negative numbers are together.

$$\underbrace{7 + 6} + \underbrace{(-2) + (-4)} = 13 + (-6) = 7.$$

More simply, we write

$$7 - 2 + 6 - 4 = 7 + 6 - 2 - 4 = 13 - 6 = 7.$$

Whenever a number is multiplied by 0, the result is 0. Thus $a \cdot 0 = 0$, where a is any real number. The next rules are for multiplication or division of positive or negative numbers.

MULTIPLICATION AND DIVISION

To multiply (or divide) two positive or two negative numbers, multiply (or divide) their absolute values. To multiply (or divide) a positive number and a negative number, multiply (or divide) their absolute values and place a minus sign in front of the result.

These rules imply that the product (or quotient) of two positive or two negative numbers is positive. The product (or quotient) of a positive number and a negative number is negative.

EXAMPLE 5

a. $(+2)(+6) = 2 \cdot 6 = 12.$

b. $\frac{-6}{-3} = \frac{6}{3} = 2.$

c. $(-2)(4) = -(2 \cdot 4) = -8.$

d. $\frac{18}{-6} = -\left(\frac{18}{6}\right) = -3.$

e. $-\frac{-36}{4} = -\left(-\frac{36}{4}\right) = -(-9) = 9$. The last step follows from the rule $-(-a) = a$.

EXAMPLE 6

a. To find $(-6)(-2)(-1)$, we shall begin by multiplying the first two numbers. This grouping will be indicated by brackets:

$$[(-6)(-2)](-1) = [12](-1) = -12.$$

In general, *the product of an* ***odd*** *number of negative numbers is negative.*

b. $(-2)(-5)(-3)(-4) = [(-2)(-5)][(-3)(-4)] = [10][12] = 120.$

The product of an ***even*** *number of negative numbers is positive.*

The result of the division $\frac{a}{b}$ is that number by which b must be multiplied to give a. For example, $\frac{6}{3}$ is 2 because $3 \cdot 2 = 6$. Division of any number by 0 *is not defined.* To see why, let us suppose that $\frac{5}{0}$ were defined. Then it would be a number such that when we multiply it by 0, we obtain 5. But any number times 0 is 0. Thus $\frac{5}{0}$ is not defined.

When you work with the numbers 0 and 1, use the following rules.

$0 + a = a.$	$0 - a = -a.$
$a - 0 = a.$	$a \cdot 0 = 0.$
$\frac{0}{a} = 0$ if $a \neq 0.$	$\frac{a}{0}$ is not defined.
$1 \cdot a = a.$	$\frac{a}{1} = a.$

EXAMPLE 7

a. $0 + (-9) = -9.$

b. $0 - 7 = -7.$

c. $(xy) - 0 = xy.$

d. $(-6)(0) = 0.$

e. $\frac{0}{12} = 0.$

f. $\frac{9}{0}$ is not defined.

g. $x = 1 \cdot x.$

h. $3xy = \frac{3xy}{1}.$

We know that the product (or quotient) of two positive or two negative numbers is positive, and the product (or quotient) of a positive number and a negative number is negative. We can generalize these results as follows.

MULTIPLICATION	DIVISION
$(a)(b) = (-a)(-b) = ab.$	$\frac{-a}{-b} = \frac{a}{b}.$
$(a)(-b) = (-a)(b) = -(ab) = -ab.$	$\frac{-a}{b} = -\frac{a}{b} = \frac{a}{-b}.$

For example, $\frac{-4}{5} = -\frac{4}{5}$ illustrates $\frac{-a}{b} = -\frac{a}{b}.$

EXAMPLE 8

a. $(-2)(-x) = 2 \cdot x = 2x.$

b. $(-2)x = -2x.$

c. $\frac{(-x)y}{z} = \frac{-xy}{z} = -\frac{xy}{z}.$

EXAMPLE 9

a. $4(-3x) = 4[(-3)x] = [4(-3)]x = -12x$. Note the use of brackets for grouping. Thus $4(-3x) = -12x$.

b. $(2x)(-3y)(-4) = [(2)(x)][(-3)(y)](-4)$

$$= [(2)(-3)(-4)](xy) = [24](xy) = 24xy.$$

c. $-\dfrac{x}{y(-z)} = -\dfrac{x}{-(yz)} = -\left(-\dfrac{x}{yz}\right) = \dfrac{x}{yz}.$

d. $-3(x-5) = (-3)(x-5) = (-3)(x) - (-3)(5)$ [distributive law]

$$= -3x - (-15) = -3x + 15.$$

In the rule $(a)(-b) = -ab$, if we let $b = 1$, then $(a)(-1) = -a \cdot 1$, or, more simply,

$$\boxed{(-1)a = -a.}$$

EXAMPLE 10

a. $(-1)x = -x$.

b. $-yz = (-1)yz$.

We follow certain rules for **order of operations**. When the order of doing operations is not clear to you, do the multiplications and divisions first. Then do the additions and subtractions. For example,

$$-4 + 2(6) = -4 + 12 = 8,$$
$$-3 - (4)(-2) = -3 - (-8) = -3 + 8 = 5.$$

To simplify $-4 + 2(6)$, don't add -4 and 2, and then multiply the result by 6.

$$-4 + 2(6) \neq -2(6).$$

In a fraction such as $\dfrac{6+2}{7-9}$, you must first simplify both the numerator and the denominator.

$$\frac{6+2}{7-9} = \frac{8}{-2} = -4.$$

When operations are inside parentheses, do those first, as Example 11 shows.

EXAMPLE 11

a. $7 - (4 - 9) = 7 - (-5) = 7 + (5) = 12.$

b. $(-4 + 3)(7 - 5) = (-1)(2) = -2.$

c. $\frac{1-(2-8)}{-3} = \frac{1-(-6)}{-3} = \frac{7}{-3} = -\frac{7}{3}.$

Exercise 1.4

In Problems **1–55**, *perform the indicated operations.*

1. $(-2)+(-3).$

2. $(+2)+(+6).$

3. $-5+0.$

4. $9+(-3).$

5. $5-7.$

6. $-5+2.$

7. $-2-8.$

8. $3-(-4).$

9. $-6-(-5).$

10. $-14-(-14).$

11. $-2-5+9-1.$

12. $3-5+6-4.$

13. $(-3)(-4).$

14. $(+3)(+4).$

15. $(-3)(7).$

16. $(-2)(-1)(-2)(-1)(-1).$

17. $(-2)(2)(-1)(-2)(-1).$

18. $\frac{10}{-2}.$

19. $\frac{+4}{+2}.$

20. $\frac{-6}{-3}.$

21. $\frac{-25}{5}.$

22. $\frac{10}{-1}.$

23. $\left(-\frac{2}{3}\right)\cdot 0.$

24. $-(-4).$

25. $-(-5xy).$

26. $-\frac{-6}{-7}.$

27. $-\frac{-14}{7}.$

28. $(-1)(-x).$

29. $(-7)(x).$

30. $(-6)(-y).$

31. $5(-3x).$

32. $(-8x)(-2).$

33. $(2x)(-3y)(-6z).$

34. $4(-5x)(-7)(-2y).$

35. $(-3x)(2-3y).$

36. $(-x+2)(-5y).$

37. $-7(y-2z+w).$

38. $(-4z)(2x-3y+6).$

39. $\frac{x}{(-y)(z)}.$

40. $-\frac{x}{(-y)z}.$

41. $-\frac{-x}{(-y)(-z)}.$

42. $\frac{(-x)(-y)}{(-w)(-z)}.$

43. $(2-3)4+5.$

44. $(-6-7)+8.$

45. $(-6)(-7)-8.$

46. $3(5+1).$

47. $(8+1)(-7).$

48. $4-8+6(-2).$

49. $(-2)(4)(-3)+6(-2).$

50. $\frac{-7-5}{4}.$

51. $\frac{-7-(-3-2)}{2}.$

52. $\frac{12}{2(-2+1)(3-4)}.$

53. $\frac{-4-(2-2)}{(-3)(2)-(-2)(4)}.$

54. $\frac{0-xy}{-(3-4)}.$

55. $\frac{0}{3-2}.$

56. On Monday, the closing price of a stock on the New York Stock Exchange was \$65. On Tuesday it declined $\$4\frac{1}{2}$, and on Wednesday it rose \$2. Use addition of signed numbers to set up an expression that gives the closing price on Wednesday and simplify your result.

57. At 3 P.M. a television weatherperson gave the current temperature as 15°C. She estimated that the low temperature that evening would be −3°C. Assuming that she is correct, set up an expression involving subtraction of signed numbers that gives the temperature drop and simplify your result.

58. The formula $d = |a - b|$ gives the distance between the numbers a and b on the number line. Find d if $a = -3$ and $b = 5$.

If your grades in two exams were 70 *and* 80, *then the average grade for both exams is* $\frac{70+80}{2} = \frac{150}{2} = 75$. *In statistics, the average,* or **mean**, *of the numbers* x_1, x_2, x_3, *and* x_4 *is given by the formula*

$$\bar{x} = \frac{x_1 + x_2 + x_3 + x_4}{4},$$

where the symbol $\bar{x}$ *(read "x-bar") represents the mean. The small numbers to the right and just below the letters are called subscripts. We read* x_1 *as "x sub one," and so on. In Problems* **59** *and* **60** *find* $\bar{x}$ *from the given information.*

59. $x_1 = -8$, $x_2 = 7$, $x_3 = 16$, $x_4 = -3$.

60. $x_1 = -3$, $x_2 = -5$, $x_3 = 12$, $x_4 = -4$.

61. A math student taking calculus must compute y' (read "y prime"), where

$$y' = \frac{(x)(1) - (x-5)(1)}{(x)(x)}$$

and $x = -2$. What answer should the student get?

62. Repeat Problem 61 if $x = 3$.

1.5 FRACTIONS

Now we turn to some rules of fractions. We assume that no denominator is 0, since division by 0 is not defined.

Rule 1. $\frac{a}{a} = 1$.

Rule 2. $\frac{a}{b} \cdot \frac{c}{d} = \frac{ac}{bd}$.

Rule 1 states that a number divided by itself is 1. Rule 2 is the rule for multiplying two fractions: Multiply their numerators and multiply their denominators.

EXAMPLE 1

a. By Rule 1,

$$\frac{4}{4} = 1 \quad \text{and} \quad \frac{-2x}{-2x} = 1.$$

b. By Rule 2, $\frac{2}{x} \cdot \frac{y}{3} = \frac{2 \cdot y}{x \cdot 3} = \frac{2y}{3x}$.

c. $\left(-\frac{2}{7}\right)\left(-\frac{x}{5}\right) = \frac{2}{7} \cdot \frac{x}{5}$ [since $(-a)(-b) = ab$]

$= \frac{2 \cdot x}{7 \cdot 5}$ [Rule 2]

$= \frac{2x}{35}$.

d. $\frac{-2}{3} \cdot \frac{-x}{-y} = \frac{(-2)(-x)}{3(-y)} = \frac{2x}{-3y} = -\frac{2x}{3y}$.

e. $3 \cdot \frac{x}{2} = \frac{3}{1} \cdot \frac{x}{2}$ $\left[\text{since } a = \frac{a}{1}\right]$

$= \frac{3 \cdot x}{1 \cdot 2} = \frac{3x}{2}$.

Generalizing Example 1e gives the rule:

Rule 3. $a \cdot \frac{b}{c} = \frac{ab}{c}$.

We can simplify the fraction $\frac{ac}{bc}$ as follows.

$$\frac{ac}{bc} = \frac{a}{b} \cdot \frac{c}{c} \qquad \text{[Rule 2]}$$

$$= \frac{a}{b} \cdot 1 = \frac{a}{b}.$$

Thus $\frac{ac}{bc} = \frac{a}{b}$. This property states that a *common factor* of *both* the numerator and denominator of a fraction may be removed (or *canceled*) and the resulting fraction is *equivalent* to the original (that is, it has the same value as the original fraction). We call this property the *cancellation property* for fractions. We can indicate the removal of the common factor by using slashes.

$$\frac{a\not{c}}{b\not{c}} = \frac{a}{b}.$$

CANCELLATION PROPERTY

Rule 4. $\frac{ac}{bc} = \frac{a}{b}$.

EXAMPLE 2

Simplify.

a. $\frac{4xy}{4yz} = \frac{\not{4}x\not{y}}{\not{4}\not{y}z} = \frac{x}{z}$.

b. $\frac{-3x}{-3}$. Since $-3x = (-3)x$ and $-3 = (-3) \cdot 1$, -3 is a common factor of the numerator and denominator. Thus

$$\frac{-3x}{-3} = \frac{\cancel{(-3)}x}{\cancel{(-3)} \cdot 1} = \frac{x}{1} = x.$$

c. $\frac{-12x}{16y}$.

$$\begin{aligned} \frac{-12x}{16y} &= -\frac{12x}{16y} \qquad \left[\text{since } \frac{-a}{b} = -\frac{a}{b}\right] \\ &= -\frac{\not{4} \cdot 3x}{\not{4} \cdot 4y} \\ &= -\frac{3x}{4y}. \end{aligned}$$

The second step may be written

$$-\frac{\overset{3}{\cancel{12}}x}{\underset{4}{\cancel{16}}y}.$$

Here we mentally canceled the common factor 4, leaving $3x$ in the numerator and $4y$ in the denominator.

d. $25 \cdot \frac{x}{5}$.

$$25 \cdot \frac{x}{5} = \frac{\overset{5}{\cancel{25}}x}{\underset{1}{\cancel{5}}} = 5x.$$

e. $\frac{3}{y} \cdot \frac{y}{6}$.

$$\frac{3}{y} \cdot \frac{y}{6} = \frac{3y}{y \cdot 6} = \frac{\cancel{3y}}{\cancel{y} \cdot \underset{2}{\cancel{6}}} = \frac{1}{2}.$$

To simplify our work, we can cancel before multiplying the fractions.

$$\frac{3}{y} \cdot \frac{y}{6} = \frac{\cancel{3}}{\cancel{y}} \cdot \frac{\cancel{y}}{\underset{2}{\cancel{6}}} = \frac{1}{2}.$$

By writing Rule 4 as $\frac{a}{b} = \frac{ac}{bc}$, you can see that if both numerator and denominator of $\frac{a}{b}$ are multiplied by c (where c is not 0), then we obtain the equivalent fraction $\frac{ac}{bc}$. It can also be shown that if both numerator and denominator of $\frac{a}{b}$ are divided by c (where c is not 0), then $\frac{a \div c}{b \div c}\left(\text{or } \frac{a/c}{b/c}\right)$ is equivalent to $\frac{a}{b}$. These results are together called the *fundamental principle of fractions.*

FUNDAMENTAL PRINCIPLE OF FRACTIONS

Rule 5. $\frac{a}{b} = \frac{ac}{bc}; \qquad \frac{a}{b} = \frac{a \div c}{b \div c}.$

EXAMPLE 3

Using the fundamental principle of fractions.

a. *Write* $\frac{5x}{8}$ *as an equivalent fraction whose denominator is* 24.

Multiplying the denominator by 3 gives 24. Thus we must also multiply the numerator by 3.

$$\frac{5x}{8} = \frac{(5x) \cdot 3}{8 \cdot 3} = \frac{15x}{24}.$$

b. *Write* $\frac{4}{10y}$ *as an equivalent fraction with denominator* $5y$.

Dividing the denominator by 2 gives $5y$. Thus we must also divide the numerator by 2.

$$\frac{4}{10y} = \frac{4/2}{10y/2} = \frac{2}{5y}.$$

Note that dividing both numerator and denominator of $\frac{4}{10y}$ by 2 is the same as cancelling the common factor 2 in the numerator and denominator.

For the division of two fractions, $\frac{a}{b} \div \frac{c}{d}$, we can apply the fundamental principle.

$$\frac{\frac{a}{b}}{\frac{c}{d}} = \frac{\frac{a}{b} \cdot \frac{d}{c}}{\frac{c}{d} \cdot \frac{d}{c}} = \frac{\frac{a}{b} \cdot \frac{d}{c}}{\frac{cd}{dc}} = \frac{\frac{a}{b} \cdot \frac{d}{c}}{1} = \frac{a}{b} \cdot \frac{d}{c}.$$

Notice that to divide $\frac{a}{b}$ by $\frac{c}{d}$, we interchange the numerator and denominator of the divisor $\frac{c}{d}$ and multiply $\frac{a}{b}$ by the result (look at last step). We say that we *inverted* the divisor and multiplied.

$$\textbf{Rule 6.}\quad \frac{a}{b} \div \frac{c}{d} = \frac{\frac{a}{b}}{\frac{c}{d}} = \frac{a}{b} \cdot \frac{d}{c}.$$

EXAMPLE 4

Division with fractions.

a. $\frac{x}{3} \div \frac{4}{5} = \frac{x}{3} \cdot \frac{5}{4} = \frac{x \cdot 5}{3 \cdot 4} = \frac{5x}{12}$.

b. $\frac{\frac{5x}{2}}{x}$. We can write the divisor x as $\frac{x}{1}$ and then apply Rule 6.

$$\frac{\frac{5x}{2}}{x} = \frac{\frac{5x}{2}}{\frac{x}{1}} = \frac{5x}{2} \cdot \frac{1}{x} = \frac{5\not{x}}{2} \cdot \frac{1}{\not{x}} = \frac{5}{2}.$$

c. $\frac{-4}{\frac{y}{3}} = -\left(4 \cdot \frac{3}{y}\right) = -\frac{12}{y}$.

Every real number except 0 has a *reciprocal*. The **reciprocal** of a is 1 divided by a.

The reciprocal of a is $\frac{1}{a}$.

Zero does not have a reciprocal because division by 0 is not defined.

EXAMPLE 5

	Number	*Reciprocal*
a.	2.	$\frac{1}{2}$.
b.	$\frac{3}{2}$.	$\frac{1}{\frac{3}{2}} = 1 \cdot \frac{2}{3} = \frac{2}{3}$.

c. $\frac{5}{x}$. $\qquad \frac{1}{\frac{5}{x}} = 1 \cdot \frac{x}{5} = \frac{x}{5}$.

Parts b and c of Example 5 show that you find the reciprocal of a fraction by simply interchanging the numerator and denominator. Thus Rule 6 states that to divide two fractions, you multiply the dividend $\frac{a}{b}$ by the reciprocal of the divisor $\frac{c}{d}$.

If we multiply a nonzero number a by its reciprocal, then—by cancellation—we have

$$a \cdot \frac{1}{a} = \cancel{a} \cdot \frac{1}{\cancel{a}} = 1.$$

Rule 7. $a \cdot \frac{1}{a} = 1.$

For example, $\frac{3}{2} \cdot \frac{2}{3} = 1$.

Because

$$a \cdot \frac{1}{b} = \frac{a \cdot 1}{b} = \frac{a}{b},$$

the division $\frac{a}{b}$ can be expressed in terms of multiplication:

Rule 8. $\frac{a}{b} = a \cdot \frac{1}{b}.$

Thus *dividing a by b is the same as multiplying a by the reciprocal of b.*

EXAMPLE 6

a. $\frac{3}{4} = 3 \cdot \frac{1}{4}$.

b. $\frac{x}{3} = x \cdot \frac{1}{3} = \frac{1}{3}x$.

c. $\frac{\frac{3}{4}}{5} = \frac{3}{4} \cdot \frac{1}{5} = \frac{3}{20}$.

d. $8 \cdot \frac{1}{5y} = \frac{8}{5y}$.

Exercise 1.5

In Problems **1–32**, *perform the indicated operations.*

1. $\frac{4w}{4w}$.

2. $-\frac{3+2x}{3+2x}$.

3. $\frac{9}{11} \cdot \frac{2}{7}$.

4. $\frac{4}{3}\cdot\frac{2}{-5}$.

5. $\frac{-3}{4}\cdot\left(-\frac{7}{2}\right)$.

6. $\frac{-6x}{5s}\cdot\frac{-3}{7t}$.

7. $-3x\cdot\frac{y}{2}$.

8. $\frac{7x}{14}$.

9. $\frac{18q}{12pq}$.

10. $\frac{-4x}{2}$.

11. $\frac{24z}{-18z}$.

12. $-\frac{-27y}{-6}$.

13. $\frac{25}{x}\cdot\frac{x}{15}$.

14. $6\cdot\frac{x}{3}$.

15. $\frac{3}{4}(-6)$.

16. $\left(-\frac{4}{5}\right)\left(\frac{-25}{8}\right)$.

17. $\frac{-2x}{21}\cdot\frac{3y}{-4x}$.

18. $\frac{1}{5}(5a)$.

19. $\frac{2}{-3}(9x)$.

20. $(-6)\left(-\frac{3}{8}x\right)$.

21. $6a\cdot\frac{1}{6a}$.

22. $-\frac{1}{a+b}\cdot(a+b)$.

23. $\frac{3}{2}\div\frac{5}{7}$.

24. $\frac{-4}{9}\div\frac{3}{8}$.

25. $\frac{5}{6}\div\frac{10}{-9}$.

26. $\left(-\frac{1}{16}\right)\div\frac{6}{-20}$.

27. $\frac{-4x}{3}\div\frac{x}{-24}$.

28. $\frac{-14y}{-6z}\div\frac{-21}{12z}$.

29. $\frac{12}{\frac{x}{6}}$.

30. $\frac{-3y}{\frac{y}{9}}$.

31. $-\frac{4x}{\frac{-y}{10}}$.

32. $\frac{-\frac{xz}{8}}{2xy}$.

33. Write $\frac{3x}{4y}$ as an equivalent fraction with denominator $24yz$.

34. Write $\frac{14zt}{36xzw}$ as an equivalent fraction with denominator $18xw$.

In Problems **35** *and* **36**, *find the reciprocals of the numbers.*

35. $\frac{2x}{3}$.

36. $\frac{5-x}{2}$.

37. The *power* of a lens (in diopters) is the reciprocal of the focal length of the lens (in meters). A doctor fitting lenses for a farsighted person finds that the lenses must have a focal length of .5 meters (m). Find the power of the prescribed lenses.

38. The formula $F = 32 + \frac{9}{5}C$ is used to convert a Celsius temperature C to a Fahrenheit temperature F. In technical situations, room temperature is often taken to be 20°C. Express this temperature in degrees Fahrenheit.

39. The formula $C = \frac{5}{9}(F - 32)$ is used to convert a Fahrenheit temperature F to a Celsius temperature C. According to the *Guinness Book of World Records*, the highest recorded temperature in the United States is 134°F (in the shade) at Death Valley, California on July 10, 1913. Express this temperature in degrees Celsius.

1.6 REVIEW

IMPORTANT TERMS AND SYMBOLS

set *(1.1)*
integer *(1.1)*
irrational number *(1.1)*
origin *(1.1)*
literal number *(1.2)*
inequality *(1.2)*
$|a|$ *(1.2)*
term *(1.3)*
factor *(1.3)*
quotient *(1.3)*
divisor *(1.3)*
denominator *(1.3)*
associative laws *(1.3)*
opposite *(1.4)*
cancellation property *(1.5)*
reciprocal *(1.5)*
positive integer *(1.1)*
rational number *(1.1)*
real number *(1.1)*
number line *(1.1)*
$a < b$, $a > b$, $a \leq b$, $a \geq b$ *(1.2)*
absolute value *(1.2)*
sum *(1.3)*
product *(1.3)*
difference *(1.3)*
dividend *(1.3)*
numerator *(1.3)*
commutative laws *(1.3)*
distributive laws *(1.3)*
equivalent fractions *(1.5)*
fundamental principle of fractions *(1.5)*

REVIEW PROBLEMS

In Problems **1–8**, *name the law used in the given statement.*

1. $8 + y = y + 8$.

2. $2(x + 3y) = 2x + 6y$.

3. $2x + (x + y) = (2x + x) + y$.

4. $2(4x) = (2 \cdot 4)x$.

5. $5(x + 4) = (x + 4)5$.

6. $5(x + 4) = 5(4 + x)$.

7. $(a - 3)b = ab - 3b$.

8. $(3x)(y) = 3(xy)$.

In Problems **9–16**, *determine whether each statement is true or false.*

9. $-3 < -2$.

10. $3 > 0$.

11. $|-6| = 6$.

12. $-5 > 1$.

13. $|-3| < |-2|$.

14. $-|-4| = 4$.

15. $2(3x) = (2 \cdot 3)(2 \cdot x)$.

16. $2(3 - x) = 6 - x$.

In Problems **17–62**, *perform the indicated operations.*

17. $\dfrac{(-2)(-4)}{-16}$.

18. $(-3)(-4 + 7)$.

19. $(3x)(y + 5)$.

20. $2(x - y)$.

21. $2(-5y)$.

22. $\frac{7}{2-2}$.

23. $-6 - (-5)$.

24. $\frac{-2}{6}$.

25. $3(-8 + 4)$.

26. $(-2)(-4)(-1)$.

27. $\frac{14}{(-2)(-7)}$.

28. $\frac{6(-3)}{-2}$.

29. $-4(5x - 3)$.

30. $8\left(\frac{-5}{12}x\right)$.

31. $(7 + x) - 8$.

32. $\frac{6-8}{-4}$.

33. $2(-6) + (-4)(-1)$.

34. $\frac{8-9}{7-6}$.

35. $-(-3xz)$.

36. $-6 + 4 - 9(2)$.

37. $\left(-\frac{3}{4}\right)(-20)$.

38. $-9(-12 - 8)$.

39. $(-8)\left(\frac{-5}{8}\right)$.

40. $(a + b)c$.

41. $(4x)(-7y)(-2z)$.

42. $-\frac{-8}{-64}$.

43. $\frac{7-9}{(7)(-9)}$.

44. $(8 - 8)(-8x)$.

45. $\frac{6(-4)}{2}$.

46. $\frac{(-2)(4)(-6)}{0-3}$.

47. $\frac{(-2) - (-1)(0)}{-2}$.

48. $\frac{8(-2) - (-2)(-8)}{3(-2) - 2}$.

49. $-4(3x)$.

50. $(3x)(1 - y + z)$.

51. $(8 - x)y$.

52. $(x + 7) - (7 - 4)$.

53. $(-2z)(4x - 3y - 8)$.

54. $-\frac{-12xy}{8xz}$.

55. $\frac{2x}{-y} \cdot \frac{3y}{14}$.

56. $\frac{6}{x} \cdot \frac{-3}{14} \cdot \frac{21x}{-5}$.

57. $15x\left(-\frac{y}{5}\right)$.

58. $\left(-\frac{18w}{15}\right) \div \frac{2w}{x}$.

59. $\frac{25}{-9x} \div \frac{15}{-3}$.

60. $\frac{5xy}{\frac{10x}{3}}$.

61. $\frac{\frac{2x}{3}}{-18x}$.

62. $1 \div \frac{1}{x+1}$.

63. Compute $\frac{a - 2b + c}{ad}$ in each case.

a. $a = 2,\ b = -3,\ c = 4,\ d = -1$.

b. $a = -3,\ b = 4,\ c = -2,\ d = -2$.

In Problems **64–65**, *for the given values of x and y, find* (a) $|x + y|$, (b) $|x - y|$, (c) $|x| + |y|$, and (d) $|x| - |y|$.

64. $x = 2,\ y = 5$.

65. $x = -3,\ y = -8$.

66. The ends of a certain copper rod are kept at different temperatures. The rate H, in calories per second, at which heat is conducted through the rod is given by

$$H = .14(T_h - T_l),$$

where T_h is the higher temperature, T_l is the lower temperature, and both T_h and T_l are in degrees Celsius. Find H for each of the following situations.

a. $T_h = 820°C$, $T_l = 200°C$.

b. $T_h = 50°C$, $T_l = -50°C$.

c. $T_h = 0°C$, $T_l = -10°C$.

d. $T_h = -10°C$, $T_l = -20°C$.

2 A First Look at Exponents and Radicals

2.1 EXPONENTS

We can use a bit of shorthand to indicate the product of four a's:

$$a \cdot a \cdot a \cdot a \quad \text{is written} \quad a^4.$$

It is read, "the fourth power of a" or "a (raised) to the fourth." We call a the **base** and 4 the **exponent**.

$$\text{base} \rightarrow a^4 \leftarrow \text{exponent}$$

In general, if a is used as a factor n times, we have:

THE nth POWER OF a

$$a^n = \underbrace{a \cdot a \cdot \ldots \cdot a}_{n \text{ factors of } a}.$$

An exponent can be 1. By a^1 we simply mean a.

$$a^1 = a.$$

For example, $5^1 = 5$.

$x^2 = x \cdot x$. Don't write $x^2 = x + x$.

EXAMPLE 1

a. $x^3 = x \cdot x \cdot x$. We read x^3 as x *cubed* or x *to the third*. The base is x; the exponent is 3.

b. $6^2 = 6 \cdot 6 = 36$. We read 6^2 as 6 *squared*.

c. $4^2x^4 = 4 \cdot 4 \cdot x \cdot x \cdot x \cdot x$.

d. $(x - 6)^1 = x - 6$. The base is $x - 6$.

e. $(1 + 3)^2 = (1 + 3)(1 + 3) = 4 \cdot 4 = 16$. Note that

$$(1 + 3)^2 \neq 1^2 + 3^2,$$

since $(1 + 3)^2 = 16$ and $1^2 + 3^2 = 1 + 9 = 10$.

EXAMPLE 2

Powers of negative numbers.

a. $(-2)^3 = (-2)(-2)(-2) = -8$.
An odd power of a negative number is negative.

b. $(-1)^4 = (-1)(-1)(-1)(-1) = 1$
An even power of a negative number is positive.

An exponent applies only to the quantity immediately to the left and below it. For example:

In $(-2)^2$ the base is -2.	$(-2)^2 = (-2)(-2) = 4.$
In -2^2 the base is 2.	$-2^2 = -(2 \cdot 2) = -4.$
In $(2x)^2$ the base is $2x$.	$(2x)^2 = (2x)(2x).$
In $2x^2$ the base is x.	$2x^2 = 2 \cdot x \cdot x.$

We can arrive at some rules of exponents. For the product $a^2 \cdot a^3$, we see that

$$a^2 \cdot a^3 = (a \cdot a)(a \cdot a \cdot a) = a \cdot a \cdot a \cdot a \cdot a$$
$$= a^5 = a^{2+3}.$$

So, $a^2 \cdot a^3 = a^{2+3}$. In general,

> *to* ***multiply*** *numbers with the same base, we may* ***add*** *the exponents and keep the base the same.*

Rule 1. $a^m a^n = a^{m+n}$.

EXAMPLE 3

a. $4^3 \cdot 4^8 = 4^{3+8} = 4^{11}$.

b. $x^5 \cdot x^2 = x^{5+2} = x^7$.

c. $(3y-1)^7(3y-1)^3 = (3y-1)^{10}$.

d. $x^3x^4x^6 = x^{3+4+6} = x^{13}$.

e. $x(x^n) = x^1x^n = x^{1+n} = x^{n+1}$.

EXAMPLE 4

a. $\dfrac{x^2x^3}{y^3y^4} = \dfrac{x^5}{y^7}$.

b. $(3x^5)(9x^2) = 3 \cdot x^5 \cdot 9 \cdot x^2 = 3 \cdot 9 \cdot x^5x^2 = 27x^7$.

When you use Rule 1, the bases must be the ***same*** *before you add exponents. For example,* $2^2 \cdot x^3 \neq (2x)^5$. *Also,*

$$(-2)^4(-2^2) \neq (-2)^6, \quad \textit{but} \quad (-2)^4(-2^2) = (16)(-4) = -64.$$

Suppose that we take a power of a number and raise it to a power, such as taking the second power of a and raising it to the third power: $(a^2)^3$. Here a^2 is the base for the exponent 3. Thus

$$(a^2)^3 = a^2 \cdot a^2 \cdot a^2$$
$$= a^6. \qquad \text{[Rule 1]}$$
$$= a^{2 \cdot 3}.$$

So $(a^2)^3 = a^{2 \cdot 3}$. In general,

> *to find a power of a power, we may* ***multiply*** *the exponents and keep the base the same.*

Rule 2. $(a^m)^n = a^{mn}$.

EXAMPLE 5

a. $(3^2)^4 = 3^{2\cdot 4} = 3^8$.

b. $(x^3)^5 = x^{3\cdot 5} = x^{15}$.

c. $(t^4)^9 = t^{36}$.

d. $[(y-2)^3]^2 = (y-2)^6$.

Don't confuse $(x^2)^3$ *with* $x^2 \cdot x^3$.

$$(x^2)^3 = x^6 \quad \textit{but} \quad x^2 \cdot x^3 = x^5.$$

Let's look now at division.* In $\frac{a^5}{a^3}$ the numerator has more factors of a than does the denominator. By cancellation,

$$\frac{a^5}{a^3} = \frac{\not{a}\cdot\not{a}\cdot\not{a}\cdot a\cdot a}{\not{a}\cdot\not{a}\cdot\not{a}} = \frac{a^2}{1} = a^2 = a^{5-3}.$$

Similarly,

$$\frac{a^3}{a^5} = \frac{\not{a}\cdot\not{a}\cdot\not{a}}{\not{a}\cdot\not{a}\cdot\not{a}\cdot a\cdot a} = \frac{1}{a^2} = \frac{1}{a^{5-3}}.$$

Notice that in both division problems, subtraction of exponents is involved. In general, we have these rules:

Rule 3. $\frac{a^m}{a^n} = a^{m-n}$ for $m > n$.

Rule 4. $\frac{a^m}{a^n} = \frac{1}{a^{n-m}}$ for $n > m$.

Rule 5. $\frac{a^n}{a^n} = 1$.

EXAMPLE 6

a. $\frac{x^{11}}{x^7} = x^{11-7} = x^4$, by Rule 3.

b. $\frac{(x^2+1)^4}{(x^2+1)^{12}} = \frac{1}{(x^2+1)^{12-4}} = \frac{1}{(x^2+1)^8}$, by Rule 4.

---

* Here, as elsewhere in this book, we assume that divisors are different from zero.

c. $\dfrac{-y^6}{y^4} = -\dfrac{y^6}{y^4} = -y^2$, by Rule 3.

d. $\dfrac{(-2)^3}{(-2)^4} = \dfrac{1}{(-2)^{4-3}} = \dfrac{1}{(-2)^1} = \dfrac{1}{-2} = -\dfrac{1}{2}$.

EXAMPLE 7

a. $\dfrac{(x^6)^4}{(x^4)^5} = \dfrac{x^{24}}{x^{20}} = x^4$, by Rules 2 and 3.

b. $\dfrac{x^2(x^4)^7}{x^{90}} = \dfrac{x^2x^{28}}{x^{90}} = \dfrac{x^{30}}{x^{90}} = \dfrac{1}{x^{60}}$, by Rules 2, 1, and 4.

Let's examine a power of a product, such as $(3 \cdot 5)^2$:

$$(3 \cdot 5)^2 = (3 \cdot 5)(3 \cdot 5) = 3 \cdot 3 \cdot 5 \cdot 5 = 3^2 5^2.$$

In general,

to raise a product to a power, we may raise each factor to that power.

Rule 6. $(ab)^n = a^n b^n$.

Now consider a quotient raised to a power, such as

$$\left(\frac{3}{5}\right)^2 = \frac{3}{5} \cdot \frac{3}{5} = \frac{3 \cdot 3}{5 \cdot 5} = \frac{3^2}{5^2}.$$

In general,

to raise a quotient to a power, we may raise both the numerator and denominator to that power.

Rule 7. $\left(\dfrac{a}{b}\right)^n = \dfrac{a^n}{b^n}$.

EXAMPLE 8

a. $(xy)^8 = x^8y^8$, by Rule 6.

b. $\left(\dfrac{x}{y}\right)^{22} = \dfrac{x^{22}}{y^{22}}$, by Rule 7.

c. $(abc)^5 = a^5b^5c^5$.

d. $\left(\frac{1}{x}\right)^4 = \frac{1^4}{x^4} = \frac{1}{x^4}$.

e. $(2x)^4 = 2^4x^4 = 16x^4$. Note that $(2x)^4 \neq 2x^4$.

EXAMPLE 9

a. $(2x^2y^3)^3 = 2^3(x^2)^3(y^3)^3 = 8x^6y^9$, by Rules 6 and 2.

b. $\left(\frac{x}{y^4}\right)^2 = \frac{x^2}{(y^4)^2} = \frac{x^2}{y^8}$, by Rules 7 and 2.

c. $\left(\frac{2b^2}{c^3d^4}\right)^6 = \frac{(2b^2)^6}{(c^3d^4)^6} = \frac{2^6(b^2)^6}{(c^3)^6(d^4)^6} = \frac{64b^{12}}{c^{18}d^{24}}$, by Rules 7, 6, and 2.

d. $\frac{(20)^3}{5^3} = \left(\frac{20}{5}\right)^3 = 4^3 = 64$, by Rule 7.

Example 10 shows you how to handle minus signs involved with powers.

EXAMPLE 10

a. $(-x)^9 = [(-1)x]^9 = (-1)^9x^9 = (-1)x^9 = -x^9$.

b. $(-x^2y)^8 = (-1 \cdot x^2 \cdot y)^8 = (-1)^8(x^2)^8y^8 = 1 \cdot x^{16}y^8 = x^{16}y^8$.

EXAMPLE 11

The area A of a circle with radius r is given by the formula $A = \pi r^2$. What happens to the area of a circle if its radius is doubled?

For a circle with radius r, the area A is πr^2. Let A_0 be the area when the radius is doubled, that is, after the radius is changed from r to $2r$. By the formula, the area is π times the square of the radius. Since the new radius is $2r$, $A_0 = \pi(2r)^2$. (In effect, we replaced A by A_0 and r by $2r$ in the formula $A = \pi r^2$.) Simplifying,

$$A_0 = \pi(2r)^2 = \pi[4r^2] = 4\pi r^2 = 4(\pi r^2).$$

But πr^2 is A, the original area. Thus

$$A_0 = 4A,$$

so the original area is multiplied by a factor of 4.

Exercise 2.1

In Problems **1–12**, *compute the numbers.*

1. 4^3.
2. $(-3)^2$.
3. $2^3 - 2^4$.
4. $(-7)^1$.
5. $-(-3)^4$.
6. $\dfrac{-2^2}{(-2)^3}$.
7. $(-2^3)(-3)^2$.
8. $(3-5)^2$.
9. $\dfrac{-(-3)^2}{(-3)^3}$.
10. $2^3 \cdot 3^2 - 6^2$.
11. $(-2^2)^3$.
12. $(-2+3^2)^2$.

In Problems **13–73**, *simplify.*

13. x^5x^8.
14. x^4x^4.
15. y^5y^4.
16. t^2t.
17. x^2x^4x.
18. x^ax^b.
19. $(x-2)^5(x-2)^3$.
20. $y^9y^{91}y^2$.
21. $\dfrac{x^5x^2}{y^2y^3y^4}$.
22. $\dfrac{x(x^2)}{y^2y^3}$.
23. $(2x^2)(7x^6)$.
24. $x^4(3x^2)$.
25. $(-3x)(4x^3)$.
26. $(-2x^5)(-3x^4)$.
27. $(x^8)^2$.
28. $(x^4)^3$.
29. $(x^3)^3$.
30. $(x^5)^7$.
31. $(t^2)^n$.
32. $(x^b)^c$.
33. $(x^4)^2(x^3)^7$.
34. $x^6(x^4)^2$.
35. $\dfrac{x^7}{x^3}$.
36. $\dfrac{x^8}{x^{12}}$.
37. $\dfrac{x^{21}}{x^{22}}$.
38. $\dfrac{(a+b)^{16}}{(a+b)^{12}}$.
39. $\dfrac{y^{14}}{-y^8}$.
40. $\dfrac{-y^2}{-y^5}$.
41. $\dfrac{x^2x^8}{x^{16}}$.
42. $\dfrac{x^{18}}{x^{10}x^{10}}$.
43. $\dfrac{(x^5)^3}{x^2}$.
44. $\dfrac{(x^4)^2}{(x^5)^3}$.
45. $\dfrac{(x^4)^2}{x(x^6)}$.
46. $\dfrac{t^{12}(t^6)}{(w^5)^3}$.
47. $\dfrac{(x^2)^4(x^4)^2}{(x^3)^7}$.
48. $\dfrac{1}{(x^4)^5}(x^4)^5$.
49. $(ab)^6$.
50. $(xy)^4$.
51. $(2x)^4$.
52. $(3x)^3$.
53. $(3y)^2(5y^4)$.
54. $(3y^2)(2y^2)^4$.
55. $\left(\dfrac{a}{b}\right)^3$.
56. $\left(\dfrac{x}{2}\right)^4$.
57. $(xy^2)^4$.
58. $(3x^2)^3$.
59. $\left(\dfrac{2y}{z}\right)^3$.
60. $\left(\dfrac{1}{x^2y^3}\right)^5$.

61. $\left(\frac{2}{3}a^2b^3c^6\right)^2.$

62. $(-4)(2x^2)^2.$

63. $\left(\frac{x^2}{y^5}\right)^3.$

64. $\left(\frac{2x^2}{w^2}\right)^4.$

65. $\left(\frac{x^2y^3}{2z^4}\right)^4.$

66. $\left(\frac{2t^4}{5x^2}\right)^3.$

67. $(-x)^{13}.$

68. $(-3x)^4.$

69. $(-2x^2y)^4.$

70. $(-3)^2(-x)^3.$

71. $\frac{(-xy)^5}{(-t)^4}.$

72. $\frac{-y^3}{(-z)^2}.$

73. $(-3x)^3(-x)^7.$

In Problems **74–76**, *find the values in an easy way.*

74. $\frac{5^{100}}{5^{99}}.$

75. $\frac{(80)^6}{(40)^6}.$

76. $(14)^5\left(\frac{1}{28}\right)^5.$

77. You might find a problem like this in calculus. Compute y' (read "y prime") if

$$y' = (x^2 - 2x)(2x) + (x^2 + 4)(2x - 2)$$

and $x = -1$.

78. Here's a problem you might find in statistics. Compute

$$\frac{(x_1 - \bar{x})^2 + (x_2 - \bar{x})^2 + (x_3 - \bar{x})^2 + (x_4 - \bar{x})^2 + (x_5 - \bar{x})^2}{4},$$

if $x_1 = 6$, $x_2 = 8$, $x_3 = 5$, $x_4 = 3$, $x_5 = 3$, and $\bar{x} = 5$. (Here $\bar{x}$ is read "x bar.")

79. The volume V of a sphere of radius r is given by $V = \frac{4}{3}\pi r^3$. What is the effect of doubling the radius?

80. The formula $I = Prt$ gives the simple interest earned when a principal of P dollars is invested for t years at an annual rate r. If person A has twice as much money invested for twice as long and at double the rate as person B, how do their amounts of interest compare?

81. A manufacturer ships a product in cylindrical cans. The volume V of each can is given by $V = \pi r^2 h$, where r is the can's radius and h is its height. (a) If the manufacturer increases the volume of each can by doubling the radius, by what factor does the volume increase? (b) If the manufacturer doubles both the radius and height, by what factor does the volume increase?

2.2 ZERO AND NEGATIVE EXPONENTS

Up to now we have dealt with exponents that are positive integers. In this section we'll define zero and negative exponents.

We begin with a^0, where $a \neq 0$.* We want to define a^0 so that the rules of exponents hold. For instance, by the rule $a^m a^n = a^{m+n}$ we must have

$$a^m a^0 = a^{m+0} = a^m.$$

That is, $a^m a^0 = a^m$. But we know that $a^m(1) = a^m$. Since a^0 appears to play the same role in multiplication as 1, it is reasonable that we make the following definition (which we shall call a rule).

Rule 8. $a^0 = 1.$

With this definition the other rules of exponents also hold if an exponent is 0. To illustrate, for the rule $(ab)^n = a^n b^n$ we have

$$(ab)^0 = 1 = 1 \cdot 1 = a^0 b^0.$$

EXAMPLE 1

Zero exponents.

a. $2^0 = 1.$

b. $(-5)^0 = 1.$

c. $3\left(\frac{4}{5}\right)^0 = 3(1) = 3.$

d. $(x^2 + 2)^0 = 1.$

We now give meaning to a negative exponent, as in a^{-n} where n is positive. Again we want the rules of exponents to hold. From the rule $a^m a^n = a^{m+n}$, we must have

$$a^n a^{-n} = a^{n+(-n)} = a^{n-n} = a^0 = 1.$$

That is, $a^n a^{-n} = 1$. But we know that multiplying a^n by its reciprocal, $1/a^n$, gives 1. Since a^{-n} appears to play the same role as $1/a^n$, it is reasonable to make the following definition for a^{-n}.

Rule 9. $a^{-n} = \dfrac{1}{a^n}.$

With this definition the other rules of exponents also hold for negative exponents. To illustrate, for the rule $(ab)^n = a^n b^n$ we have

$$(ab)^{-n} = \frac{1}{(ab)^n} = \frac{1}{a^n b^n} = \frac{1}{a^n} \cdot \frac{1}{b^n} = a^{-n} b^{-n}.$$

---

* 0^0 has no meaning.

We can obtain another rule by applying Rule 9 to $\frac{1}{a^{-n}}$.

$$\frac{1}{a^{-n}} = \frac{1}{\frac{1}{a^n}} = 1 \cdot \frac{a^n}{1} = a^n.$$

Rule 10. $\frac{1}{a^{-n}} = a^n.$

EXAMPLE 2

Negative exponents.

a. $5^{-2} = \frac{1}{5^2} = \frac{1}{25}$, by Rule 9.

b. $-x^{-6} = -(x^{-6}) = -\frac{1}{x^6}$.

c. $\frac{1}{2^{-4}} = 2^4 = 16$, by Rule 10.

d. $\frac{1}{x^2} = x^{-2}$, by Rule 9.

e. $2x^{-3} + (2x)^{-3} = 2\left(\frac{1}{x^3}\right) + \frac{1}{(2x)^3} = \frac{2}{x^3} + \frac{1}{8x^3}$.

f. $\left(\frac{a}{b}\right)^{-2} = \frac{1}{\left(\frac{a}{b}\right)^2} = \frac{1}{\frac{a^2}{b^2}} = 1 \cdot \frac{b^2}{a^2} = \frac{b^2}{a^2}$.

Generalizing part f of Example 2 gives the following rule.

Rule 11. $\left(\frac{a}{b}\right)^{-n} = \frac{b^n}{a^n} = \left(\frac{b}{a}\right)^n.$

a. $\frac{1}{(-2)^{-3}} = (-2)^3 = -8$. *We change the sign of the exponent only, not the sign of the base.* ***Don't write*** $\frac{1}{(-2)^{-3}} = (2)^3$.

b. $x^{-2} + y^{-2} = \frac{1}{x^2} + \frac{1}{y^2}$. ***Don't write*** $x^{-2} + y^{-2} = \frac{1}{x^2 + y^2}$.

Using Rules 8–10, we can replace Rules 3–5 by the following:

Rule 12. $\frac{a^m}{a^n} = a^{m-n} = \frac{1}{a^{n-m}},$

regardless of the values of m and n. That is, we subtract the exponent n in the denominator from the exponent m in the numerator to obtain a^{m-n}, or vice versa to obtain $\frac{1}{a^{n-m}}$. For example, since $\frac{a^m}{a^n} = a^{m-n}$, we can write

$$\frac{x^3}{x^6} = x^{3-6} = x^{-3} = \frac{1}{x^3}.$$

More simply, since $\frac{a^m}{a^n}$ is also $\frac{1}{a^{n-m}}$, we have

$$\frac{x^3}{x^6} = \frac{1}{x^{6-3}} = \frac{1}{x^3},$$

which is the same result that you would get by applying Rule 4 to x^3/x^6.

Using negative exponents we can manipulate factors in a fraction. Observe that

$$\frac{x^3y^{-4}}{z^5} = x^3 \cdot y^{-4} \cdot \frac{1}{z^5} = x^3 \cdot \frac{1}{y^4} \cdot z^{-5} = \frac{x^3z^{-5}}{y^4}.$$

Compare the factors in the first and last fractions. You can see that *a **factor** of the numerator can be moved to the denominator if you **change** the sign of its exponent. Similarly,* a ***factor** of the denominator can be moved to the numerator by **changing** the sign of its exponent.* Thus

$$\frac{x^{-7}y^6}{3z^2} = \frac{y^6}{3x^7z^2}.$$

EXAMPLE 3

Simplify and give all answers with positive exponents.

a. $x^{-2}y^{-2} = \frac{1}{x^2y^2}.$

b. $\frac{16x^{-4}}{32x^7} = \frac{1}{2x^7x^4} = \frac{1}{2x^{11}}.$

c. $\frac{-x^{-2}}{y^{-3}z^2} = -\frac{x^{-2}}{y^{-3}z^2} = -\frac{y^3}{x^2z^2}.$

d. $\frac{x^{-7}y^6}{x^9y^{-2}} = \frac{y^6y^2}{x^9x^7} = \frac{y^8}{x^{16}}.$

Table 2-1 gives a summary of rules of exponents. *These rules are true for positive, negative, and zero exponents.* The next example shows how these rules are used to simplify expressions involving negative exponents.

TABLE 2-1

$$a^0 = 1.$$

$$a^{-n} = \frac{1}{a^n}.$$

$$a^m a^n = a^{m+n}.$$

$$(a^m)^n = a^{mn}.$$

$$\frac{a^m}{a^n} = a^{m-n} = \frac{1}{a^{n-m}}.$$

$$(ab)^n = a^n b^n.$$

$$\left(\frac{a}{b}\right)^n = \frac{a^n}{b^n}.$$

$$\left(\frac{a}{b}\right)^{-n} = \left(\frac{b}{a}\right)^n.$$

EXAMPLE 4

Perform the operations and write the answers with positive exponents only.

a. $x^{11}x^{-5} = x^{11+(-5)} = x^6.$

b. $2x^{-7}x^{-6} = 2x^{-7-6} = 2x^{-13} = \frac{2}{x^{13}}.$

c. $(3x^7y^{-4})^2 = 3^2(x^7)^2(y^{-4})^2 = 9x^{14}y^{-8} = \frac{9x^{14}}{y^8}.$

d. $(x^{-3}y^5)^{-4} = (x^{-3})^{-4}(y^5)^{-4} = x^{12}y^{-20} = \frac{x^{12}}{y^{20}}.$

e. $\left(\frac{x^{-2}y^3}{z^{-4}}\right)^{-6} = \frac{(x^{-2}y^3)^{-6}}{(z^{-4})^{-6}} = \frac{x^{12}y^{-18}}{z^{24}} = \frac{x^{12}}{y^{18}z^{24}}.$

Another way to do this problem is

$$\left(\frac{x^{-2}y^3}{z^{-4}}\right)^{-6} = \left(\frac{y^3z^4}{x^2}\right)^{-6} = \left(\frac{x^2}{y^3z^4}\right)^6 = \frac{x^{12}}{y^{18}z^{24}}.$$

EXAMPLE 5

When a number is written as the product of a number between 1 *and* 10 *and some integral power of* 10, *then the number is said to be in* ***scientific notation****. For example,* 832 *in scientific notation is* 8.32×10^2. *Given the formula (from physics)*

$$F = \frac{(9 \times 10^9)q_1q_2}{r^2},$$

find F if $q_1 = 1.6 \times 10^{-19}$, $q_2 = 1.6 \times 10^{-19}$, *and* $r = 5.3 \times 10^{-11}$. *Give the answer in scientific notation.*

Let's handle the powers of 10 separately.

$$\begin{aligned} F &= \frac{(9 \times 10^9)(1.6 \times 10^{-19})(1.6 \times 10^{-19})}{(5.3 \times 10^{-11})^2} \\ &= \frac{(9 \times 1.6 \times 1.6)(10^9 \times 10^{-19} \times 10^{-19})}{(5.3)^2(10^{-22})} \\ &= \left[\frac{9 \times 1.6 \times 1.6}{(5.3)^2}\right] 10^{9-19-19+22}. \end{aligned}$$

The first factor is approximately .82. Thus

$$F = .82 \times 10^{-7}.$$

To write the answer in scientific notation, we want the .82 to be 8.2 (which is a number between 1 and 10). Thus we multiply by 10×10^{-1} (which is 1):

$$\begin{aligned} F &= .82 \times 10^{-7} = (.82 \times 10) \times (10^{-1} \times 10^{-7}) \\ &= 8.2 \times 10^{-8}. \end{aligned}$$

Exercise 2.2

In Problems **1–15**, *find the values of the numbers.*

1. 7^0. **2.** $\left(\frac{3}{5}\right)^0$. **3.** 2^{-3}.

4. 3^{-2}. **5.** $\frac{1}{3^{-3}}$. **6.** $\frac{2}{4^{-2}}$.

7. $2x^0 + (2x)^0$. **8.** $2(-3)^0$. **9.** $\frac{-1^0}{4^{-1}}$.

10. $-5^{-2}(25)$. **11.** $\frac{1}{(-3)^{-3}}$. **12.** $2^{-1} + 3^{-1}$.

13. $2^{-2} + 9(3^{-2})$. **14.** $4^{-2}(2^{-4})$. **15.** $\frac{(3^{-2})^0}{1^{-1}}$.

In Problems **16–30**, *write the expression using positive exponents only. Simplify.*

16. x^{-2}. **17.** x^{-6}. **18.** $2^{-1}x$.

19. $\frac{1}{x^{-3}}$. **20.** $\frac{1}{3x^{-2}}$. **21.** $3y^{-4}$.

22. $2^{-2}x^{-4}$.

23. $\dfrac{x}{4^{-2}}$.

24. $\dfrac{7^0}{x^{-1}yz^{-2}}$.

25. $x^{-5}y^{-7}$.

26. $x^{-1}y^{-2}z^4$.

27. $\dfrac{2a^2b^{-4}}{c^{-5}}$.

28. $\dfrac{a^5b^{-4}}{c^{-3}d}$.

29. $\dfrac{x^9y^{-12}}{w^2z^{-4}}$.

30. $\dfrac{(x^2+4x^4)^0}{x^{-2}}$.

In Problems **31–56**, *perform the operations and simplify. Give all answers with positive exponents only.*

31. x^8x^{-7}.

32. $x^{-7}x$.

33. $x^{-2}x^{-3}$.

34. $x^2x^{-4}x^9$.

35. $(3x^{-4})(2x)$.

36. $(3x^{-7})(4x^{-5})$.

37. $(xy^{-5})^{-4}$.

38. $(2x^2y^{-1})^2$.

39. $2(x^{-1}y^2)^2$.

40. $(x^{-5}y^6)^{-1}$.

41. $(3t)^{-2}$.

42. $(x^{-4}y^{-4})^4$.

43. $\dfrac{t^{-8}}{t^{-12}}$.

44. $\dfrac{-2b^{-30}}{4b^{-5}}$.

45. $\dfrac{x^{-2}y^4}{x^6y^{-1}}$.

46. $\dfrac{x^{-7}x^9}{x^2x^{-3}}$.

47. $\dfrac{x^{-2}}{(x^{-3})^2}$.

48. $\dfrac{x^2y^2}{x^{-2}y^{-2}}$.

49. $\left(\dfrac{x}{y}\right)^{-4}$.

50. $\left(\dfrac{y}{z^{-1}}\right)^{-1}$.

51. $\left(\dfrac{8x^2}{5y^2}\right)^{-1}$.

52. $\dfrac{1}{(3x^{-1})^{-1}}$.

53. $\left(-\dfrac{z^{-1}}{x}\right)^{-1}$.

54. $\left(\dfrac{x^{-1}y^4}{z^2}\right)^{-5}$.

55. $\left(\dfrac{y^{-6}z^2}{2x}\right)^{-2}$.

56. $\left[\left(\dfrac{x}{y}\right)^{-2}\right]^{-4}$.

In Problems **57–62**, *perform the indicated operations and give your answer in scientific notation.*

57. $(1.3 \times 10^{-4})(2.0 \times 10^6)$.

58. $\dfrac{9.3 \times 10^{-1}}{3.1 \times 10^5}$.

59. $\dfrac{(3.0 \times 10^{11})(4.2 \times 10^{-4})}{2 \times 10^{13}}$.

60. $\dfrac{(4.8 \times 10^{-1})(5.0 \times 10^{-2})}{(3.2 \times 10^{-3})(3.0 \times 10^{-4})}$.

61. $\dfrac{(1.0 \times 10^4)^3}{2.5 \times 10^5}$.

62. $\dfrac{(1.2 \times 10^{-2})^2}{2.88 \times 10}$.

63. Given the formula $F = IBL$, find F if $I = 10^3$, $B = 5 \times 10^{-5}$, and $L = 4$. Give your answer in scientific notation.

64. Given the formula $F = \dfrac{a}{b}$, find F if $a = 3 \times 10^{10}$ and $b = 1.5 \times 10^8$. Give your answer in scientific notation.

2.3 RADICALS

Since $2^4 = 16$, we say that 2 is a *fourth root* of 1. More generally,

> r is an **nth root of a** means $r^n = a$.*

Thus 5 is a *cube* (or third) *root* of 125 because $5^3 = 125$.

Similarly, since $3^2 = 9$ and $(-3)^2 = 9$, both 3 and -3 are *square* (or second) *roots* of 9. If a number a has a *positive* nth root, we call that root the **principal nth root** of a. For example, 3 is the principal square root of 9. Sometimes the principal nth root is simply called **the** nth root. Thus

3 is *the* square root of 9.

When a number has only a negative nth *root, that root is the principal* nth *root.* Thus *the* (principal) cube root of -8 is -2 because $(-2)^3 = -8$ and there are no positive cube roots of -8. We define any root of 0 to be 0.

To refer to the principal nth root of a, we use the symbol $\sqrt[n]{a}$ (called a **radical**), where a is the **radicand** and n is the **index**.

PRINCIPAL nth ROOT OF a

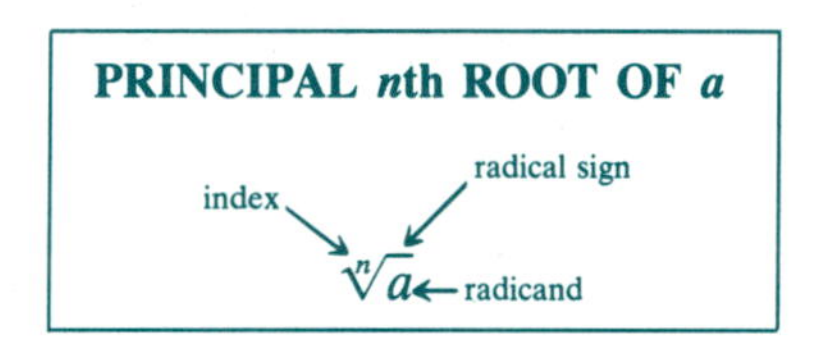

Thus $\sqrt[3]{-8} = -2$ and $\sqrt[2]{9} = 3$ (**not** -3). In $\sqrt[3]{-8}$, the index is 3 and the radicand is -8. For a principal square root we omit the index 2; thus $\sqrt{9} = 3$. In Appendix A there is a table of square roots and cube roots.

There are some troubles with even roots of negative numbers. For example, there is no real number whose square is -4. This means that the symbol $\sqrt{-4}$ does not represent a real number. For the present we shall avoid such roots.

EXAMPLE 1

Some principal roots.

a. $\sqrt{25} = 5$. *The* square root of 25 is 5, since $5^2 = 25$ and 5 is positive.

b. $\sqrt[4]{16} = 2$. *The* fourth root of 16 is 2, since $2^4 = 16$ and 2 is positive.

c. $\sqrt[3]{-1} = -1$, because $(-1)^3 = -1$ and no positive number has -1 for its cube.

d. $\sqrt[5]{0} = 0$, because any root of 0 is 0.

---

* n is a positive integer.

EXAMPLE 2

a. $\sqrt[3]{\frac{1}{125}} = \frac{1}{5}$, because $\left(\frac{1}{5}\right)^3 = \frac{1}{125}$.

b. $\sqrt{.01} = .1$, because $(.1)^2 = .01$.

c. $-\sqrt{81} = -(\sqrt{81}) = -(9) = -9$.

A root of a number may be irrational. For example, $\sqrt{2}$ is irrational. Although an approximate value of $\sqrt{2}$ is 1.414, we'll usually keep $\sqrt{2}$ in radical form rather than give a decimal form. Similarly, we'll keep $\sqrt[3]{5}$ and other irrational roots in radical form.

Since $\sqrt{9}$ is a number whose square is 9, we have $(\sqrt{9})^2 = 9$. Similarly,

$$\left(\sqrt{5}\right)^2 = 5 \quad \text{and} \quad \left(\sqrt[3]{4}\right)^3 = 4.$$

A general rule is:

Rule 13. $\left(\sqrt[n]{a}\right)^n = a.$

Moreover, since a to the nth power is a^n, a is an nth root of a^n. In fact, we have the following rule.*

Rule 14. $\sqrt[n]{a^n} = a.$

For example, $\sqrt[3]{6^3} = 6$.

The next two rules involve roots of products or quotients.

Rule 15. $\sqrt[n]{ab} = \sqrt[n]{a} \cdot \sqrt[n]{b}$.†

Rule 16. $\sqrt[n]{\frac{a}{b}} = \frac{\sqrt[n]{a}}{\sqrt[n]{b}}$.

---

* Here a must be nonnegative if n is even.

† To see why $\sqrt[n]{ab} = \sqrt[n]{a}\sqrt[n]{b}$, note that

$$\left(\sqrt[n]{a} \cdot \sqrt[n]{b}\right)^n = \left(\sqrt[n]{a}\right)^n\left(\sqrt[n]{b}\right)^n = ab.$$

Since $\left(\sqrt[n]{a} \cdot \sqrt[n]{b}\right)^n = ab$, then $\sqrt[n]{a} \cdot \sqrt[n]{b}$ is the nth root of ab. That is, $\sqrt[n]{ab} = \sqrt[n]{a} \cdot \sqrt[n]{b}$. In a similar way we could show that $\sqrt[n]{\frac{a}{b}} = \frac{\sqrt[n]{a}}{\sqrt[n]{b}}$.

These rules let us simplify many radicals by writing them as a product or quotient of radicals so that one of them is easy to compute.

EXAMPLE 3

Simplifying radicals.

a. $\sqrt{20}$.
We factor 20 so that one factor is the square of an integer. Then we use Rule 15.

$$\sqrt{20} = \sqrt{4 \cdot 5} = \sqrt{4} \cdot \sqrt{5} = 2\sqrt{5}.$$

Here we wrote $\sqrt{20}$ as a product of two radicals, one of which we could easily compute. Writing $\sqrt{20} = \sqrt{10} \cdot \sqrt{2}$ would not have helped us.

b. $\sqrt{18} = \sqrt{9 \cdot 2} = \sqrt{9} \cdot \sqrt{2} = 3\sqrt{2}$.

c. $\sqrt[3]{54} = \sqrt[3]{27 \cdot 2} = \sqrt[3]{27} \cdot \sqrt[3]{2} = 3\sqrt[3]{2}$.

d. $\sqrt{\frac{7}{4}} = \frac{\sqrt{7}}{\sqrt{4}} = \frac{\sqrt{7}}{2}$ by Rule 16. Here we wrote $\sqrt{\frac{7}{4}}$ as a quotient of two radicals, since we could compute $\sqrt{4}$ easily.

In part d of Example 3, we simplified a square root that had a fraction under the radical sign. It was simple to do because we can easily find the square root of the denominator, a perfect square. The radical $\sqrt{\frac{5}{3}}$ is harder to simplify because the denominator, 3, is not a perfect square. Example 4 shows how to handle this kind of situation.

EXAMPLE 4

Simplify $\sqrt{\frac{5}{3}}$.

By multiplying the numerator and denominator of $\frac{5}{3}$ by 3, we make the denominator a perfect square. Thus

$$\sqrt{\frac{5}{3}} = \sqrt{\frac{5 \cdot 3}{3 \cdot 3}} = \sqrt{\frac{15}{3^2}} = \frac{\sqrt{15}}{\sqrt{3^2}} = \frac{\sqrt{15}}{3}.$$

Sometimes a product or quotient of radicals can be written as a single radical and simplified, as Example 5 shows.

EXAMPLE 5

a. *Simplify* $\sqrt[3]{2}\sqrt[3]{4}$.

By Rule 15,

$$\sqrt[3]{2}\sqrt[3]{4} = \sqrt[3]{2 \cdot 4} = \sqrt[3]{8} = 2.$$

b. *Simplify* $\frac{\sqrt{20}}{\sqrt{5}}$.

By Rule 16,

$$\frac{\sqrt{20}}{\sqrt{5}} = \sqrt{\frac{20}{5}} = \sqrt{4} = 2.$$

EXAMPLE 6

Given the formula

$$d = \sqrt{(x_2 - x_1)^2 + (y_2 - y_1)^2},$$

find d if $x_1 = 1$, $x_2 = -7$, $y_1 = 2$, *and* $y_2 = 8$.

$$\begin{aligned} d &= \sqrt{(x_2 - x_1)^2 + (y_2 - y_1)^2} \\ &= \sqrt{(-7 - 1)^2 + (8 - 2)^2} \\ &= \sqrt{(-8)^2 + (6)^2} \\ &= \sqrt{64 + 36} = \sqrt{100}. \\ d &= 10. \end{aligned}$$

This formula, called the *distance formula*, is used in analytic geometry.

In Example 6, note that $\sqrt{64 + 36} = \sqrt{100} = 10$. ***Don't write*** $\sqrt{64 + 36}$ *as* $\sqrt{64} + \sqrt{36} = 8 + 6 = 14$. *In general,* $\sqrt{a + b}$ *is* ***not*** $\sqrt{a} + \sqrt{b}$. *Moral: Don't try fancy tricks of your own. Stick to basics.*

Exercise 2.3

In Problems **1–32**, *compute the numbers.*

1. $\sqrt{36}$.

2. $\sqrt{81}$.

3. $\sqrt[3]{8}$.

4. $\sqrt[3]{125}$.

5. $\sqrt{49}$.

6. $\sqrt{100}$.

7. $\sqrt[3]{-27}$.

8. $\sqrt[3]{-8}$.

9. $\sqrt[3]{-64}$.

10. $\sqrt[4]{1}$.

11. $\sqrt[4]{16}$.

12. $\sqrt[5]{-1}$.

13. $\sqrt[5]{0}$.

14. $\sqrt[3]{64}$.

15. $\sqrt[3]{-125}$.

16. $\sqrt[5]{32}$.

17. $\sqrt[6]{64}$.

18. $\sqrt{12 \cdot 3}$.

19. $\sqrt{.04}$.

20. $\sqrt{.25}$.

21. $\sqrt{\frac{1}{16}}$.

22. $\sqrt{\frac{1}{100}}$.

23. $-\sqrt{25}$.

24. $-\sqrt[3]{-1}$.

25. $\sqrt{81}-\sqrt[3]{-8}$.

26. $\sqrt{64}-\sqrt[3]{-8}$.

27. $\dfrac{\sqrt{64}+\sqrt[3]{-64}}{\sqrt{81}+\sqrt[4]{81}}$.

28. $\sqrt[3]{|-8|}$.

29. $\sqrt{5}\cdot\sqrt{5}$.

30. $\sqrt[3]{7}\cdot\sqrt[3]{7}\cdot\sqrt[3]{7}$.

31. $\left(\sqrt[4]{4}\right)^4+\sqrt[4]{4^4}$.

32. $\left(\sqrt{3}\right)^2+\sqrt[5]{2^5}$.

In Problems **33–59**, *simplify the numbers.*

33. $\sqrt{50}$.

34. $\sqrt{75}$.

35. $\sqrt{12}$.

36. $\sqrt{32}$.

37. $\sqrt{8}$.

38. $\sqrt{400{,}000}$.

39. $\sqrt{14{,}400}$.

40. $\sqrt[3]{24}$.

41. $\sqrt[4]{48}$.

42. $\sqrt[4]{162}$.

43. $\sqrt[5]{64}$.

44. $\sqrt[3]{-54}$.

45. $\sqrt[3]{-500}$.

46. $\sqrt{\frac{5}{4}}$.

47. $\sqrt{\frac{14}{9}}$.

48. $\sqrt{\frac{2}{25}}$.

49. $\sqrt{\frac{1}{15}}$.

50. $\sqrt{\frac{3}{2}}$.

51. $\sqrt[3]{\frac{10}{27}}$.

52. $\dfrac{\sqrt{90}}{\sqrt{10}}$.

53. $\dfrac{\sqrt{50}}{\sqrt{2}}$.

54. $\dfrac{\sqrt[4]{64}}{\sqrt[4]{4}}$.

55. $\dfrac{\sqrt[3]{-2}}{\sqrt[3]{16}}$.

56. $\dfrac{\sqrt[5]{-4}}{\sqrt[5]{-128}}$.

57. $\sqrt{8}\cdot\sqrt{2}$.

58. $\sqrt{27}\cdot\sqrt{3}$.

59. $\sqrt[3]{4}\cdot\sqrt[3]{16}$.

In Problems **60** *and* **61**, *use the distance formula* $d=\sqrt{(x_2-x_1)^2+(y_2-y_1)^2}$, *as given in Example* 6, *to find d with the given information.*

60. $x_1=10,\ y_1=2,\ x_2=7,\ y_2=-2$.

61. $x_1=-1,\ y_1=1,\ x_2=-6,\ y_2=13$.

In the right triangle shown in Fig. 2-1, *the length c of the hypotenuse is given by*

$$c=\sqrt{a^2+b^2}.$$

In Problems **62** *and* **63**, *find c from the given information.*

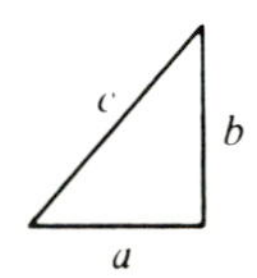

FIGURE 2-1

62. $a=5,\ b=12$.

63. $a=7,\ b=24$.

The expression $\dfrac{-b+\sqrt{b^2-4ac}}{2a}$ *is used in solving a certain type of equation. In Problems* **64–67**, *evaluate this expression for the given values of a, b, and c.*

64. $a = 4, b = 5, c = 0.$

65. $a = 1, b = 2, c = -15.$

66. $a = 1, b = -9, c = 14.$

67. $a = 4, b = -12, c = 9.$

68. In statistics, the standard deviation of the numbers x_1 and x_2 is given by

$$\sqrt{\frac{(\overline{x}-x_1)^2+(\overline{x}-x_2)^2}{2}},$$

where $\overline{x} = \dfrac{x_1+x_2}{2}$. Find the standard deviation of $x_1 = 0$ and $x_2 = 4$.

69. Compute y' (read "y prime") where

$$y' = \frac{\dfrac{x+5}{2\sqrt{x+1}} - \sqrt{x+1}}{(x+5)^2},$$

if $x = 3$.

2.4 REVIEW

IMPORTANT TERMS AND SYMBOLS

base *(2.1)*

exponent *(2.1)*

zero exponent *(2.2)*

negative exponent *(2.2)*

*n*th root *(2.3)*

principal root *(2.3)*

radical *(2.3)*

radicand *(2.3)*

index *(2.3)*

$\sqrt[n]{a}$ *(2.3)*

REVIEW PROBLEMS

In Problems **1–24**, *determine whether each statement is true or false.*

1. $a^3a^2 = a^6.$

2. $\left(\dfrac{1}{x}\right)^3 = \dfrac{1}{x^3}.$

3. $2^2 + 3^2 = 5^2.$

4. $(-a)^4 = a^4.$

5. $3 + 3 + 3 = 3^3.$

6. $(x^2y)^2 = x^4y.$

7. $(2x^2)^3 = 2^3x^5.$

8. $\dfrac{x^5}{x^3} = x^2.$

9. $(-y)^3 = -y^3.$

10. $2^2 \cdot 2^4 = 4^8.$

11. $3x = xxx.$

12. $(-3)^3 = -3^3.$

13. $(-2)^2 = -2^2$.
14. $(2+5)^2 = 2^2 + 5^2$.
15. $a^3a^7 = a^7a^3$.

16. $(x^2)^y = x^{2y}$.
17. $(2x)^2 = 2x^2$.
18. $\sqrt{16} = -4$.

19. $\sqrt{5}\sqrt{5} = \sqrt{25}$.
20. $\sqrt{4 \cdot 2} = 4\sqrt{2}$.
21. $\sqrt[3]{-\frac{1}{8}} = -\frac{1}{2}$.

22. $\sqrt[3]{0} = 0$.
23. $\sqrt{1} + \sqrt{1} = \sqrt{2}$.
24. $\left(\sqrt[3]{-1}\right)^2 = 1$.

In Problems **25–30**, *evaluate the expressions.*

25. 3^0.
26. $2\left(-\frac{2}{3}\right)^0$.
27. 5^{-1}.

28. $(-3)^{-1}$.
29. $4\left(-\frac{2}{3}\right)^{-2}$.
30. $\frac{2^{-1}}{4^{-2}}$.

In Problems **31–71**, *simplify. Use only positive exponents in your answers.*

31. $x^6x^4x^3$.
32. $\frac{x^2x^{20}}{y^5y^2}$.
33. $\frac{(x^2)^5}{(y^5)^{10}}$.

34. $(5x^2)(2x^3)$.
35. $(-2xy^4)^5$.
36. $\left(\frac{2xy^3}{z^2}\right)^2$.

37. $\frac{(x^3)^6}{x(x^3)}$.
38. $\frac{(xy^2)^2}{(t^2w)^3}$.
39. $(-x)^2(-x)^3$.

40. $\frac{(x^2)^3(x^4)^5}{(x^3)^8}$.
41. $\frac{(5^7)^2}{(5^4)^4}$.
42. $\frac{2^6 2^{11}}{(2^5)^3}$.

43. $-\sqrt{36} + \sqrt{81}$.
44. $\frac{-\sqrt{144}}{|-12|}$.
45. $|-2| - \sqrt[4]{16}$.

46. $\left(\sqrt[3]{2}\right)^3$.
47. $\sqrt{13} \cdot \sqrt{13}$.
48. $\frac{(-1)^2 - 3\sqrt{9}}{|-4| + \sqrt{16}}$.

49. $\frac{1}{3}\sqrt[3]{\frac{1}{27}}$.
50. $\frac{\sqrt[5]{-32}}{\sqrt{9}}$.
51. $\sqrt{.01} + \sqrt{.0025}$.

52. $\frac{\sqrt{4} - \sqrt[3]{8}}{\sqrt{6}}$.
53. $\sqrt{45}$.
54. $\sqrt{72}$.

55. $\sqrt[4]{810}$.
56. $\sqrt[3]{40}$.
57. $\sqrt{\frac{7}{5}}$.

58. $\sqrt{\frac{10}{9}}$.
59. $\frac{\sqrt[3]{24}}{\sqrt[3]{3}}$.
60. $\frac{\sqrt{72}}{\sqrt{2}}$.

61. $\sqrt{18} \cdot \sqrt{2}$.
62. $\sqrt[3]{8} \cdot \sqrt[3]{16}$.
63. $2x^{-1}x^{-3}$.

64. xy^{-1}.
65. $(4t^2)^{-2}$.
66. $(-2z)^{-3}$.

67. $(x^{-2}y^2)^3$.
68. $\left(\frac{x^{-2}}{w^2y^{-2}}\right)^{-4}$.
69. $\frac{x^{-2}y^{-6}z^2}{xy^{-1}}$.

70. $(2x^{-1}y)^{-2}$.

71. $\dfrac{x^3y^{-2}}{x^5z^2}$.

72. The area of the right triangle in Fig. 2-2 is given by

$$\frac{1}{2}a\sqrt{c^2 - a^2}.$$

If $a = 3$ and $c = 5$, find the area of the triangle.

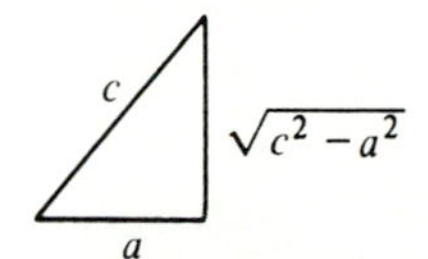

FIGURE 2-2

73. The area A of a circle of radius r is given by $A = \pi r^2$. By what factor does the area increase if the radius is (a) tripled, and (b) halved?

3 Operations with Algebraic Expressions

3.1 ADDITION AND SUBTRACTION OF EXPRESSIONS

An **algebraic expression** is either a number, including a literal number, or a combination of numbers obtained by one or more of the operations of addition, subtraction, multiplication, division, raising to powers, or taking of roots. For example, some algebraic expressions are 7, $2x^2$, $3 + 4\sqrt{y}$, and x/y.

In an algebraic expression that is a sum or difference, each part connected by a plus sign or minus sign, together with its sign, is called a **term** of the expression. For example, let's take a close look at the expression $2x^2 - 7y + 6$.

> Terms of $2x^2 - 7y + 6$ are
>
> $+2x^2, \quad -7y, \quad +6.$

In the *first* term $2x^2$, we say that the numerical factor 2 is the **coefficient*** of the term and the literal factor x^2 is the **literal part.** The coefficient of the *second* term $-7y$ is -7. The literal part is y. The *third* term 6 is called a **constant term.** It has a fixed value. We say that 6 is the coefficient of this term.

---

* Sometimes it is called the *numerical* coefficient.

EXAMPLE 1

a. The expression $t - t^2$ has two terms, t and $-t^2$. Since $t = 1 \cdot t$, the coefficient of t is 1 and the literal part is t. The coefficient of $-t^2$ (or $-1 \cdot t^2$) is -1 and the literal part is t^2.

b. $\frac{x}{10}$ consists of one term. Since $\frac{x}{10} = \frac{1}{10} \cdot x$, the coefficient is $\frac{1}{10}$ and the literal part is x.

c. $\frac{x+3}{x-4}$ has one term. However, both the numerator and denominator have two terms.

Just as

$$4 \text{ apples} + 5 \text{ apples} = 9 \text{ apples},$$

we have $4x + 5x = 9x$. This is a result of the distributive law because

$$4x + 5x = (4+5)x = 9x.$$

We say that $4x$ and $5x$ are **similar** (or *like*) **terms** because they have the *same literal part*, x. We can combine similar terms under addition (or subtraction), but we **cannot** combine nonsimilar terms such as those in the expression

$$2x^2 + 4x \qquad [\text{not similar terms, since } x^2 \neq x].$$

EXAMPLE 2

a. $2ab$ and $-4ab$ are similar terms because they have the same literal part, ab.

b. $3xy^2$ and $2x^2y$ are *not* similar terms because $xy^2 \neq x^2y$.

EXAMPLE 3

Combining similar terms by using the distributive law.

a. $2x^2 + 7x^2 + x^2 = 2x^2 + 7x^2 + 1 \cdot x^2 = (2 + 7 + 1)x^2 = 10x^2$.

b. $9a - 14a = (9 - 14)a = -5a$.

c. $-4a^2b + a^2b = -4a^2b + 1 \cdot a^2b = (-4 + 1)a^2b = -3a^2b$.

d. $2xy + 3yx = 2xy + 3xy = 5xy$; $2xy$ and $3yx$ are similar terms because $xy = yx$.

To simplify

$$3x^2 - 2x + 4 + 12x^2 - 9x - 3,$$

we "shuffle" the terms until similar terms are together.

$$\underbrace{3x^2 + 12x^2}_{\text{similar terms}} \underbrace{- 2x - 9x}_{\text{similar terms}} \underbrace{+ 4 - 3}_{\text{similar terms}}$$

Then we combine:

$$15x^2 - 11x + 1.$$

EXAMPLE 4

Simplify.

$$\begin{aligned} & a^2 - 2ab + b^2 - 3a^2 + 4b^2 \\ &= a^2 - 3a^2 - 2ab + b^2 + 4b^2 \quad \text{[getting similar terms together]} \\ &= -2a^2 - 2ab + 5b^2 \quad \text{[combining similar terms].} \end{aligned}$$

At times an expression contains grouping symbols, such as parentheses ( ), brackets [ ], or braces { }. These can be removed by the use of the distributive law.

EXAMPLE 5

Removing grouping symbols by using the distributive law.

a. $3a + 2(b - c) = 3a + 2b - 2c$. We multiplied *each* enclosed term by 2.

b. $-2(-3 + 4x)$.

We have to multiply *each* enclosed term by -2.

$$\begin{aligned} -2(-3 + 4x) &= (-2)(-3) + (-2)(4x) \\ &= 6 + (-8x) \\ &= 6 - 8x. \end{aligned}$$

c. $-8[x^2 - 2x + 1] = -8x^2 + 16x - 8$. We multiplied *each* enclosed term by -8.

*Be careful when you are using the distributive law. For example, $x + 3(y + 2) \neq x + 3y + 2$, because the 2 on the right side was not multiplied by 3. You must multiply **each** term inside the parentheses by 3. Thus $x + 3(y + 2) = x + 3y + 6$.*

We can find rules for removing grouping symbols preceded by plus or minus signs. Notice that

$$2 + (x - 2y) = 2 + (1)(x - 2y)$$
$$= 2 + x - 2y \qquad \text{[distributive law]},$$

but

$$2 - (x - 2y) = 2 + (-1)(x - 2y)$$
$$= 2 + (-1)(x) - (-1)(2y) \qquad \text{[distributive law]}$$
$$= 2 - x + 2y.$$

These examples illustrate the following rules.

Rule 1. If a plus sign is immediately in front of grouping symbols, then you can remove these symbols without changing the signs of the enclosed terms.

Rule 2. You can remove a minus sign immediately in front of grouping symbols by removing the grouping symbols **and** changing the sign of each of the enclosed terms.

EXAMPLE 6

Removing grouping symbols.

a. $5a + (3 - 4b + 3c) = 5a + 3 - 4b + 3c$ by Rule 1.

b. $(5x - 2y) + 3 = 5x - 2y + 3$. Here we assume that a plus sign is in front of the parentheses.

c. $-[-2 - 4a + 3b] = +2 + 4a - 3b = 2 + 4a - 3b$ by Rule 2.

d. $-(3a - 2b) = -3a + 2b$. Note that the first term inside the parentheses is $+3a$.

EXAMPLE 7

Simplify.

a. $(3x - y) - (-4y - 2x + 3)$

$$= 3x - y + 4y + 2x - 3 \qquad \text{[removing parentheses]}$$
$$= 5x + 3y - 3 \qquad \text{[combining similar terms]}.$$

b. $x^2 + 3[-6x + x^2] = x^2 - 18x + 3x^2$ [removing brackets]

$$= 4x^2 - 18x \qquad \text{[combining similar terms]}.$$

c. $3(x^2 - 5) - 4[-a^2 + 2x^2] + 5\{x^2 + 3\}$

$= 3x^2 - 15 + 4a^2 - 8x^2 + 5x^2 + 15$ [removing grouping symbols]

$= 4a^2$ [combining similar terms].

Note that we removed the brackets by multiplying each term within the brackets by -4.

When grouping symbols appear within other grouping symbols, as in

$$-2[x - (5x - 3)],$$

it's best to remove first the innermost symbols, which in this case are parentheses.

$-2[x - (5x - 3)] = -2[x - 5x + 3]$ [removing parentheses]

$= -2[-4x + 3]$ [combining within brackets]

$= 8x - 6$ [removing brackets].

EXAMPLE 8

Simplify $2x - 3[-2(x - 1) - x]$.

Here we want to remove all symbols of grouping and combine similar terms.

$2x - 3[-2(x - 1) - x]$

$= 2x - 3[-2x + 2 - x]$ [removing parentheses]

$= 2x - 3[-3x + 2]$ [combining within brackets]

$= 2x + 9x - 6$ [removing brackets]

$= 11x - 6$ [combining].

EXAMPLE 9

Simplify.

$-2\{2 - 3[-(x + 1)]\} - x$

$= -2\{2 - 3[-x - 1]\} - x$ [removing parentheses]

$= -2\{2 + 3x + 3\} - x$ [removing brackets]

$= -2\{3x + 5\} - x$ [combining within braces]

$= -6x - 10 - x$ [removing braces]

$= -7x - 10$ [combining].

Exercise 3.1

In Problems **1–36,** *simplify.*

1. $8x^2 + 3x^2 + 4x^2$.

2. $5t^3 + 4t^3 - t^3$.

3. $2y - 10y - 4y$.

4. $-5x^2 + 7x^2 - 2x^2$.

5. $5x - 8 + 2x + 3$.

6. $9x^2 - 3x - 4x^2 - 5x$.

7. $4(9x + 3y + x)$.

8. $3(a - 5b) - a$.

9. $-2(7x - 8y + 2y)$.

10. $5(a - 5b - a)$.

11. $4y + (4y - 8)$.

12. $2 - (6x + 7) - 2x^2$.

13. $3x + 5y - (8y - x)$.

14. $2(x^2 + 5) - 8x$.

15. $(3a - 2b) + (4a - 6b + c)$.

16. $(6x + 5) - (8x + 3)$.

17. $2(x^2 - 3) + 3(x^2 + 7)$.

18. $4(3 - a) - 2(a - 1)$.

19. $5(xy + z) - (3xy + 7)$.

20. $2[x + 3z] - 4(z + x) + 3\{x - 4z\}$.

21. $6(x^2 - 2x) + (3x - 5) - 4(x^2 + 2)$.

22. $2[(3x - 5) + 4]$.

23. $2[-(5 - x) + x]$.

24. $3[4x + (2 + 5x)]$.

25. $5[3x + 2(5 - x)]$.

26. $7[4 - (y + 8)]$.

27. $-\{4a - (6 + 3a)\}$.

28. $-2\{9(z^2 - 1)\}$.

29. $5x^2 + 3[8(x^2 - 1)] + 2$.

30. $[2a + 3(b - a)] + 7a$.

31. $\{9a - (3b + a + c)\} - 4\{2b + 3c\}$.

32. $4x^2 - [8x + 2(x + x^2)]$.

33. $3\{4x - 2[5 - (x + 1)]\}$.

34. $-\{2[3(2x + 5) + 6x]\}$.

35. $9 - 2\{8x - 3[4(y - 2x) - 6(x - y)]\}$.

36. $3\{5 - [xy - 2(xy + x)]\} - 4x$.

37. Subtract $6x - 4$ from $4x - 8$.

38. Subtract $2(1 - x)$ from $9x^2 - 3x$.

39. In a physical study of a steel beam, the following equation occurred:

$$M = 1970x - 1200(x - 3) - 1000(x - 9).$$

Find a simplified expression for M.

40. The voltage, in volts, measured across the terminals of a battery is found to be

$$12.0[1 - .0004(T - 20)],$$

where T is the surrounding temperature in degrees Celsius. (a) Simplify the expression. (b) Find the voltage if $T = 22°C$.

41. In a discussion of the motion of a disc, the expression

$$10[(10 - 100\theta) - (10 + 100\theta)]$$

occurs, where θ is the Greek letter theta. Simplify the expression.

3.2 MULTIPLICATION OF EXPRESSIONS

An expression with exactly one term, such as $3x^2$, is called a **monomial.** If an expression has more than one term, such as $x^2y + 5$, it is called a **multinomial.**

EXAMPLE 1

Monomials	*Multinomials*
$4xy$.	$7x - 3$.
-3.	$a + b - c$.
$\frac{10}{x^3}$.	$-2x^3 - 3x^2 + 4x - 5$.

To multiply the monomials $3x^2$ and $4xy$, we have

$$\begin{aligned}(3x^2)(4xy) &= 3 \cdot x^2 \cdot 4 \cdot x \cdot y \\ &= 3 \cdot 4 \cdot x^2 \cdot x \cdot y && \text{[rearranging factors by commutative law]} \\ &= 12x^3y && \text{[rule of exponents].}\end{aligned}$$

In short, *to find the product of monomials, multiply their coefficients and multiply their literal parts.*

EXAMPLE 2

Multiplying monomials.

a. $(2ab^2)(-4a^2b^2) = (2)(-4)(a \cdot a^2)(b^2 \cdot b^2) = -8a^3b^4$.

b. $(2x)(3xy)(-2y^2) = (2)(3)(-2)(x \cdot x)(y \cdot y^2) = -12x^2y^3$. Another way to find the product is to begin by multiplying the first two factors.

$$(2x)(3xy)(-2y^2) = (6x^2y)(-2y^2) = -12x^2y^3.$$

c. $-4x^{-2}y^3(-x^4y^{-4})$. The coefficient of $-x^4y^{-4}$ is -1. Thus

$$-4x^{-2}y^3(-x^4y^{-4}) = (-4)(-1)(x^{-2}x^4)(y^3y^{-4}) = 4x^2y^{-1} = \frac{4x^2}{y}.$$

Note that we wrote our answer with positive exponents only. We shall follow this practice.

EXAMPLE 3

a. *Find* $(3a^3bc)(ab)^2$.

Here we multiply $3a^3bc$ **not** by ab, but by $(ab)^2$, which is a^2b^2. The squaring *must* be done first.

$$(3a^3bc)(ab)^2 = (3a^3bc)(a^2b^2) = 3a^5b^3c.$$

b. *Find* $(-2x^2y)^3(-xy^2)^2$.

$$(-2x^2y)^3(-xy^2)^2 = (-8x^6y^3)(x^2y^4) = -8x^8y^7.$$

In the product $(2x)(x + 3)$ the first factor, $2x$, is a monomial and the second factor, $x + 3$, is a multinomial. To find the product, we use the distributive law $a(b + c) = ab + ac$ by *multiplying each term in the multinomial by the monomial*. Then we simplify.

$$\underbrace{(2x)}_{\substack{\uparrow \\ a}}(\underset{\substack{\uparrow \\ b}}{x} + \underset{\substack{\uparrow \\ c}}{3}) = \underbrace{(2x)}_{\substack{\uparrow \\ a}}\underset{\substack{\uparrow \\ b}}{(x)} + \underbrace{(2x)}_{\substack{\uparrow \\ a}}\underset{\substack{\uparrow \\ c}}{(3)} = 2x^2 + 6x.$$

EXAMPLE 4

Multiplying monomials and multinomials.

a. $4ab(a^2 - 2b + 1) = (4ab)(a^2) - (4ab)(2b) + (4ab)(1)$

$$= 4a^3b - 8ab^2 + 4ab.$$

b. $-3x(x - 4) = (-3x)(x) - (-3x)(4)$

$$= -3x^2 - (-12x) = -3x^2 + 12x.$$

c. $(x^2y)^2(x - y + 2) = (x^4y^2)(x - y + 2)$

$$= (x^4y^2)(x) - (x^4y^2)(y) + (x^4y^2)(2)$$

$$= x^5y^2 - x^4y^3 + 2x^4y^2.$$

d. $5(3x^2 + 3) - 2x(x - 4) = 15x^2 + 15 - 2x^2 + 8x$

$$= 13x^2 + 8x + 15 \qquad \text{[combining similar terms].}$$

e. $2a[x(x + 1) - 7] = 2a[x^2 + x - 7] = 2ax^2 + 2ax - 14a.$

Let's now find the product of two multinomials, such as $(x+2)(x+3)$. We use the distributive law $(a+b)c = ac + bc$, where $x+2$ matches $a+b$ and $x+3$ plays the role of c.

$$(x+2)\underbrace{(x+3)} = x\underbrace{(x+3)} + 2\underbrace{(x+3)} = x^2 + 3x + 2x + 6 = x^2 + 5x + 6.$$

$$\begin{matrix} \uparrow \quad \uparrow & \uparrow & \uparrow & \uparrow & \uparrow & \uparrow \\ (a+b) & c & = a & c & +b & c \end{matrix}$$

That is, *to find the product of two multinomials, multiply each term in the first factor by the entire second factor* and then simplify.

EXAMPLE 5

Multiplying multinomials.

a. $(x^2-2)(x-3) = x^2(x-3) - 2(x-3) = x^3 - 3x^2 - 2x + 6.$

b. $x - (x+1)(x-2).$

The first minus sign applies to the *product* of the multinomials and must be kept until *after* that product is found.

$$\begin{aligned} x - (x+1)(x-2) &= x - [x(x-2) + 1(x-2)] = x - [x^2 - 2x + x - 2] \\ &= x - [x^2 - x - 2] = x - x^2 + x + 2 \\ &= -x^2 + 2x + 2. \end{aligned}$$

c. $$\begin{aligned} 2x(x+1)(x-1) &= [2x(x+1)](x-1) = [2x^2 + 2x](x-1) \\ &= 2x^2(x-1) + 2x(x-1) = 2x^3 - 2x^2 + 2x^2 - 2x \\ &= 2x^3 - 2x. \end{aligned}$$

d. $$\begin{aligned} &(x^2 + x - 1)(x^2 - 2x + 1) \\ &= x^2(x^2 - 2x + 1) + x(x^2 - 2x + 1) - 1(x^2 - 2x + 1) \\ &= x^4 - 2x^3 + x^2 + x^3 - 2x^2 + x - x^2 + 2x - 1 \\ &= x^4 - x^3 - 2x^2 + 3x - 1. \end{aligned}$$

Do not confuse $x^2 - 2(x-3)$ with $(x^2-2)(x-3)$. The first expression is the difference of x^2 and $2(x-3)$, while the second expression is a product. That is,

$$x^2 - 2(x-3) = x^2 - 2x + 6,$$

$$\textit{but} \qquad (x^2-2)(x-3) = x^3 - 3x^2 - 2x + 6,$$

as we showed in Example 5a. Similarly,

$$(x^2-2)x - 3 \neq (x^2-2)(x-3).$$

Sometimes it is convenient to use a vertical arrangement to do a multiplication problem. For example, let's redo Example 5d. As shown below, we multiply the first term (x^2) in the second row by *each* term in the first row. This gives $x^4 - 2x^3 + x^2$. This is repeated with the second term and then with the third term in the second row. Then we add the results.

$$
\begin{array}{rrrrr}
 & & x^2 & -2x & +1 \\
 & & x^2 & +x & -1 \\
\hline
x^4 & -2x^3 & +x^2 & & \\
 & x^3 & -2x^2 & +x & \\
 & & -x^2 & +2x & -1 \\
\hline
x^4 & -x^3 & -2x^2 & +3x & -1
\end{array}
$$

(The three middle rows are marked "add".)

Notice that to simplify the addition, we placed similar terms in the same column.

Exercise 3.2

In Problems **1–62,** *perform the indicated operations.*

1. $(2x)(3xy)$.

2. $-a(ac)$.

3. $3ab(a^2b)$.

4. $xy(xz)(xw)$.

5. $2xy^2(-4x^3y^2)$.

6. $-a^2b(-ac)$.

7. $x^{-2}(3x^3y)$.

8. $(xy^5)y^{-3}$.

9. $(3x^{-4}y^5)(2xy^{-1})$.

10. $(3x^{-7}y^{-3})(x^2y^2)(4x)$.

11. $ab(a^2b)(bc^2)$.

12. $x(xy^2)(y^2z)$.

13. $2x^2yz^2(\frac{1}{2}xz^2)$.

14. $(2xy^2)^3(2y)^2$.

15. $(-3x)(4xy^2)(-2x^2y^3)$.

16. $x(x^3y^2)^2(-2xy^2)^3$.

17. $(10x^4)(3x^3)^2$.

18. $(3y^3)^2(-y)^3$.

19. $(ab^2c^3)^{-2}(a^2b^3)^5$.

20. $(x^{-2}y^{-3})(x^2y)^3$.

21. $a(-bc)^2(cd)^2$.

22. $(-xy)^3(-x)^2$.

23. $x(x^2-4x+7)$.

24. $-2(x-2y^2+7xy)$.

25. $a^2b(-3+ab-a)$.

26. $2x(-x+2y+1)$.

27. $x^{-2}(x^3+4x^4)$.

28. $x^5(1-x^{-3})$.

29. $-5xy(x^2-y^2+xy)$.

30. $a(b-a)(ac)$.

31. $(2x^2y)^2(x+2y^2-3x^2)$.

32. $(ab)^2(a^2-b^2c+ac^2)$.

33. $(x+2)(x+5)$.

34. $(x+4)(x+5)$.

35. $(y-2)(3y+2)$.

36. $(y-2)(y+2)$.

37. $(3x-1)(3x-1)$.

38. $(2x+3)(3x+2)$.

39. $(4x+y)^2$.

40. $(x^2+4)(x^4+4)$.

41. $(3x^2)(2xy)(x+y)$.

42. $(x+3)(5x^2)$.

43. $(t-2)(t^2+2t+4)$.

44. $(x+5)(x^2-x-1)$.

45. $(x^2-2)(x^2-5x+1)$.

46. $(2+x-y)(x-y)$.

47. $(x+y+1)(x+y-1)$.

48. $(2x-1)(2x^3-3x+1)$.

49. $4x(2x-1)(2x+1)$.

50. $-3x^2(x-2)(x-1)$.

51. $xy+y(y+x)$.

52. $(x-3)x-4$.

53. $(x^2+1)(2x)-(x^2-2)(2x)$.

54. $x^2y-2x-x^2y(1+x)$.

55. $x(x-1)-2(3-x)$.

56. $2xy(x^2y)-y^2(2x^3-2xy)$.

57. $3x(x^2y-xy^2)-3y(x^3+4x^2y)$.

58. $2(x^3y-2x^2)-2x^2(x^2-2xy)$.

59. $3(x-x^2)+(x+1)(x-1)$.

60. $(x+1)(x^2-x+1)-1-x^3$.

61. $x^3+y^3-(x+y)(x^2-xy+y^2)$.

62. $(-x-2)(x+2))-(-1+x)x$.

63. A muscle shortens when a load, such as a weight, is imposed upon it. The equation

$$(P+a)(v+b)=k$$

is called the *fundamental equation of muscle contraction*. Here P is the load, v is the velocity of the shortening of the muscle fibers, and a, b, and k are constants. Find the product of the expressions on the left side of the equation.

64. Given the formula $m=ab$, find m if $a=2t-1$ and $b=t+4$.

3.3 DIVISION

We now turn to division. To divide a monomial by a monomial, we use ordinary arithmetic and the rules of exponents. For example,

$$\frac{2x^3y^6}{8x^2y^8}=\frac{2}{8}\cdot\frac{x^3}{x^2}\cdot\frac{y^6}{y^8}=\frac{1}{4}\cdot x\cdot\frac{1}{y^2}=\frac{x}{4y^2}.$$

You may even think of this as cancelling common factors:

$$\frac{2x^3y^6}{8x^2y^8}=\frac{\overset{}{\cancel{2}}\overset{x}{\cancel{x^3}}\cancel{y^6}}{\underset{4}{\cancel{8}}\cancel{x^2}\underset{y^2}{\cancel{y^8}}}=\frac{x}{4y^2}.$$

In fact, after some practice you may be able to do some steps in your head. Try this one.

$$\frac{-42x^3y^4}{7xy^5} = \frac{-6x^2}{y} = -\frac{6x^2}{y}.$$

EXAMPLE 1

Dividing monomials by monomials.

a. $\dfrac{(3a^2b^3)(2a^2c^4)}{-2a^3b^7c^4d} = \dfrac{6a^4b^3c^4}{-2a^3b^7c^4d}$ [simplifying numerator]

$$= \frac{6}{-2}\cdot\frac{a^4}{a^3}\cdot\frac{b^3}{b^7}\cdot\frac{c^4}{c^4}\cdot\frac{1}{d}$$

$$= -3\cdot a\cdot\frac{1}{b^4}\cdot 1\cdot\frac{1}{d} = -\frac{3a}{b^4d}.$$

b. $\left(\dfrac{x^2y^3}{2xy^2}\right)^3 = \left(\dfrac{xy}{2}\right)^3 = \dfrac{x^3y^3}{8}$. Here we chose to simplify *before* cubing.

c. $\dfrac{(2a^2x)^3(-2a^4y)}{(2xy^2)^2} = \dfrac{(8a^6x^3)(-2a^4y)}{4x^2y^4} = \dfrac{-16a^{10}x^3y}{4x^2y^4} = -\dfrac{4a^{10}x}{y^3}.$

$\dfrac{(ax)^2}{ay} \neq \dfrac{(\not{a}x)^2}{\not{a}y}$. *You must square ax before cancelling. Thus*

$$\frac{(ax)^2}{ay} = \frac{a^2x^2}{ay} = \frac{ax^2}{y}.$$

When the numerator of a fraction has more than one term, we can break up the fraction into simpler fractions, as follows.

$$\boxed{\frac{a+b+c}{d} = \frac{a}{d} + \frac{b}{d} + \frac{c}{d}.}$$

That is, *to divide a multinomial by a monomial, divide **each** term of the multinomial by the monomial.* Thus

$$\frac{8x^6 + x^4}{2x} = \frac{8x^6}{2x} + \frac{x^4}{2x} = 4x^5 + \frac{x^3}{2}.$$

A monomial divided by a multinomial ***cannot****, as a rule, be broken into simpler fractions. That is,*

$$\frac{a}{b+c} \neq \frac{a}{b} + \frac{a}{c}.$$

EXAMPLE 2

Dividing multinomials by monomials.

a. $\dfrac{4x^2+2x+3}{2}=\dfrac{4x^2}{2}+\dfrac{2x}{2}+\dfrac{3}{2}=2x^2+x+\dfrac{3}{2}.$

b. $\dfrac{2(x+y)}{xy}$. We first multiply 2 by $x+y$ to obtain a multinomial in the numerator.

$$\frac{2(x+y)}{xy}=\frac{2x+2y}{xy}=\frac{2x}{xy}+\frac{2y}{xy}=\frac{2}{y}+\frac{2}{x}.$$

Note that $\dfrac{a+ax}{a} \neq \dfrac{\not{a}+ax}{\not{a}}$, *but* $\dfrac{a+ax}{a}=\dfrac{a}{a}+\dfrac{ax}{a}=1+x$. *You may cancel only common **factors**, not terms.*

If some terms in the numerator have minus signs, we use the same procedure as in Example 2. Thus

$$\frac{a-b-c}{d}=\frac{a}{d}-\frac{b}{d}-\frac{c}{d}.$$

EXAMPLE 3

a. $$\frac{12a^2b^3+4ab-6}{4ab}=\frac{12a^2b^3}{4ab}+\frac{4ab}{4ab}-\frac{6}{4ab}$$
$$=3ab^2+1-\frac{3}{2ab}.$$

b. $$\frac{5x^2y-x^3z}{-4xyz^2}=\frac{5x^2y}{-4xyz^2}-\frac{x^3z}{-4xyz^2}$$
$$=-\frac{5x}{4z^2}-\left(-\frac{x^2}{4yz}\right)=-\frac{5x}{4z^2}+\frac{x^2}{4yz}.$$

c. $$\frac{3x-(2x^2y)^2-7x^2(x^2y^2)}{3x(2y^2)}=\frac{3x-4x^4y^2-7x^4y^2}{6xy^2}$$
$$=\frac{3x-11x^4y^2}{6xy^2}=\frac{3x}{6xy^2}-\frac{11x^4y^2}{6xy^2}$$
$$=\frac{1}{2y^2}-\frac{11x^3}{6}.$$

EXAMPLE 4

$\dfrac{4+y}{y^{-6}} = \dfrac{4}{y^{-6}} + \dfrac{y}{y^{-6}} = 4y^6 + y^7$. Another way to handle this problem is

$$\frac{4+y}{y^{-6}} = (4+y)y^6 = 4y^6 + y^7.$$

Don't write $\dfrac{4+y}{y^{-6}} = 4 + y(y^6) = 4 + y^7$.

Exercise 3.3

Do the divisions. Give all answers with positive exponents only.

1. $\dfrac{2ab}{4a}$.
2. $\dfrac{-3x^2y}{xy}$.
3. $\dfrac{-14ab^2}{7a^2}$.
4. $\dfrac{35x^3y^2}{-7xy}$.
5. $\dfrac{6abc}{-6ab}$.
6. $\dfrac{-ax^2y^5}{-x^3y^3}$.
7. $\dfrac{-16x^2yz}{-32xy^2z^3}$.
8. $\dfrac{-25ab^3c^2}{5ab^2}$.
9. $\dfrac{-a^2b(abc^2)}{a^2b^4c}$.
10. $\dfrac{(abc^2)^2(ab)}{2(ab)^2}$.
11. $\dfrac{(3xy)^2}{y}$.
12. $\dfrac{2xy^2}{(2xy)^2}$.
13. $\left(\dfrac{8x^2y}{24xy^2}\right)^2$.
14. $\left(\dfrac{3xy}{6y}\right)^2$.
15. $\dfrac{-(-2xy)^2}{14x^2y}$.
16. $\dfrac{(-3x^2y)^3}{(-2x)^2(9x^3y^2)^2}$.
17. $\dfrac{4x^2-6x+8}{2}$.
18. $\dfrac{3x^2-8y+xy}{xy}$.
19. $\dfrac{5-x^2+2x}{x}$.
20. $\dfrac{9x^2-15}{-3x}$.
21. $\dfrac{10x^3-15x+2}{5x}$.
22. $\dfrac{3xy-2y}{6}$.
23. $\dfrac{4x+2y}{-2x}$.
24. $\dfrac{2x^2-y^2x}{x^2}$.
25. $\dfrac{xy+x^2}{xy}$.
26. $\dfrac{x-xy}{x}$.
27. $\dfrac{3x^2y-x^3y^2+1}{x^2y^2}$.
28. $\dfrac{2+2y}{2y}$.
29. $\dfrac{6x^2y-2y^2+7x-4}{-3x}$.
30. $\dfrac{2x^2y^2-6xz^2+4x}{2xy^2z}$.
31. $\dfrac{-20x^2y^2+5xy^2-2x}{2xy}$.
32. $\dfrac{-6a^4b^2c^3+3a^5b^2c-12a^6bc^2}{-6a^4b^2c^2}$.
33. $\dfrac{2xy-(2xy^2)^2-3x^3y^3}{(xy)^2}$.
34. $\dfrac{x(xy)+y(-x^2)+x^3y^3}{-x(xy)}$.

35. $\dfrac{2x(x^3y^2)^2 + 3x^2(2y^2) - x}{-xy^4}$.

36. $\dfrac{x(x-y) - y(2x)}{xy}$.

37. $\dfrac{x(x^2 - 2x + 1)}{x^2}$.

38. $\dfrac{(x-2)^2}{x}$.

39. $\dfrac{x^{-2}(yz)^2 w}{x^{-3}y^5}$.

40. $\dfrac{x^6}{(x^{-4}y^8)y^{-8}}$.

41. $\dfrac{2(2x^2y)^2}{3y^{-13}z^{-2}}$.

42. $\dfrac{x^2y^{-5}}{(x^{-8}y^6)^{-3}}$.

43. $\left(\dfrac{x^{-3}y^{-6}z^2}{2xy^{-1}}\right)^{-2}$.

44. $\left[\dfrac{x^{-1}y^4}{(z^2)^{-2}}\right]^{-5}$.

3.4 DIVISION OF POLYNOMIALS

In this section we'll divide special types of expressions called *polynomials*. The expression $3x^2 - 8x + 9$ is a polynomial in x. In general, a **polynomial** in x is an expression in which *each term is either a constant or of the form* ax^n; here a is a constant (a fixed number) and n is a positive integer. In $3x^2 - 8x + 9$, the term with the greatest power of x is $3x^2$. We say that the exponent 2 in this term is the **degree** of the polynomial. The coefficient 3 of this term is the **leading coefficient** of the polynomial.

$3x^2 - 8x + 9$ is a polynomial of degree 2
and has leading coefficient 3.

Constants, such as 8 and 0, are also polynomials and are called **constant polynomials.** The degree of a nonzero constant, such as 8, is zero. The constant 0 has no degree assigned to it.

EXAMPLE 1

a. $y^6 + 7y^4 - 4y - 5$ is a polynomial in y.

degree: 6. *leading coefficient*: 1 [since $y^6 = 1 \cdot y^6$].

b. $3x^3 - 7x^8$ is a polynomial in x. The term with the greatest power of x is $-7x^8$, so $3x^3 - 7x^8$ has

degree: 8. *leading coefficient*: -7.

c. $\dfrac{3}{x^2}$ is *not* a polynomial, because it is not a constant or of the form ax^n where n is a positive integer.

d. $6x$ is a polynomial in x.

degree: 1. *leading coefficient*: 6.

In Example 1, the polynomials in parts a and b are also multinomials. In parts c and d, both expressions are monomials, but only d is a polynomial.

To divide polynomials we use so-called long division. Before we show you how, let's look at long division with numbers. In this way we can recall some words used in describing division.

Let's find $\frac{25}{6}$ (25 divided by 6):

$$\begin{array}{r} 4 \\ 6\overline{)25} \\ \underline{24} \\ 1 \end{array}$$

6 is the *divisor*;
25 is the *dividend*;
4 is the *quotient*;
1 is the *remainder.*

We usually write the answer as $4\frac{1}{6}$ or $4 + \frac{1}{6}$. Thus

$$\frac{25}{6} = 4 + \frac{1}{6} = \text{quotient} + \frac{\text{remainder}}{\text{divisor}}.$$

You can check a division by making sure that

$$(\text{quotient})(\text{divisor}) + \text{remainder} = \text{dividend}.$$

$$(4)(6) + 1 = 25,$$

$$25 = 25. \quad \checkmark$$

Now we're ready for division of polynomials, such as

$$\frac{6x^2 - 13x + 10}{3x - 2}.$$

We'll go through it step by step.

$$\begin{array}{r} 2x \\ 3x - 2\overline{)6x^2 - 13x + 10} \end{array}$$

divide $6x^2$
by $3x$

Divide $6x^2$ by $3x$ and get $2x$.

multiply

$$\begin{array}{r} 2x \\ 3x - 2\overline{)6x^2 - 13x + 10} \\ \ominus \quad \oplus \\ \underline{+6x^2 - 4x} \\ -9x + 10 \end{array}$$

Multiply $2x$ by $3x - 2$ and get $6x^2 - 4x$. Then *subtract* this from $6x^2 - 13x$ by changing signs and proceeding as in addition. The circles show the sign changes. We get $-9x$. Then we bring down the $+10$.

$$\begin{array}{r} 2x - 3 \\ 3x - 2\,\overline{)\,6x^2 - 13x + 10} \\ \ominus \quad \oplus \qquad \\ +6x^2 - 4x \qquad \\ \hline -9x + 10 \end{array}$$

divide $-9x$ by $3x$

Repeat the process again. Divide $-9x$ by $3x$ and get -3.

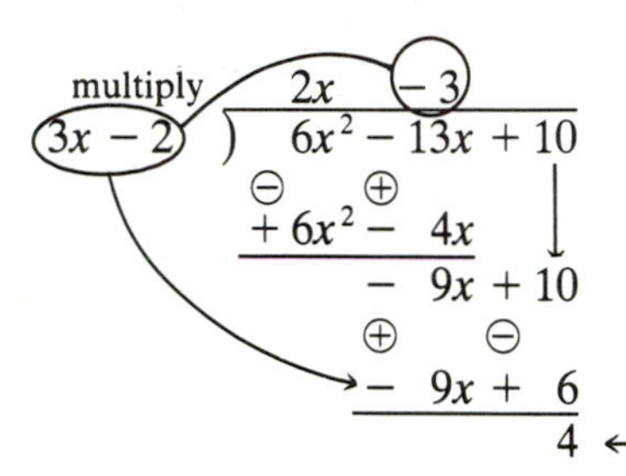

Multiply -3 by $3x - 2$ and get $-9x + 6$. Then *subtract* this from $-9x + 10$ and get 4, the remainder.

Thus

$$\frac{6x^2 - 13x + 10}{3x - 2} = \text{quotient} + \frac{\text{remainder}}{\text{divisor}} = 2x - 3 + \frac{4}{3x - 2}.$$

Just as before, we can check our division by making sure that

$$(\text{quotient})(\text{divisor}) + \text{remainder} = \text{dividend}.$$

$$\begin{aligned}(2x - 3)(3x - 2) + 4 &= 2x(3x - 2) - 3(3x - 2) + 4 \\ &= 6x^2 - 13x + 10. \quad \checkmark\end{aligned}$$

Notice that the divisor $3x - 2$ has degree one and the remainder 4 has degree zero. In general, we stop the long-division process when the remainder is zero or has degree less than the degree of the divisor.

EXAMPLE 2

Do the division $\dfrac{2 + 9x - 5x^2 - 2x^3}{x + 3}$.

When you write the divisor and dividend in long division, *make sure that the powers of x decrease from left to right and that the constant term is to the right.* So, write $2 + 9x - 5x^2 - 2x^3$ as $-2x^3 - 5x^2 + 9x + 2$.

$$\begin{array}{r} -2x^2 + x + 6 \\ x + 3\,\overline{)\,-2x^3 - 5x^2 + 9x + 2} \\ \oplus \quad \oplus \qquad\qquad \\ -2x^3 - 6x^2 \qquad\qquad \\ \hline x^2 + 9x \qquad \\ \ominus \quad \ominus \qquad \\ +\ x^2 + 3x \qquad \\ \hline 6x + 2 \\ \ominus \quad \ominus \\ +6x + 18 \\ \hline -16 \end{array}$$

Thus

$$\frac{2+9x-5x^2-2x^3}{x+3} = -2x^2+x+6-\frac{16}{x+3}.$$

EXAMPLE 3

Divide $2y^4 - 5y^3 - y - 4$ *by* $2y - 1$.

In the dividend $2y^4 - 5y^3 - y - 4$, the powers of y decrease from left to right and the constant term is to the right. But there is a term "missing": the y^2-term. When powers of y or the constant term are missing, we fill in with zeros. Thus we write the dividend as $2y^4 - 5y^3 + 0y^2 - y - 4$. This is done to make the long division process more orderly.

$$\begin{array}{r|l}
 & y^3 - 2y^2 - y - 1 \\
\hline
2y - 1 & 2y^4 - 5y^3 + 0y^2 - y - 4 \\
 & \underline{\overset{\ominus}{+2y^4} \overset{\oplus}{-} y^3} \\
 & \quad -4y^3 + 0y^2 \\
 & \quad \underline{\overset{\oplus}{-}4y^3 \overset{\ominus}{+} 2y^2} \\
 & \qquad -2y^2 - y \\
 & \qquad \underline{\overset{\oplus}{-}2y^2 \overset{\ominus}{+} y} \\
 & \qquad\quad -2y - 4 \\
 & \qquad\quad \underline{\overset{\oplus}{-}2y \overset{\ominus}{+} 1} \\
 & \qquad\qquad -5
\end{array}$$

Thus

$$\frac{2y^4-5y^3-y-4}{2y-1} = y^3-2y^2-y-1-\frac{5}{2y-1}.$$

Exercise 3.4

In Problems **1–10**, *determine if the given expression is a polynomial in* x. *If it is, find the* (*a*) *degree and* (*b*) *leading coefficient.*

1. $-4x + 2$.

2. $\frac{x^2}{2}$.

3. $7x^2 + 6x - 2$.

4. $3x^3 - 7x$.

5. $x + 2x^{-2}$.

6. $-x^2 + 1$.

7. $\frac{18 - 2x^4}{3}$.

8. $x + \frac{1}{x}$.

9. $3x^3 - 2x^5 + x^2$.

10. $x^6 - x^5 + x^4 + 1$.

In Problems **11–20,** *do the divisions and check your answers.*

11. $\frac{2x^2 + 3x - 4}{x - 2}$.

12. $\frac{9x^2 - 6x - 6}{3x - 1}$.

13. $\frac{x + 3}{x + 2}$.

14. $\frac{x}{x + 1}$.

15. $\frac{4x^2 - 7x - 5}{4x + 1}$.

16. $\frac{5x^2 + 26x + 8}{x + 5}$.

17. $\frac{4x^2 - 5}{2x + 3}$.

18. $\frac{3 + 6x - 10x^2}{5x - 3}$.

19. $\frac{x^3 + 2x^2 - 5x + 2}{x - 1}$.

20. $\frac{3x^3 + 2x^2 + 3x + 1}{3x + 2}$.

In Problems **21–30,** *do the divisions.*

21. $\frac{1 + 2x - x^2 - 2x^3 - 3x^4}{3x + 5}$.

22. $\frac{8x^4 - 4x^2 + x}{2x + 1}$.

23. $\frac{8x^3 - 2x^2 + 4x - 3}{4x - 1}$.

24. $\frac{3x^3 + x^2 - 6x + 1}{3x + 1}$.

25. $\frac{x^4 - 2x^2 + 1}{x - 1}$.

26. $\frac{x^4 - 8x^2 + 16}{x + 2}$.

27. $\frac{81 - x^4}{x + 3}$.

28. $\frac{x^3}{x - 4}$.

29. $\frac{x^3 + x^2 + x + 3}{x^2 - x + 1}$.

30. $\frac{x^3 + 3x^2 - 2x - 12}{x^2 - 4}$.

31. In the division of polynomials, if the remainder is zero, then the divisor is a factor of the dividend. Explain why this is true.

3.5 REVIEW

IMPORTANT TERMS AND SYMBOLS

algebraic expression (*3.1*)
terms of an expression (*3.1*)
coefficient of a term (*3.1*)
literal part of a term (*3.1*)
similar terms (*3.1*)
grouping symbols (*3.1*)
parentheses, ( ) (*3.1*)
brackets, [ ] (*3.1*)
braces, { } (*3.1*)
monomial (*3.2*)
multinomial (*3.2*)
polynomial (*3.4*)
degree of polynomial (*3.4*)
leading coefficient (*3.4*)

REVIEW PROBLEMS

In Problems **1–52,** *perform the operations and simplify.*

1. $(3x + 2y - 5) + (8x - 4y + 2)$.

2. $7x + 5(4x + 3)$.

3. $6(a + 3b) - (8a - b - 4)$.

4. $2(a + 4b) - 3(3b - 2a)$.

5. $2[3(xy - 5) + 7(4 - xy)]$.

6. $(4xy + 7) - 5[xy - (4 - 3xy)]$.

7. $4x + [-5(2 - x) - 8]$.

8. $\{1 - 2[x - (x - 1)]\} + 1$.

9. $-3\{x^2 - [3(2x - 4) + 2x^2]\}$.

10. $2\{x^2 + 3[x - (x^2 + 4)]\} + 7$.

11. $(2x^2yz)(xy^3z^6)$.

12. $3xy(-2xy^2)$.

13. $8ab^2(3a^2b)^2$.

14. $-xy^3(-3xz)$.

15. $(2x^2y)^2(xy)^3(xy^2)$.

16. $(xy)^2(xz)^2(yz)^2$.

17. $2x^{10}y^{-3}x^{-5}$.

18. $(4x^{-2})^{-2}(-2xy^{-1})^2$.

19. $(2x^{-3}y^2)(x^4y^{-1})$.

20. $x^{-3}(2x^4y^{-2})$.

21. $x(x^2 - 2x + 4)$.

22. $x^2y(xy - xz + yz)$.

23. $a^2b(-2a^2b + 2ab - 3)$.

24. $(2xy)^2(x - y + 2xy)$.

25. $(x + 3)(x - 4)$.

26. $(y - 4)(y^2 - 3y + 5)$.

27. $(x + 2)(x - 2)$.

28. $(3 - x)(3 + x)$.

29. $(x - 3)(x - 2)$.

30. $(x - 3)^2$.

31. $(x + 2y)^2$.

32. $(x^2 + 1)(x^2 - 2)$.

33. $(x^2 + 3)(x - 4)$.

34. $(x - 3y)^2$.

35. $(3x - 1)(2x^3 - 3x^2 + 5)$.

36. $(y - 4)(y^2 - 3y + 5)$.

37. $(x + y + 2)(2x - 3y + 1)$.

38. $(1 + x + y)(1 - x - y)$.

39. $3x(x - 4) - 2(x^2 - 9)$.

40. $2x^2(x^2 - xy) + 2(x^4 - x^3y)$.

41. $\dfrac{ax^2y^5}{x^3y^3}$.

42. $\dfrac{2xy^2}{4y^3}$.

43. $\dfrac{(x^2y^{-1}z)^{-2}}{(xy^2)^{-4}}$.

44. $\dfrac{(ab)^2(2ab^2)}{2ab^3}$.

45. $\dfrac{(-3x^2y)(2xy)}{(4x^2y)^2}$.

46. $\dfrac{(xy^2)^2(-3x)^3}{9x(xy)}$.

47. $\dfrac{x^2 - 5x + 7}{x}$.

48. $\dfrac{-3x^3 - 5x^2 + 6}{30x}$.

49. $\dfrac{x^2y - 5xy^3 + 7xy}{xy^2}$.

50. $\dfrac{6x^2y - 3y^2 + 2x - 2}{-4x}$.

51. $\dfrac{2x^2y + (2xy^2w)^2 - 4x^3y^3w}{-2xy}$.

52. $\dfrac{3xy^2 - xy^2 + 3}{xy^2}$.

In Problems **53–58**, *do the long divisions.*

53. $\dfrac{6x^3 + 3x^2 - 5x - 1}{2x - 1}$.

54. $\dfrac{3x^3 - 5x^2 + x + 1}{3x + 1}$.

55. $\dfrac{x^4 + x^3 + 8x - 30}{x + 3}$.

56. $\dfrac{5 - x + 4x^2 - 3x^3 - 4x^4}{4x + 3}$.

57. $\dfrac{3x + 3x^3 - 2x^4}{2x + 1}$.

58. $\dfrac{9 - x^3}{x - 2}$.

4
Equations and Inequalities

4.1 LINEAR EQUATIONS

In mathematics, a literal symbol that represents more than one number is called a **variable.** Letters commonly used for variables are x, y, z, s, and t. A symbol that represents exactly one number is called a **constant.** Common literal symbols for constants are a, b, c, and k. For example, in the expression $2x + 3a$ we would usually assume that x is a variable and a is a constant. Specified numbers, such as 2 or π, are also constants.

One important use of mathematics is to solve equations. An **equation** is a statement that two expressions are equal. For example,

$$3x - 1 = 14$$

is an equation. Here

$3x - 1$ is the *left side*,

14 is the *right side*,

$3x$, -1, and 14 are *terms*, and

x is a *variable* or *unknown*.

There may be values of x that make the previous equation a true statement. Other values may make it false. For example, if x is 0, then $3x - 1 = 14$ becomes

$$3(0) - 1 = 14,$$

or simply

$$-1 = 14, \quad \text{a } \textbf{false} \text{ statement.}$$

However, if x is 5, then $3x - 1 = 14$ becomes

$$3(5) - 1 = 14,$$

or simply

$$14 = 14, \quad \text{a } \textbf{true} \text{ statement.}$$

Any value of a variable for which an equation is true is called a **solution,** or **root,** of that equation. Thus 5 is a solution of $3x - 1 = 14$.

Solving an equation means to find *all* of its solutions. This is done by means of *equivalent equations*. Two or more equations are **equivalent** if they have exactly the same solutions. To solve an equation, we apply rules that give us other equations equivalent to the given equation. We apply these rules until we have an equation whose solutions are *obvious*. For the present, this means an equation in which the variable is isolated on one side. The solutions of this equation are the same as the solutions of the original equation. Here are the rules that we shall apply and that guarantee equivalence.*

RULES THAT GUARANTEE EQUIVALENT EQUATIONS

1. If the same number is added to (or subtracted from) **both** sides of a given equation, then the resulting equation is equivalent to the given one.
2. If **both** sides of a given equation are multiplied (or divided) by the same number—**except** 0—then the resulting equation is equivalent to the given one.

Let's use these rules to solve our original equation,

$$3x - 1 = 14.$$

Since we want to isolate the variable x on one side, first we look for the term involving x:

$$\underset{\text{involves } x}{\text{this term}} \rightarrow \textcircled{3x} - 1 = 14.$$

Next, we use the rules to get this term, $3x$, by itself. To get rid of the 1 that is *subtracted* from $3x$, we *add* 1 to **both** sides (Rule 1) and simplify.

---

* In later chapters you will see that there are other rules that may be applied when solving an equation. But they do not necessarily guarantee that you obtain equivalent equations.

$$3x - 1 + 1 = 14 + 1,$$
$$3x + 0 = 15,$$
$$3x = 15. \qquad (1)$$

Now $3x$ is by itself. But we want x by itself. Since x is *multiplied* by 3, we'll *divide* **both** sides of Eq. 1 by 3 (Rule 2) and simplify.

$$\frac{3x}{3} = \frac{15}{3},$$

$$\frac{\overset{1}{\cancel{3}}x}{\underset{1}{\cancel{3}}} = \frac{\overset{5}{\cancel{15}}}{\underset{1}{\cancel{3}}},$$

$$\boxed{x = 5.}$$

Here x is isolated and the solution is obviously 5. Since this equation was obtained from the original equation by applying Rules 1 and 2, both equations are equivalent. Thus the solution of $3x - 1 = 14$ is 5. We write our solution as $x = 5$. It may also be written as $\{5\}$, which is called the *solution set*. It's a good idea to put a box (or circle) around a solution as we did. It shows at a glance the result of your hard work.

*Be sure you do the same thing to **both** sides of an equation. If*

$$3x = 15,$$

***don't write** $x = 15 - 3$. Here the left side was **divided** by 3, but on the right side 3 was **subtracted** from 15. You should divide both sides of $3x = 15$ by 3, as we did above.*

The equation $3x - 1 = 14$ that we solved above is an example of a *linear equation*. A **linear,** or **first-degree, equation** in x is an equation that can be expressed in the form

$$ax + b = 0, \qquad (1)$$

where a and b are constants and $a \neq 0$. [Note in Eq. (1) that x appears to the first power only.] For example, $3x - 1 = 14$ is a linear equation because it can be expressed as $3x + (-15) = 0$. Here a is 3 and b is -15. However, $2x^2 + 5 = 0$ is not a linear equation. For the time being, we shall be concerned with solving linear equations.

EXAMPLE 1

Solve $-\frac{3}{8}u + 1 = -2.$

We want to isolate the unknown (or variable) u. Here 1 is *added* to $-\frac{3}{8}u$. So to get the term involving u by itself, we *subtract* 1 from both sides (Rule 1).

$$-\frac{3}{8}u + 1 = -2,$$

$$-\frac{3}{8}u + 1 - 1 = -2 - 1 \qquad \text{[subtracting 1 from both sides]},$$

$$-\frac{3}{8}u = -3.$$

Let's clear of fractions by multiplying both sides by the denominator 8 (Rule 2).

$$8\left(-\frac{3}{8}u\right) = 8(-3) \qquad \text{[multiplying both sides by 8]},$$

$$-3u = -24 \qquad \left[\text{since } 8\left(-\frac{3}{8}u\right) = -\left(\cancel{8}\cdot\frac{3}{\cancel{8}}\right)u = -3u\right],$$

$$\frac{\cancel{(-3)}u}{\cancel{-3}} = \frac{-24}{-3} \qquad \text{[dividing both sides by } -3],$$

$$\boxed{u = 8.}$$

Here is a guide that you may follow to solve more complicated linear equations:

STEPS TO SOLVE EQUATIONS

1. Remove symbols of grouping.
2. Clear of fractions.
3. Add or subtract to get the terms involving x (or other unknown) on one side and the constants on the other.
4. Multiply or divide so that the value of x is obvious.

We shall use this four-step procedure in the next example.

EXAMPLE 2

Solve $6x - 3(4x - 5) = 5(9 - x)$.

Step 1.

$$6x - 12x + 15 = 45 - 5x \qquad \text{[removing parentheses]}.$$

Step 2.
Here there are no fractions to clear.

Step 3.

$$-6x + 15 = 45 - 5x \quad \text{[combining on left side]},$$

$$-6x + 15 + 5x = 45 - 5x + 5x \quad \text{[adding } 5x \text{ to both sides]},$$

$$-x + 15 = 45 \quad \text{[combining]},$$

$$-x + 15 - 15 = 45 - 15 \quad \text{[subtracting 15 from both sides]},$$

$$-x = 30 \quad \text{[combining]}.$$

Step 4.
If we multiply (or divide) both sides by -1, we will then have x by itself on the left.

$$(-1)(-x) = (-1)30 \quad \text{[multiplying both sides by } -1\text{]},$$

$$\boxed{x = -30} \quad \text{[simplifying]}.$$

EXAMPLE 3

Solve $3\left(\frac{2x}{5} - \frac{1}{2}\right) = 2x$.

Step 1.

$$3 \cdot \frac{2x}{5} - 3 \cdot \frac{1}{2} = 2x \quad \text{[removing parentheses]},$$

$$\frac{6x}{5} - \frac{3}{2} = 2x.$$

Step 2.
When two or more terms of an equation have fractions, to clear of fractions you may multiply both sides by the least common denominator (L.C.D.).* Here the L.C.D. is 10.

$$10\left(\frac{6x}{5} - \frac{3}{2}\right) = 10(2x) \quad \text{[multiplying both sides by 10]},$$

$$12x - 15 = 20x \quad \text{[removing parentheses]}.$$

Step 3.

$$12x - 15 - 12x = 20x - 12x \quad \text{[subtracting } 12x \text{ from both sides]},$$

$$-15 = 8x \quad \text{[combining]}.$$

---

* The *least common denominator* of two or more fractions is the smallest number with all the denominators as factors. That is, the L.C.D. is the least common multiple of all the denominators.

Step 4.

$$-\frac{15}{8} = x \qquad \text{[dividing both sides by 8]},$$

or

$$\boxed{x = -\frac{15}{8}.}$$

Our four step procedure to solve equations is *only* a guide. For instance, in Example 3, instead of first removing parentheses, we could first clear of fractions by multiplying both sides by 10.

$$10 \cdot 3\left(\frac{2x}{5} - \frac{1}{2}\right) = 10 \cdot 2x,$$

$$30\left(\frac{2x}{5} - \frac{1}{2}\right) = 20x,$$

$$\overset{6}{\cancel{30}} \cdot \frac{2x}{\cancel{5}} - \overset{15}{\cancel{30}} \cdot \frac{1}{\cancel{2}} = 20x \qquad \text{[removing parentheses]},$$

$$12x - 15 = 20x.$$

Proceeding as in Steps 3 and 4, we can now solve for x. In both approaches to the problem, remember that our ultimate aim is to obtain an equivalent equation in which the solution is obvious.

Formulas are equations that involve two or more literal numbers. Let's look at some.

EXAMPLE 4

The formula for the circumference of a circle is

$$C = 2\pi r,$$

where C is the circumference and r is the radius. If the circumference is 16 centimeters (cm), find the radius.

Letting $C = 16$, we have

$$16 = 2\pi r.$$

Now we solve for r. Since $2\pi r$ is just $(2\pi)r$, we divide both sides by 2π.

$$\frac{16}{2\pi} = \frac{2\pi r}{2\pi},$$

$$\boxed{\frac{8}{\pi} = r.}$$

Thus the radius is $\frac{8}{\pi}$ cm.

EXAMPLE 5

The formula for the perimeter P of a rectangle is $P = 2l + 2w$. Here l is the length and w is the width. Solve for w.*

Since we want to solve for w, we think of w as the unknown and begin to isolate it.

$$P = 2l + 2w,$$
$$P - 2l = 2l + 2w - 2l \quad \text{[subtracting } 2l \text{ from both sides]},$$
$$P - 2l = 2w,$$
$$\frac{P-2l}{2} = \frac{2w}{2} \quad \text{[dividing both sides by 2]},$$
$$\frac{P-2l}{2} = w.$$

Our answer is

$$\boxed{w = \frac{P-2l}{2}.}$$

Note that if we were given values for P and l, it would be easy to find the **width** w with this equation.

*In the **answer** to Example 5, we **cannot** cancel the 2's. That is, $w \neq \frac{P - \not{2}l}{\not{2}}$. In a fraction we can cancel only **factors** of the numerator with **factors** of the denominator. In $\frac{P-2l}{2}$, 2 is a factor of a **term** of the numerator; it is not a factor of the numerator. Now, **if** w were equal to $\frac{2(P-2l)}{2}$ (which it isn't)*, then we could *cancel the factor 2 of the numerator with the factor 2 of the denominator:*

$$\frac{\not{2}(P-2l)}{\not{2}}.$$

Exercise 4.1

In Problems **1–44**, *solve the equations.*

1. $x + 3 = 0.$

2. $-6x = 18.$

3. $\frac{y}{5} = -6.$

4. $6x + 7 = 7.$

5. $13 - 3x = 19.$

6. $\frac{x}{3} = \frac{7}{2}.$

---

* The perimeter of a geometric figure is the length of its boundary.

7. $\frac{2x}{5} = -\frac{3}{8}$.

8. $\frac{u}{4} - \frac{4}{3} = 8$.

9. $\frac{5}{3}S - 6 = -1$.

10. $\frac{13}{9} = 1 - \frac{4}{3}x$.

11. $\frac{3}{4} = \frac{2 - x}{3}$.

12. $\frac{3}{8} = \frac{2x - 2}{6}$.

13. $5y - 2y + 3 = 12$.

14. $2t + 6 = 6t + 2$.

15. $6x = 4x$.

16. $x = 3 - 2x$.

17. $8x = 4x - 12$.

18. $3x + 2x = 4x + 6$.

19. $3x + 6 = 7x - 2$.

20. $9x - 4 = 9 - 4x$.

21. $2(y - 5) = y + 1$.

22. $-3(y - 1) = 4y + 17$.

23. $7(3 - 2z) = 3 - 5z$.

24. $8 - 6z = 4(3z + 5)$.

25. $(4x + 3) - (7 - x) = 7x$.

26. $2(x - 1) - (3x + 7) = x$.

27. $17x - 3(x - 5) = 2 + 4(3x + 2)$.

28. $6 = -[3 - (x - 2)]$.

29. $4[2 - 3(r + 6)] = 10r$.

30. $2(3r + 2) = -[1 + (2 - r)]$.

31. $t + \frac{t}{2} = 6$.

32. $\frac{x}{2} + 1 = \frac{x}{3}$.

33. $\frac{x}{3} = \frac{10 + x}{2}$.

34. $\frac{z}{2} = \frac{z}{3}$.

35. $3\left(\frac{x}{2} - 4\right) = 5x$.

36. $\frac{2}{3}(x - 5) = 7$.

37. $\frac{3}{4}(x - 1) = 7 + x$.

38. $2\left(x - \frac{1}{5}\right) = 3x$.

39. $\frac{2}{3}(x - 5) = 4x + \frac{1}{2}$.

40. $\frac{x}{3} + 1 = 4\left(x - \frac{1}{2}\right)$.

41. $2[3z + 4(z - 1)] = -(7 - z)$.

42. $2\{4x - [8 + 2(5x - 4)]\} = 0$.

43. $\frac{1}{2}\left(x + \frac{4}{3}\right) = 3 - (x + 1)$.

44. $\frac{7}{2} - (x + 5) = \frac{x}{4}$.

45. A gas has a volume of 200 cubic centimeters (cm^3) at a temperature of 273 K (the Kelvin scale). Under certain conditions, its volume (in cubic centimeters) at 373 K is given by v, where

$$\frac{v}{200} = \frac{373}{273}.$$

Find v. You may round your answer to the nearest cubic centimeter.

46. In a certain chemical solution, the hydrogen ion concentration, written $[H^+]$, in moles per liter is given by

$$1 \times 10^{-4} = [H^+] \times 1 \times 10^{-2}.$$

Solve for $[H^+]$ and give your answer in scientific notation.

47. Use the formula $P = 2l + 2w$ to find the width w of a rectangle whose perimeter P is 960 m and whose length l is 360 m.

48. Use the formula $A = \frac{1}{2}bh$ to find the height h of a triangle whose area A is 75 square centimeters (cm^2) and whose base b is 15 cm.

49. Solve $C = 2\pi r$ for r.

50. $F = ma$ is a formula found in physics. It is one of Newton's laws. Here F is force, m is mass, and a is acceleration. Solve for m.

51. The formula for simple interest I is $I = Prt$. Here P is principal, r is rate of interest, and t is time. Solve for r. *Hint*: First write Prt as $(Pt)r$.

52. Solve $I = Prt$ for P.

53. The amount A to which a principal P grows at a simple rate of interest r after t years is given by $A = P + Prt$. Solve for t.

54. The total surface area A of a right circular cylinder (think of a can of soda with a top and bottom) is given by $A = 2\pi r^2 + 2\pi rh$. Here r is the radius of the cylinder and h is the height. Solve for h.

55. When radar is used on a highway to determine the speed of a car, a radar beam is sent out and reflected from the moving car. The difference F (in cycles per second) in frequency between the original and reflected beams is given by

$$F = \frac{vf}{334.8},$$

where v is the speed of the car in miles per hour (mi/h) and f is the frequency of the original beam (in megacycles per second).

Suppose you are driving along a highway with a speed limit of 55 mi/h. A police officer aims a radar beam with a frequency of 2450 (megacycles per second) at your car, and the officer observes the difference in frequency to be 420 (cycles per second). Can the officer claim that you were speeding?

56. The beam in Fig. 4-1 with forces F_1 and F_2 acting on it will balance on the pivot if $F_1 d_1 = F_2 d_2$. Solve this equation for F_1.

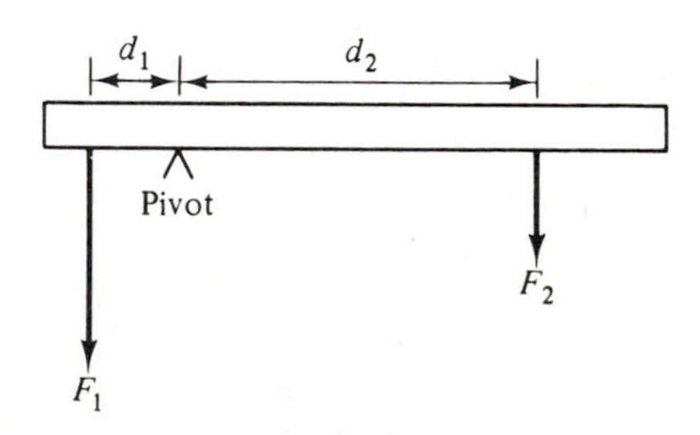

FIGURE 4-1

57. In Problem 56, suppose $F_1 = 20$ lb, $d_1 = 2$ ft, and $F_2 = 8$ lb. Find d_2 (in feet).

58. Use the formula in Problem 56 to find the force F (in pounds) in Fig. 4-2 so that the beam balances.

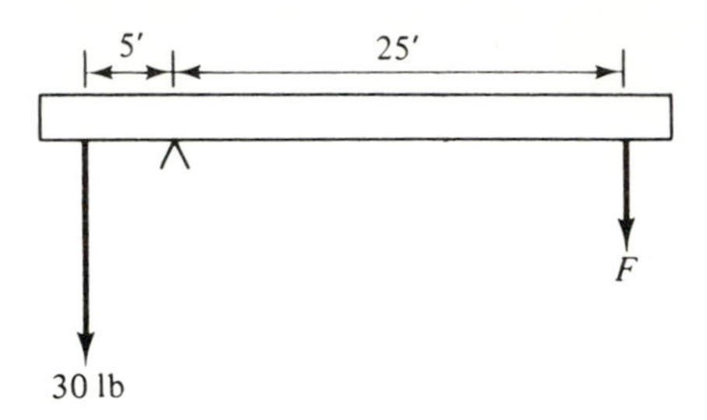

FIGURE 4-2

59. At time t, the height s of an object thrown up from the ground with a velocity v_0 is given by

$$s = v_0 t - 16t^2.$$

If an object is to have a height of 16 ft after 2 seconds, find v_0 (in feet per second).

60. Boyle's law for gases is given by $V_1 P_1 = V_2 P_2$. Here the V's are volumes and the P's are pressures. If $V_1 = 100$ cubic centimeters (cm^3), $P_1 = 756$ millimeters (mm) of mercury, and $P_2 = 720$ mm of mercury, find V_2 (in cubic centimeters).

61. If you purchase an item for business use, in preparing your income tax you may be able to spread out its expense over the life of the item. This is called *depreciation*. One method of depreciation is *straight-line depreciation*, in which the annual depreciation is computed by dividing the cost of the item, less its estimated salvage value, by its useful life. Suppose the cost is C dollars, the useful life is N years, and there is no salvage value. Then the value V (in dollars) of the item at the end of n years is given by

$$V = C\left(1 - \frac{n}{N}\right).$$

Suppose new office furniture is purchased for \$1600, has a useful life of 8 years, and has no salvage value. After how many years will it have a value of \$1000?

62. $F = \frac{9}{5}C + 32$ is a formula relating Fahrenheit and Celsius temperatures. Solve for C.

63. $V = E(x_b - x_a)$ is a formula that can be used in a certain situation involving electric intensity. Solve for x_b.

64. When solid objects are heated, they expand in length—which is why expansion joints are placed in bridges and pavements. Generally, when the temperature of a solid body of length l_0 is increased from T_0 to T, the body's length l is given by

$$l = l_0[1 + \alpha(T - T_0)],$$

where α (α is the Greek letter *alpha*) is called the *coefficient of linear expansion.*

Suppose a metal rod 1 m long at 0°C expands .001 m when it is heated from 0°C to 100°C. Find the coefficient of linear expansion.

65. In analyzing an electric circuit, the following equation occurred.

$$\frac{V-100}{40}+\frac{V}{80}+\frac{V-150}{60}=0.$$

Solve for V (to one decimal place).

4.2 TRANSLATING ENGLISH TO ALGEBRA

In the next section you will begin to solve word problems. There you will have to translate situations, stated in words, into algebraic statements. The following examples will show you some translations.

EXAMPLE 1

Translate each of the following word statements into algebraic statements.

a. *x is 15 more than y.*

The phrase "15 more than y" translates into $15 + y$ (or $y + 15$). Since x is (or equals) this expression, our complete translation is

$$x = 15 + y.$$

This situation could also be stated as *y is 15 less than x.* The phrase "15 less than x" translates into $x - 15$, **not** $15 - x$. Thus

$$y = x - 15.$$

b. *Twice a certain number is 14.*

Let the number be x. Then

$$2x = 14.$$

c. *The difference of x and the sum of 3 and y.*

$$x - (3 + y).$$

d. *The sum of y, and 3 times x, is $\frac{1}{2}$.*

$$y + 3x = \tfrac{1}{2}.$$

e. *Four times a number is 5 less than 8 times the number.*

Let the number be x. Then

$$4x = 8x - 5.$$

EXAMPLE 2

A bank pays 6% interest per year on savings accounts. How much interest is earned after 1 year on a deposit of x dollars?

Since 6% in decimal form is .06, the interest (in dollars) is

$$.06x.$$

EXAMPLE 3

Translate each of the following word statements into algebraic statements.

a. *Mary is twice as old as Alice.*

Let m be Mary's age and a be Alice's age (in years). The statement means that Mary's age is twice Alice's age. Thus

$$m = 2a.$$

b. *There are three times as many dimes* (d) *as quarters* (q).

Here d is the number of dimes and q is the number of quarters. You may think that the statement translates into $3d = q$, but it **does not.** Whatever is the number of quarters, the number of dimes is three times as many. Since there are q quarters, there are $3q$ dimes. Thus

$$d = 3q.$$

The order of the words in an English statement may not be the same as the order of numbers and letters in its algebraic translation. In the statement of Example 3b,

there are three times as many dimes as quarters,

the order of key words is

three, times, dimes, as, quarters.

This may lead you to think that the algebraic translation is $3d = q$. But that is wrong, as Example 3b shows. When translating from English to algebra, you have to analyze *carefully* what the given statement *means*.

Exercise 4.2

In Problems **1–26,** *translate the word statement into an algebraic statement.*

1. The sum of x and y is 8.

2. The difference of 6 and w is 2.

3. 13 is 4 more than x.

4. x is 5 less than y.

5. z is 2 less than the difference of x and y.

6. x is 3 more than the product of y and z.

7. The distance s traveled by a particle is equal to the product of its velocity v and the time t it travels.

8. The product of the pressure P of a gas and the volume V of the gas is equal to a constant k.

9. Three times a number x is y.

10. The sum of 4 and the product of 6 and w is 24.

11. Ten times a number x is 6 more than 3 times the number.

12. Tom is 4 years older than Jack. (Let T be Tom's age and J be Jack's age.)

13. Mary's age is half of Joan's age. (Let M be Mary's age and J be Joan's age.)

14. In a factory there are twice as many male workers (m) as female workers (f).

15. In a contest there are 100 times as many losers (l) as winners (w).

16. The sum of twice a number x and 5.

17. Twice the difference of a number x and 12.

18. A number x is 5 less than twice the number z.

19. A rectangle has a length (l) that is 4 centimeters less than 6 times its width (w).

20. The reciprocal of t_1 plus the reciprocal of t_2 is equal to the reciprocal of T.

21. If a mass were converted into energy, the amount E of energy liberated is equal to the product of the mass m and the square of c, where c is the speed of light.

22. The product of x and the sum of a and b is equal to the sum of b times x and a times x.

23. If 3 is decreased by 15 times a number x, the result is 3, plus 4 times the number.

24. The sum of a number n and 2 is multiplied by the number decreased by 2. The result is equal to -4 plus the square of the number.

25. Fahrenheit temperature F is equal to $\frac{9}{5}$ Celsius temperature C, plus 32.

26. At a certain college bookstore, for every two fiction books (f) that are sold, seven nonfiction books (n) are sold.

In Problems **27–39,** *answer each question with an algebraic expression.*

27. A person deposits x dollars into a savings account that earns $8\frac{1}{2}\%$ interest per year. How much interest (in dollars) is earned at the end of 1 year?

28. If Sam is now x years old, what was his age 3 years ago?

29. A tank contains 100 gallons of an alcohol and water solution. If x gallons are alcohol, how many gallons are water?

30. A person has $5000 and invests x dollars of it. How much money is not invested?

31. A solution is 25% acid, by volume. How much acid is there in x milliliters of the solution?

32. A certain coffee costs $3.50 per pound. How many dollars do x pounds cost?

33. In a bottling plant, a machine fills q bottles per minute. In 1 hour, how many bottles does it fill?

34. The average temperature today is 22°C. The average temperature tomorrow is expected to be x°C lower. What is the expected average temperature tomorrow?

35. A stock selling for $100 on Monday decreased by x dollars on Tuesday and then increased by y dollars on Wednesday. What is its price (in dollars) on Wednesday?

36. The cost of renting a boat for a n-person fishing trip is d dollars. What is each person's equal share of the cost?

37. In one semester a student purchased x paperback books selling for $3 each, y books at $3.50 each, and z books at $4 each. (a) What is the total cost (in dollars) of the books? (b) What is the average cost (in dollars) per book? (*Hint*: The average cost per book is the total cost divided by the total number of books.)

38. The cost of a dress to a retailer is $40, and the dress is sold at a profit of p dollars. What is the selling price of the dress?

39. A spring stretches 5.3 cm for each newton of weight that is suspended from it. (In physics, a *newton* is a unit of force.) If the initial length of the spring is 20 cm, what equation relates the length L (in centimeters) of the stretched spring to the load W (in newtons) suspended from it?

40. Symbolize the following. The sum of three consecutive integers is 36. Simplify your answer. (*Hint*: If the first integer is n, then the next consecutive integer is $n + 1$.)

41. A typewriter selling for p dollars is given a 15% discount. One week later, the price is further reduced by 10% of the first discounted price. (a) Find the price after the discount of 15%. (b) Find the price after the 10% discount. (c) Considering the discounts above of 15% and 10%, would you prefer a straight 25% discount?

4.3 WORD PROBLEMS

Now let's turn to "word" problems. With a word problem an equation is not handed to you. You have to set it up by translating verbal statements into an equation. This is called *mathematical modeling*. After modeling, you solve the equation. Here are some steps that you can follow as a guide.

GUIDE TO SOLVING WORD PROBLEMS

1. Read the problem more than once so that you clearly understand what facts are given and what you are asked to find.
2. Choose a letter to represent the unknown quantity that you want to find.
3. Use the relationships and facts given in the problem and translate them into an equation involving the letter.
4. Solve the equation. Check your solution to see if it answers what was asked. Sometimes the solution to the *equation* will not be the answer to the *problem*, but it may be useful in obtaining that answer.

It takes a lot of practice to get a good feeling for word problems. Read the following examples carefully.

EXAMPLE 1

A student installed a tape player in his car. He has \$50 *and wants to spend it all on tapes of his favorite country group. If tapes sell for* \$7.25 *each, how many tapes can he buy?*

You might be able to solve this problem in your head. But let's model it. First, read the problem again so that you understand it. Then say to yourself, "What is the problem asking?"

How many tapes can the student buy?

Next, choose a letter to represent a quantity that you want to find. Write down exactly what this letter stands for. We'll let

t = number of tapes the student can buy.

We chose the letter t as the unknown because it reminds us of what we want to find.

Now let's set up an equation. Since each tape costs \$7.25, then t tapes cost $7.25t$.

$7.25t$ = total cost of t tapes.

The total cost of these t tapes equals the amount he will spend, \$50. So the equation to solve is

$$7.25t = 50.$$

This equation is our model. Solving, we obtain

$$\frac{7.25t}{7.25} = \frac{50}{7.25},$$

$$t = \frac{50}{7.25} = \boxed{6.9} \qquad \text{(approximately)}.$$

The number 6.9 is the solution to the *equation*, but it is not a solution to the *problem*. Buying 6.9 tapes makes no sense. The student has only enough money to buy six tapes.

You are probably familiar with percentage. For example, to find 20% of 50, we first write 20% as a decimal, .20. Then we multiply 50 by .20:

$$(.20) \cdot 50 = 10.$$

Thus 20% of 50 is 10.

Notice the connection between the phrase "20% *of* 50 *is* 10" and the equation $(.20) \cdot 50 = 10$:

$$\begin{array}{cccc} 20\% & \text{of} & 50 \text{ is} & 10 \\ \updownarrow & \updownarrow & \updownarrow & \\ (.20) & \cdot & 50 = & 10 \end{array}$$

The 20% is replaced by its decimal form. The word *of* is replaced by the multiplication symbol. And the word *is* is replaced by the symbol = . This connection will help you set up different types of percentage problems.

EXAMPLE 2

70 *is* 20% *of what number?*

Let n be the number for which we're looking. The decimal form of 20% is .20.

$$\begin{array}{ccccc} 70 & \text{is} & 20\% & \text{of} & \underbrace{\text{what number}} \\ \downarrow & \downarrow & \downarrow & \downarrow & \downarrow \\ 70 & = & (.20) & \cdot & n \end{array}$$

The equation to solve is $70 = (.20) \cdot n$, or

$$.20n = 70,$$

$$\frac{.20n}{.20} = \frac{70}{.20},$$

$$\boxed{n = 350.}$$

Thus 70 is 20% of 350.

EXAMPLE 3

At a certain college there are 2100 *freshmen, and* 1365 *of them take a math course. What percentage of the freshmen class is taking math?*

Here we want to find a percentage. Let P be the percentage of the freshmen class taking math. Since P will occur in an equation, P will be expressed in decimal form.

$$P = \begin{cases}\text{percentage of freshmen class} \\ \text{taking math (expressed as a decimal)}\end{cases}$$

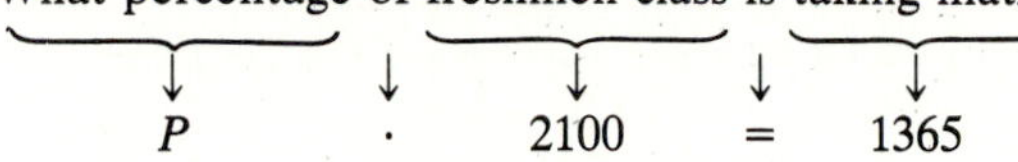

The equation we want to solve is $P \cdot (2100) = 1365$, or

$$2100P = 1365,$$

$$\frac{2100P}{2100} = \frac{1365}{2100},$$

$$\boxed{P = .65.}$$

But $P = .65$ is the decimal form of percentage. So, 65% of the freshmen take math.

EXAMPLE 4

A chemist must prepare 350 ml *of a chemical solution. It is to be made up of* 2 *parts alcohol and* 3 *parts acid. How much of each should be used?*

Let n = number of milliliters in each part. Figure 4-3 shows the situation.

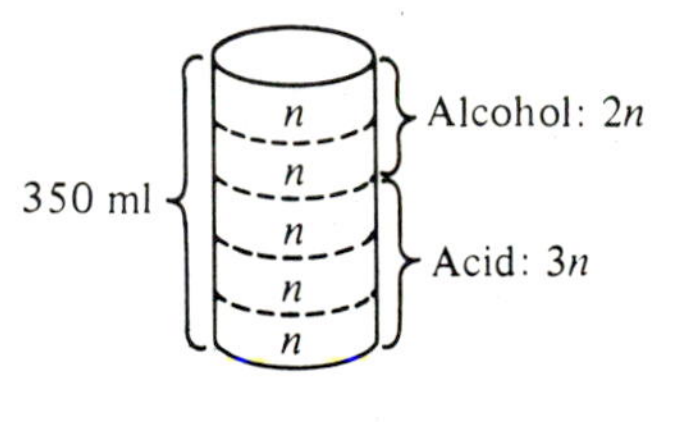

FIGURE 4-3

From the diagram we have

$$2n + 3n = 350,$$

$$5n = 350,$$

$$n = \frac{350}{5} = 70.$$

But $n = 70$ *is not* the answer to the original problem. *Each part* (n) has 70 ml. The amount of alcohol is $2n = 2(70) = 140$, and the amount of acid is $3n = 3(70) = 210$. Thus the chemist should use 140 ml of alcohol and 210 ml of acid. This example shows how helpful a diagram can be in setting up a word problem.

EXAMPLE 5

A company produces pencil sharpeners and has **fixed costs** *that total* \$20,000 *per year. Fixed costs are costs that the company is charged whether or not any product is produced. These costs include such things as rent, light, heat, insurance, and maintenance. There is also the manufacturing cost (labor, material) of* \$3.50 *for each sharpener produced. Next year the company wants the* **total cost** *of making sharpeners to be* \$48,000. *How many pencil sharpeners should be made?*

Did you reread the problem? If not, do it now. Here the question is "How many pencil sharpeners should be made?" Let

$$s = \text{number of sharpeners that should be made.}$$

The total cost is made up of two parts.

$$\text{total cost} = (\text{fixed costs}) + \begin{pmatrix}\text{manufacturing cost}\\ \text{for } s \text{ sharpeners}\end{pmatrix}.$$

We are given that

$$\text{total cost} = 48{,}000$$

$$\text{and} \qquad \text{fixed costs} = 20{,}000.$$

Since the manufacturing cost for each sharpener is \$3.50, then

$$\begin{pmatrix}\text{manufacturing cost}\\ \text{for } s \text{ sharpeners}\end{pmatrix} = 3.50s.$$

The equation to solve is

$$48{,}000 = 20{,}000 + 3.50s.$$

Solving, we obtain

$$48{,}000 - 20{,}000 = 20{,}000 + 3.50s - 20{,}000,$$

$$28{,}000 = 3.50s,$$

$$\frac{28{,}000}{3.50} = \frac{3.50s}{3.50},$$

$$\boxed{8000 = s.}$$

Thus 8000 sharpeners should be made next year.

When a company sells its product, it receives money. The total of all money received is called the **total revenue**. The items that are sold are called *units* of output and

$$\text{total revenue} = (\text{price per unit})(\text{number of units sold}).$$

The company's **profit** is given by

$$\text{profit} = \text{total revenue} - \text{total cost}.$$

EXAMPLE 6

A company produces a product for which the fixed costs are \$80,000 *and the manufacturing cost per unit is* \$6. *It sells each unit for* \$10. *How many units must the company sell so that it earns a profit of* \$60,000?

Let q be the number of units that must be sold. (In many business problems, q represents quantity.) The total revenue is then $10q$ (dollars):

$$\text{total revenue} = 10q.$$

The total cost is the fixed cost, or \$80,000, plus the total manufacturing cost for q units, or $6q$:

$$\text{total cost} = 80{,}000 + 6q.$$

Since

$$\text{profit} = \text{total revenue} - \text{total cost},$$

our model is

$$60{,}000 = 10q - (80{,}0000 + 6q). \quad (1)$$

Solving gives

$$60{,}000 = 10q - 80{,}000 - 6q,$$

$$140{,}000 = 4q,$$

$$\frac{140{,}000}{4} = q,$$

$$\boxed{35{,}000 = q.}$$

Thus 35,000 units must be sold to earn a profit of \$60,000.

In Eq. (1) *the parentheses are crucial because we must subtract* ***all*** *of* $80{,}000 + 6q$ *from* $10q$. ***Don't write*** $60{,}000 = 10q - 80{,}000 + 6q$.

Exercise 4.3

Model each problem and solve.

1. If 4 less than twice a number is 16, what is the number?
2. If one-third of the sum of a number and 20 is 5, what is the number?
3. A city set aside \$26,000 to purchase parking meters. If each meter costs \$70, how many meters can be purchased?

4. A company makes car stereos. The manufacturing cost for each stereo is \$45. The company has fixed costs of \$4150 per month. How many stereos can it make next month for a total cost of \$10,000?

5. The Ace Wallet Co. has fixed costs of \$50,000 per year. The manufacturing cost for each wallet is \$5. How many wallets can the company make next year for a total cost of \$125,000?

6. A farmer can harvest 45 bushels (bu) of corn per acre. How many acres should be planted for a total harvest of 3600 bu?

7. A certain machine can perform 34 chemical analyses per day. But a lab technician can perform only 7. A laboratory must make 110 analyses tomorrow, and it has only two machines. How many technicians will be needed to complete the job?

8. A hospital has only private and semiprivate rooms. It has 90 private rooms (one bed to a room). However, the total number of beds in the hospital is 318. How many semiprivate rooms (two beds to a room) does it have?

9. The *current ratio* of a company is the value of its current assets divided by its current liabilities. A company has a current ratio of 2.5 and current liabilities of \$80,000. What are its current assets?

10. The IQ (intelligence quotient) of a person is found by dividing his or her mental age by his or her chronological age and then multiplying that result by 100. For example, a person with a mental age of 11 and a chronological age of 10 has an IQ of $\frac{11}{10} \cdot 100 = 110$. Find the mental age of a person with chronological age 12 and IQ of 125.

11. 90 is 30% of what number?

12. 72 is 45% of what number?

13. 75 is 12% of what number?

14. 34 is 85% of what number?

15. What percentage of 200 is 8?

16. What percentage of 60 is 9?

17. What percentage of 50 is 55?

18. What percentage of 1400 is 70?

19. A college dormitory houses 210 students. This fall, rooms are available for 76 freshmen. On the average, 95% of those freshmen who request room applications actually reserve a room. How many room applications should the college send out if it wants to receive 76 reservations?

20. A group of people were polled and 20%, or 700, of them favored a new product over the best-selling brand. How many people were polled?

21. Approximately 21% of the air we breathe is oxygen. To the nearest milliliter, how many milliliters of air contain 1 milliliter (ml) of oxygen?

22. On a square centimeter of a leaf of a certain corn plant, there are 8928 stomata (small pores) on the upper surface and 15,872 on the under surface. What percentage of the total stomata is on the upper surface?

23. A team of biologists studying ABO blood groups tested 2000 people. They found 850 had antigen A, 780 had antigen B, and 370 had no antigen. What percentage of the

people tested had antigen A? (Antigens are foreign substances involved with the production of antibodies.)

24. A few years ago, cement drivers were on strike for 46 days. Before the strike, these drivers earned \$7.50 per hour and worked 260 eight-hour days a year. What percentage increase is needed in yearly income to make up for the lost time within 1 year?

25. It was reported that in a certain women's jail, women prison guards, called matrons, received 30% (or \$200) a month less than their male counterparts, deputy sheriffs. Find the yearly salary of a deputy sheriff. Give your answer to the nearest dollar.

26. The cost of making a pair of jeans is c dollars, and each pair sells for p dollars. If n pairs of jeans are sold, what is the profit?

27. One of the most important defoliating insects is the gypsy moth caterpillar, which feeds on foliage of shade, forest, and fruit trees. A homeowner lives in an area in which the gypsy moth has become a problem. She wishes to spray the trees on her property before more defoliation occurs. She needs 128 oz of a solution made up of 3 parts of insecticide A and 5 parts of insecticide B. The solution is then mixed with water. How many ounces of each insecticide should be used?

28. A builder makes a certain type of concrete by mixing together 1 part cement, 3 parts sand, and 5 parts stone (by volume). If he wants 585 ft^3 of concrete, how many cubic feet of each ingredient does he need?

29. According to *The Consumer's Handbook* (Paul Fargis, ed., (New York: Hawthorn, 1974)), a good oiled furniture finish contains two parts boiled linseed oil and one part turpentine. If you need a pint (16 fluid oz) of this furniture finish, how many fluid ounces of turpentine are needed?

30. A student scored an 82 and an 88 on her first two math exams. What score does she have to get on her next exam so that the average of the three exams will be 90? (*Hint*: To find the average of three numbers, add the numbers and divide this sum by 3.)

31. A student has two exam grades of 74 and one of 76 in his sociology course. What score must he get in his next exam so that the average of the four exams is 80? (*Hint*: To find the average of four numbers, add the numbers and divide this sum by 4.)

32. The Geometric Products Co. manufactures product Z. The manufacturing cost for each unit is \$2.20, and fixed costs are \$95,000. Each unit sells for \$3. How many units must be sold for the company to have a profit of \$50,000?

33. The Clark Co. manufactures hair dryers. The manufacturing cost for each dryer is \$15, and fixed costs are \$600,000. A dryer sells for \$20. If the company wants to earn a profit of \$100,000, how many dryers must be sold?

34. A company manufactures two types of prefabricated houses: ranch and colonial. Last year they sold three times as many ranch models as they did colonial models. If a total of 2640 houses were sold last year, how many of each model were sold?

35. As a fringe benefit for its employees, a company established a vision-care plan. Under this plan, each year the company will pay the first $10 of an employee's vision-care expenses and 80% of all additional vision-care expenses up to a maximum *total* benefit payment of $60. For an employee, find the total annual vision-care expenses covered by this program.

36. Over a period of time, the manufacturer of a caramel-center candy bar found that 2% of the bars were rejected for imperfections.

(a) If c candy bars are made in a year, how many would the manufacturer expect to be rejected?

(b) This year, annual consumption of this candy is projected to be 2,000,000 bars. Approximately how many bars will have to be made if rejections are taken into consideration?

37. An electric utility company is going to locate its new power plant along a road connecting the towns of Exton and Whyton, which are 10 miles apart. See Fig. 4-4.

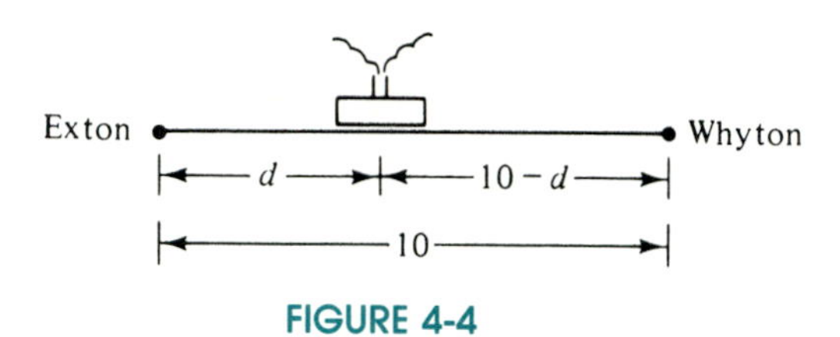

FIGURE 4-4

For political reasons, the utility company will buy coal from both towns. The price of coal per ton from Exton is $72.65 plus $0.45 per mile for delivery. The price per ton from Whyton is $72.25 plus $0.25 per mile for delivery. How far from Exton should the plant be located if the price of coal per ton delivered from Exton is to be equal to that from Whyton? (*Hint*: If d is the distance of the plant from Exton, then $10 - d$ is the distance from Whyton.)

38. An investment club bought a bond of an oil corporation for $5000. The bond yields 8% per year. The club now wants to buy shares of stock in a utility company. The stock sells at $20 per share and earns a dividend of $0.50 per share per year. How many shares should the club buy so that its total investment in stocks and bonds yields 5% per year?

39. A manufacturer of video-game cartridges sells each cartridge for $19.95. The manufacturing cost of each cartridge is $14.95. Monthly fixed costs are $8000. During the first month of sales of a new game, how many cartridges must be sold in order for the manufacturer to break even (that is, in order that total revenue equal total cost)?

40. The owner of a 20-room motel, which is 70% occupied, decides to charge $8 more than the single occupancy rate if two or more people occupy the room. This situation occurs in 75% of the occupied rooms, on the average. What should the two rates be so that the owner receives $160,000 in annual income to cover expenses and yield a reasonable profit? (Assume that a year has 365 days and give your answer to the nearest dollar.)

4.4 RATE AND MIXTURE PROBLEMS

We continue our discussion of word problems with examples of "rate" and "mixture" type problems.

EXAMPLE 1

*On the moon a lunar rover traveled from point A to point B at the rate of 5 kilometers per hour (km/h). It returned to A at the rate of 15 km/h. The **total** traveling time was 2 h. Find the distance from A to B.*

You may recall that

$$\text{distance} = (\text{rate})(\text{time}).$$

Two other forms of this are

$$\text{time} = \frac{\text{distance}}{\text{rate}}, \qquad \text{rate} = \frac{\text{distance}}{\text{time}}.$$

Now, let d be the distance (in kilometers) from A to B (see Fig. 4-5). Then the time to go from A to B at 5 km/h is $\frac{\text{distance}}{\text{rate}}$, or $\frac{d}{5}$. From B to A the distance is also d,

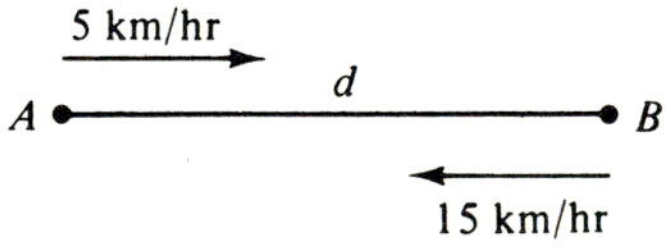

$$A \bullet \xrightarrow{5 \text{ km/hr}} \quad d \quad \xleftarrow{15 \text{ km/hr}} \bullet B$$

FIGURE 4-5

but the time at 15 km/h is $\frac{\text{distance}}{\text{rate}} = \frac{d}{15}$. Thus

$$\begin{pmatrix}\text{time}\\ \text{from}\\ A \text{ to } B\end{pmatrix} + \begin{pmatrix}\text{time}\\ \text{from}\\ B \text{ to } A\end{pmatrix} = \text{total time},$$

$$\frac{d}{5} + \frac{d}{15} = 2,$$

$$15\left[\frac{d}{5} + \frac{d}{15}\right] = 15 \cdot 2 \qquad \text{[multiplying both sides by 15, the L.C.D.]},$$

$$15 \cdot \frac{d}{5} + 15 \cdot \frac{d}{15} = 30,$$

$$3d + d = 30,$$

$$4d = 30,$$

$$d = \frac{30}{4} = \frac{15}{2} = 7\frac{1}{2}.$$

Thus the distance from A to B is $7\frac{1}{2}$ km.

Here's another way to do the problem. Instead of finding the distance right away, we first find the *time* it takes to go from A to B.

Let t = time (in hours) to go from A to B at 5 km/h. Because the total time traveled is 2 h, then $2 - t$ is the time to go from B to A at 15 km/h. See Fig. 4-6.

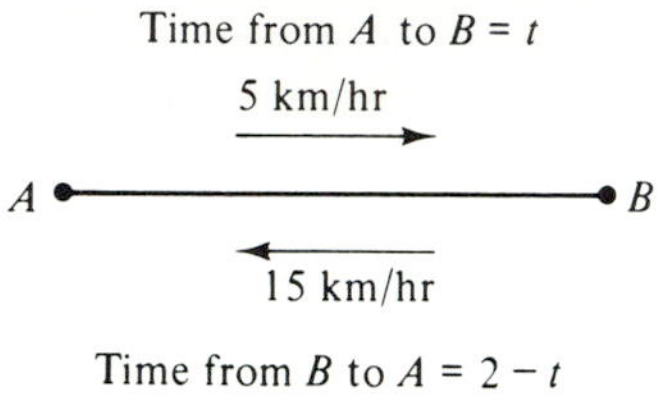

Time from A to $B = t$
5 km/hr
A —— B
15 km/hr
Time from B to $A = 2 - t$

FIGURE 4-6

Now, distance = (rate)(time) and

$$\begin{pmatrix}\text{distance}\\ \text{from}\\ A \text{ to } B\end{pmatrix} = \begin{pmatrix}\text{distance}\\ \text{from}\\ B \text{ to } A\end{pmatrix}.$$

Thus we have

$$\begin{aligned}(\text{rate})(\text{time}) &= (\text{rate})(\text{time}),\\ 5t &= 15(2 - t),\\ 5t &= 30 - 15t,\\ 20t &= 30,\\ t &= \frac{30}{20} = \frac{3}{2}.\end{aligned}$$

Thus the distance from A to B is (rate)(time) = $(5)(\frac{3}{2}) = \frac{15}{2} = 7\frac{1}{2}$ km.

EXAMPLE 2

A new insect spray, Bug Off, is in the experimental stages. It contains the remarkable new "killer" ingredient K-57. A lab assistant has available two spray formulas: formula A, of which 10% *is K-57; and formula B, of which* 16% *is K-57. So far, formula A has proved too weak. On the other hand, formula B seems too strong to be used near house pets. The lab assistant is told to mix formula A with* 400 ml *of formula B so that the result is* 14% *K-57. How many milliliters of formula A should be used?*

Let a be the number of milliliters of formula A to be added to the 400 ml of formula B. Then we end up with $a + 400$ ml, of which 14% must be K-57. That is,

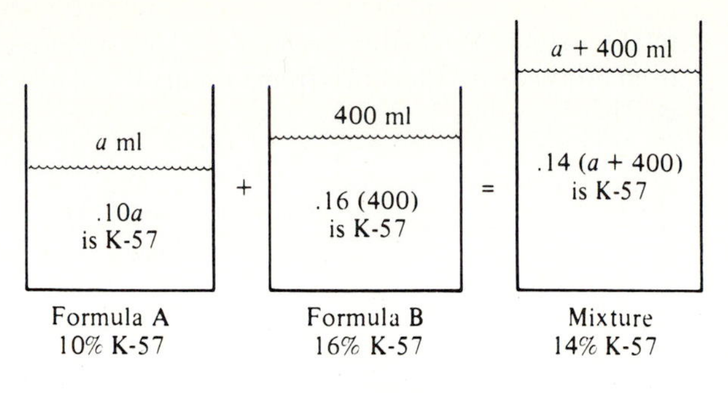

FIGURE 4-7

$.14(a+400)$ is K-57. See Fig. 4-7. This K-57 comes from two sources: $.10a$ comes from formula A and $.16(400)$ comes from formula B. Thus

$$.10a + .16(400) = .14(a+400),$$
$$.10a + 64 = .14a + 56,$$
$$8 = .04a,$$
$$a = \frac{8}{.04} = \boxed{200.}$$

Therefore 200 ml of formula A must be used.

EXAMPLE 3

Suppose that the lab assistant in Example 2 had needed exactly 500 ml of a 14% K-57 solution. How much of each formula would be used?

Let a be the number of milliliters of formula A to be used. Then to get a total of 500 ml of a 14% solution, there must be $500 - a$ ml of formula B. See Fig. 4-8. The

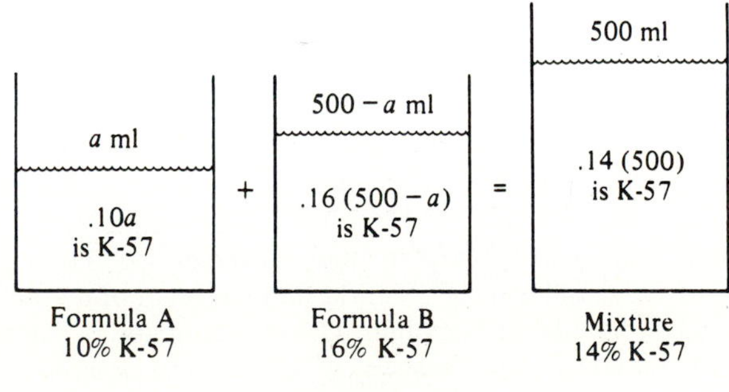

FIGURE 4-8

total amount of K-57 in the 500 ml of the 14% solution is .14(500). This K-57 comes from two sources: $.10a$ comes from formula A and $.16(500-a)$ comes from formula B. Thus

$$
\begin{aligned}
.10a + .16(500 - a) &= .14(500), \\
.10a + 80 - .16a &= 70, \\
-.06a &= -10, \\
a &= \frac{-10}{-.06} = \boxed{166\tfrac{2}{3}.}
\end{aligned}
$$

Therefore $500 - a = 500 - 166\frac{2}{3} = 333\frac{1}{3}$. The lab assistant should mix $166\frac{2}{3}$ ml of formula A with $333\frac{1}{3}$ ml of formula B.

Exercise 4.4

1. Suppose that the lunar rover in Example 1 traveled from A to B at 6 km/h and returned at 10 km/h. If the total time was 3 h, find the distance from A to B.

2. Suppose that the total time for the trip in Example 1 was 3 h. Based on the rates given in that example, find the distance from A to B.

3. A traveling salesperson drove 100 mi from Exton to Whyton in 2 h. At first he averaged 55 mi/h. But then he ran into bumpy road conditions for the rest of the trip. On that part he averaged 40 mi/h. For how long a time was he on the *bumpy* part of the road?

4. From two airports that are 300 mi apart, two airplanes leave at the same time and fly toward each other. One flies at 275 mi/h and the other at 325 mi/h. How long will it take for the planes to pass each other? (*Hint*: When they pass, the sum of the distances traveled by the planes is 300 mi.)

5. The water level in a certain reservoir is 6 ft deep, but the level is sinking at the rate of 4 in. a day. The water in another reservoir is 2 ft 9 in. deep and is rising $5\frac{3}{4}$ in. a day. When will the depths of the two reservoirs be the same? What will this depth be?

6. A pilot, flying against a headwind, traveled from A to B at 250 mi/h. She flew back, with the wind, at 300 mi/h. Her trip from B to A took 1 hour less than the trip from A to B. Find the distance from A to B.

7. For an airplane flight, the *point of no return* is the point on the flight where it will take as much *time* to fly on to the destination as to fly back to the starting point. Suppose an airplane is to fly from Honolulu to San Francisco, for which the air distance is 2397 statute miles. If the airplane has an average speed of 350 mi/h in still air and there is an average tail wind of 50 mi/h, find the point of no return. Give your answer to the nearest mile from Honolulu. (*Hint*: The speed of the plane to San Francisco is its speed in still air *plus* the speed of the wind. To Honolulu, it is its speed in still air *minus* the speed of the wind.)

8. A sharpshooter heard his bullet strike a target 2 seconds after firing his rifle. If the

bullet travels 580 m/s and sound travels 331 m/s, find the distance to the target. Give your answer to the nearest meter.

9. The manager of a fast-food restaurant finds that his "soyburgers" aren't selling too well. Soyburgers are a blend of 70% hamburger meat and 30% soy protein. To boost sales he decides to increase the percentage of hamburger meat to 80%. He has 40 lb of raw soyburger meat. How much pure hamburger must be added to it so that the resulting blend is 80% hamburger?

10. A 6-gal truck radiator is two-thirds full of water. How much of a 90% antifreeze solution (90% is antifreeze by volume) must be added to it to make a 10% antifreeze solution in the radiator?

11. A chemical manufacturer mixes a 20% acid solution (20% is acid by volume) with a 30% acid solution to get 700 gal of a 24% acid solution. How many gallons of each are used?

12. A chemical manufacturer wants to fill an order for 500 gal of a 25% acid solution (25% is acid by volume). Solutions of 30% and 18% are available in stock. How many gallons of each must he mix to fill the order?

13. A student in Maine commutes by car between her home and college. She doesn't want to miss any classes on cold winter days, so she decides to protect the car's cooling system from freeze-ups down to −34°F. She checks the antifreeze solution in the car's radiator and finds that it will give protection down to −12°F. The radiator has a 10-qt capacity. It contains 4 qt of pure antifreeze and 6 qt of water. For the added protection, 50% of the solution must be pure antifreeze. How many quarts of the present solution must she drain and replace with pure antifreeze so that she has the protection she wants?

14. How many milliliters of water must be evaporated from 80 ml of a 12% salt solution (12% is salt by volume) so that what remains is a 20% salt solution?

15. In a manufacturing process, a solution which is 50% chemical A and 50% chemical B is to be added to 3000 gal of chemical A so that after 30 minutes chemical B is 20% of the mixture. At what constant rate, in gal/min, should the solution be added?

4.5 LINEAR INEQUALITIES

An inequality looks like an equation, except that the symbol $=$ is replaced by either $<$, $>$, $\leq$, or $\geq$. We'll look at inequalities involving one variable: for example, $2x + 3 < 13$. Our goal is to find the values of the variable that make the inequality a true statement.

There are certain rules that we use to solve inequalities. The first is:

Rule 1. If $a < b$, then $a + c < b + c$ and $a - c < b - c$.

Rule 1 states that you can add (or subtract) the same number to (or from) both sides, or *members*, of an inequality and the inequality symbol remains the same. This rule is also true if you replace $<$ by $\leq$, $>$, or $\geq$.

EXAMPLE 1

Use of Rule 1.

a. Since $7 < 10$, then $7 + 3 < 10 + 3$. Thus $10 < 13$.

b. If $x + 2 > 9$, then $x + 2 - 2 > 9 - 2$. Thus $x > 7$.

Rule 2. If $a < b$ and c is a positive number, then

$$ac < bc \quad \text{and} \quad \frac{a}{c} < \frac{b}{c}.$$

Rule 2 states that if you multiply (or divide) both sides of an inequality by a *positive* number, the inequality symbol remains the same. This rule is also true for the other inequality symbols.

EXAMPLE 2

Use of Rule 2.

a. Since $4 < 9$, then $4 \cdot 2 < 9 \cdot 2$. Thus $8 < 18$.

b. If $2x \leq -4$, then $\frac{2x}{2} \leq \frac{-4}{2}$. Thus $x \leq -2$.

Rule 3. If $a < b$ and c is a positive number, then

$$a(-c) > b(-c) \quad \text{and} \quad \frac{a}{-c} > \frac{b}{-c}.$$

Rule 3 states that if you multiply (or divide) both sides of an inequality by a **negative** number, then you must **change** the direction (or *sense*) of the inequality symbol. Thus

$$< \text{ changes to } >.$$

If you started with the symbol $\leq$, it would change to $\geq$:

$$\leq \text{ changes to } \geq.$$

Similarly,

$$> \text{ changes to } < \quad \text{and} \quad \geq \text{ changes to } \leq.$$

EXAMPLE 3

Use of Rule 3.

a. Since $-2 < 100$, then $-2(-1) > 100(-1)$. Thus $2 > -100$.

b. If $-3x \geq 6$, then $\frac{-3x}{-3} \leq \frac{6}{-3}$. Thus $x \leq -2$.

Whenever one of Rules 1–3 is applied to an inequality, the resulting inequality is equivalent to the given one. (Equivalent inequalities, like equivalent equations, have the same solutions.) Thus to solve an inequality, we apply Rules 1–3 until the values of the variable are obvious.

EXAMPLE 4

Solve $2x + 3 < 13$.

$$2x + 3 < 13,$$

$$2x + 3 - 3 < 13 - 3 \quad \text{[subtracting 3 from both sides (Rule 1)]},$$

$$2x < 10 \quad \text{[simplifying]},$$

$$\frac{2x}{2} < \frac{10}{2} \quad \text{[dividing both sides by 2 (Rule 2)]},$$

$$\boxed{x < 5.}$$

Thus *all* numbers less than 5 are solutions to $2x + 3 < 13$. For example, since $1 < 5$, then 1 is a solution. Let's check it:

$$2x + 3 < 13,$$

$$2(1) + 3 < 13,$$

$$5 < 13. \ \checkmark$$

We can geometrically represent all the solutions by a bold line segment on a number line. See Fig. 4-9. The hollow dot means that 5 is **not included**. Notice that infinitely many values make the inequality true.

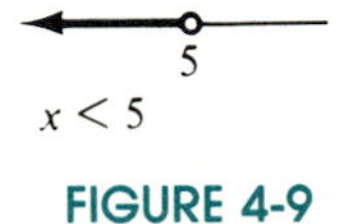

FIGURE 4-9

In Example 4 we call $2x + 3 < 13$ a *linear inequality* in x. Any inequality that can be put in the form $ax + b < 0$ is called a **linear inequality** in the variable x.* If the symbol $<$ in $ax + b < 0$ were replaced by $\leq$, $>$, or $\geq$, we would still have a linear inequality.

---

* We assume $a \neq 0$.

EXAMPLE 5

Solve $5x - 2(7x + 3) \leq 12$.

$$5x - 2(7x + 3) \leq 12,$$

$$5x - 14x - 6 \leq 12,$$

$$-9x - 6 \leq 12,$$

$$-9x \leq 18$$ [adding 6 to both sides (Rule 1)],

$$\frac{-9x}{-9} \geq \frac{18}{-9}$$ [dividing both sides by -9 and changing the direction of inequality symbol (Rule 3)],

$$\boxed{x \geq -2.}$$

Thus *all* numbers greater than or equal to -2 are solutions. We can represent them geometrically by the bold line segment in Fig. 4-10. The solid dot means that -2 **is included**.

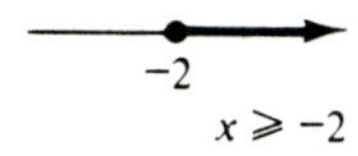

FIGURE 4-10

EXAMPLE 6

Solve $\frac{3}{2}t - 1 < \frac{5}{3}(-3 + t)$.

To clear of fractions, we multiply both sides by the L.C.D., 6.

$$6\left(\frac{3}{2}t - 1\right) < 6 \cdot \frac{5}{3}(-3 + t),$$

$$9t - 6 < 10(-3 + t),$$

$$9t - 6 < -30 + 10t,$$

$$-6 < -30 + t$$ [subtracting $9t$ from both sides],

$$24 < t$$ [adding 30 to both sides],

$$\boxed{t > 24}$$ [rewriting].

See Fig. 4-11.

24

$t > 24$

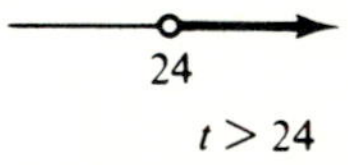

FIGURE 4-11

EXAMPLE 7

Solve $2(x-4)-3 \geq 2x+1$.

$$2(x-4)-3 \geq 2x+1,$$
$$2x-8-3 \geq 2x+1,$$
$$-11 \geq 1 \qquad \text{[subtracting } 2x \text{ from both sides]}.$$

Since $-11 \geq 1$ is never true, there is **no solution**.

EXAMPLE 8

Solve $3z+7>8z-(5z-2)$.

$$3z+7>8z-(5z-2),$$
$$3z+7>3z+2,$$
$$7>2 \qquad \text{[subtracting } 3z \text{ from both sides]}.$$

Since $7>2$ is always true (true for any value of z), the solution is **all real numbers**. See Fig. 4-12.

FIGURE 4-12

EXAMPLE 9

A builder must decide whether to rent or buy an excavating machine. If he were to rent the machine, the rental fee would be \$600 *per month (on a yearly basis), and the daily cost (gas, oil, and driver) would be* \$60 *for each day it is used. If he were to buy it, his fixed annual cost would be* \$4000, *and daily operating and maintenance costs would be* \$80 *for each day the machine is used. What is the least number of days each year that he would have to use the machine to justify renting it rather than buying it?*

Let d be the number of days each year that the machine is used. If the machine is rented, the total yearly cost consists of rental fees, which are (12)(600), and daily charges of $60d$. If the machine is purchased, the cost per year is $4000+80d$. We want

$$\text{cost}_{\text{rent}} < \text{cost}_{\text{purchase}},$$
$$12(600)+60d < 4000+80d,$$
$$7200+60d < 4000+80d,$$
$$3200 < 20d,$$
$$\frac{3200}{20} < \frac{20d}{20},$$
$$160 < d.$$

Thus the builder must use the machine at least 161 days to justify renting it.

Exercise 4.5

In Problems **1–32**, *solve each inequality and indicate your answer on a number line.*

1. $4x > 8$.

2. $8x < -2$.

3. $3x - 4 \leq 5$.

4. $5x \geq 0$.

5. $-4x \geq 2$.

6. $6 \leq 5 - 3y$.

7. $3 - 5s > 5$.

8. $4s - 1 < -5$.

9. $6x - 15 < 2x - 3$.

10. $3x + 1 > 2(x + 4)$.

11. $2x - 3 \leq 4 + 7x$.

12. $-3 \geq 8(2 - x)$.

13. $3 < 2y + 3$.

14. $2y + 3 \leq 1 + 2y$.

15. $3(2 - 3x) > 4(1 - 4x)$.

16. $8(x + 1) + 1 < 3(2x) + 1$.

17. $2(3x - 2) > 3(2x - 1)$.

18. $3 - 2(x - 1) \leq 2(4 + x)$.

19. $\frac{5}{3}x < 10$.

20. $-\frac{1}{2}x > 6$.

21. $\dfrac{9y + 1}{4} \leq 2y - 1$.

22. $\dfrac{4y - 3}{2} \geq \dfrac{1}{3}$.

23. $4x - 1 \geq 4(x - 2) + 7$.

24. $0x \leq 0$.

25. $\dfrac{1 - t}{2} < \dfrac{3t - 7}{3}$.

26. $\dfrac{3(2t - 2)}{2} > \dfrac{6t - 3}{5} + \dfrac{t}{10}$.

27. $2x + 3 \geq \frac{1}{2}x - 4$.

28. $4x - \frac{1}{2} \leq \frac{3}{2}x$.

29. $\frac{2}{3}r < \frac{5}{6}r$.

30. $\frac{7}{4}t > -\frac{2}{3}t$.

31. $\dfrac{y}{2} + \dfrac{y}{3} > y + \dfrac{y}{5}$.

32. $\dfrac{5y - 1}{-3} < \dfrac{7(y + 1)}{-2}$.

33. A businessperson wants to determine the difference between the costs of owning and renting an automobile. He can rent a small car for \$135 per month (on an annual basis). Under this plan, his cost per mile (gas and oil) is \$0.05. If he were to purchase the car, his fixed annual expense would be \$1000 and other costs would amount to \$0.10 per mile. What is the least number of miles per year he would have to drive to make renting no more expensive than purchasing?

34. Suppose a company offers you a sales position with your choice of two methods of determining your yearly salary. One method pays \$12,600 plus a bonus of 2% of your yearly sales. The other method pays a straight 8% commission of your sales. For what yearly sales' level is it better to choose the first method?

35. Painters are often paid either by the hour or on a per-job basis. The rate they receive can affect their working speed. For example, suppose they can work for \$8.50 per hour, or for \$300 plus \$3 for each hour less than 40 if they complete the job in less than 40 h. Suppose the job will take t hours. If $t \geq 40$, clearly the hourly rate is better. If $t < 40$, for what value of t is the hourly rate the better pay scale?

36. A company produces alarm clocks. During the regular work week, the labor cost for producing one clock is \$2.00. However, if a clock is produced in overtime, the labor cost is \$3.00. Management has decided to spend no more than a total of \$25,000 per week for labor. The company must produce 11,000 clocks this week. What is the minimum number of clocks that must be produced during the regular work week?

37. The dean of student affairs of a college is arranging for a rock group to perform a concert on campus. Their charge is a flat fee of \$2440 or, instead, a fee of \$1000 plus 40% of the gate. It is likely that 800 students will attend. At most, how much could the dean charge for a ticket so that the second arrangement is no more costly than the flat fee? If this maximum is charged, how much money will be left over to pay for publicity, guards, and other concert expenses?

4.6 REVIEW

IMPORTANT TERMS

variable (*4.1*)
constant (*4.1*)
equation (*4.1*)
unknown (*4.1*)
solution of an equation (*4.1*)
root of an equation (*4.1*)
equivalent equations (*4.1*)
linear equation (*4.1*)
total cost (*4.3*)
total revenue (*4.3*)
profit (*4.3*)
linear inequality in one variable (*4.5*)

REVIEW PROBLEMS

In Problems **1–18**, *solve the equations.*

1. $4x + 1 = 3$.

2. $9x - 7 = 11$.

3. $5 = 8 - 2y$.

4. $6y = 0$.

5. $8 - \frac{4x}{3} = 10$.

6. $6x - \frac{1}{3} = 5$.

7. $\frac{3}{4}z + 2 = \frac{1}{3}$.

8. $\frac{1}{10} - \frac{2z}{5} = 4$.

9. $\frac{4x + 3}{4} = 7$.

10. $\frac{2}{5} = \frac{5 - 3x}{10}$.

11. $9(3u + 2) = 3 - (u + 7)$.

12. $4\left(u - \frac{5}{7}\right) = -3u$.

13. $\frac{3}{2}(x - 8) = 2x + 4$.

14. $3\{2x + 4[(7 - 2x) - 5x]\} = 0$.

15. $5[3 - 2(3x - 4)] = 18 - 9x$.

16. $7x - [8x + 4(x + 2)] = -(1 + x)$.

17. $\frac{2y - 7}{3} + \frac{8y - 9}{14} = \frac{3y - 5}{21}$.

18. $y - \frac{y}{2} + \frac{y}{3} - \frac{y}{4} = \frac{y}{5}$.

In Problems **19–26**, *solve the inequalities.*

19. $3x + 5 < 6$.

20. $3(x + 1) > 9$.

21. $4 - 2x \geq 8$.

22. $-2(x + 6) \leq x + 4$.

23. $3(t + 4) < 9 + 6t$.

24. $5(s + 3) > 2(s - 1)$.

25. $\frac{1}{3}(x + 2) \geq \frac{1}{4}x + 4$.

26. $\frac{x + 1}{5} \leq \frac{x}{10} + 2$.

27. The formula for the velocity v of a certain object moving along a straight line is $v = v_0 + at$. The initial velocity is v_0, the acceleration is a, and t is time. Solve for a.

28. $E = \frac{1}{2}mv^2$ is a formula for kinetic energy E. Here m is mass and v is velocity. Solve for m.

29. The formula $V = \frac{\pi}{4}d^2 l$ gives the volume (in cubic millimeters) of fluid flowing per second from a capillary. Here d is the diameter (in millimeters) of the capillary and l is the linear velocity (in millimeters per second) of the fluid. If $V = .000015\pi$ and $d = .01$, find l.

30. Symbolize the statement: In a hospital there are four times as many nurses (n) as physicians (p).

31. 96 is 80% of what number?

32. 28 is 35% of what number?

33. 9 is what percentage of 75?

34. 162 is what percentage of 900?

35. In a certain city, 40%, or 3300, of the registered voters went to the polls. How many registered voters were there?

36. A scientist proposed a new theory. Testing it with 65 experiments, she was successful in getting a desired result 52 times. What was the percentage of successful experiments?

37. The Sparkle Lamp Company will have fixed costs of $26,500 this year. If the manufacturing cost of each lamp is $11, how many lamps can be made this year for a total cost of $76,000?

38. A builder has a client who wants an L-shaped living and dining area in his new house. See Fig. 4-13. The area is to be a total of 385 square feet. What should the length l of the living area be?

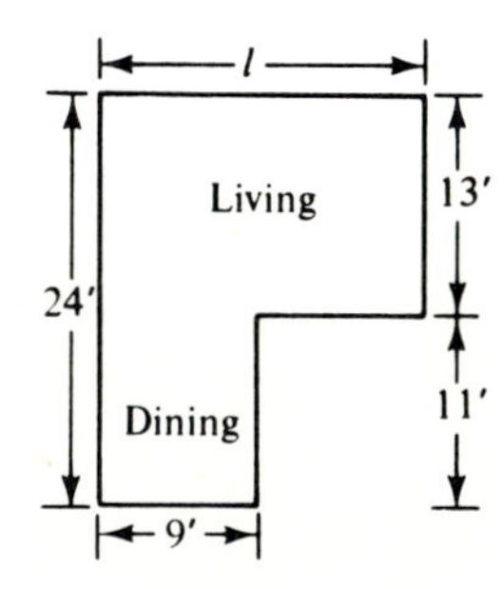

FIGURE 4-13

39. A certain alloy is made up of 8 parts of metal A, 3 parts of metal B, and 1 part of metal C by weight. How much of each metal is needed to make 168 tons of the alloy?

40. In a certain heat experiment, 2 kg of water at 79°C is mixed with 4 kg of water at 40°C. The final temperature of the mixture is given by t, where

$$2000(1)(79 - t) = 4000(1)(t - 40).$$

Solve for t.

41. A company manufactures a product at a cost of $8 per unit. Fixed costs are $15,000 and each unit sells for $10. How many units must be sold to earn a profit of $20,000?

42. In a poll of 360 people, four times as many favored a new product as compared to those who favored the best-selling brand. How many people favored the new product?

43. After you stain furniture, *The Consumer's Handbook** urges that you protect the color by sealing the stain. "For a homemade sealer, mix one part shellac and five parts denatured alcohol." If you need a pint (16 fluid ounces) of sealer, how many fluid ounces of shellac and how many ounces of denatured alcohol are needed?

44. An office has two photocopy machines. One produces 10 copies a minute and the other produces 25 copies a minute. If both machines are used together, approximately how many minutes would it take to produce 500 copies? Give your answer to the nearest minute.

45. A man can row 7 mi/h in still water. The current of a stream is 2 mi/h. How far upstream can he row if he is to be back at his starting point in 2 h? *Hint*: Upstream he goes 5 mi/h (which is rate − current), and downstream he goes 9 mi/h (which is rate + current).

46. How many gallons of antifreeze that is 70% alcohol (by volume) must be added to 10 gallons of a 35% solution to get a 50% solution?

47. A company manufactures a drain cleaner. The cleaner consists of a chemical compound and metal shavings. The chemical compound not only loosens grease, but when dissolved in water it gives off heat, which speeds up the reaction, and it reacts with the metal to generate hydrogen, which also loosens dirt and grease. The company markets two forms of the cleaner: industrial strength, of which 9% is metal shavings (by weight); and household strength, of which 6% is metal shavings. A motel chain has placed an order with the company to supply them with 12,000 kg of a new form of the cleaner, which is 8% metal shavings. To fill the order, the company will mix the industrial and household forms. How many kilograms of each should go into the mixture?

48. Suppose that consumers will purchase q units of a product at a price of $\frac{100}{q} + 1$ dollars per unit. What is the minimum number of units that must be purchased in order that sales revenue be greater than $5000?

49. A manufacturer has 2500 units of product in stock at present. The product is now selling at $4 per unit. Next month the unit price will increase by $0.50. The manufacturer wants the total revenue received from the sale of the 2500 units to be no less than $10,750. What is the maximum number of units that can be sold this month?

---

* Paul Fargis, ed. (New York: Hawthorn, 1974), p. 90.

5 Special Products and Factoring

5.1 SPECIAL PRODUCTS

In mathematics certain products occur so often that we find it worthwhile to memorize their patterns. But just knowing these *special products* is not enough. You have to be so familiar with them that you can recognize them in any form.

Each special product is found by applying the distributive law. We'll show this in the first special product. You can check the others for yourself.

To begin, let's look at the square of $a + b$. We call $a + b$ a **binomial** because it has exactly *two* terms.

$$(a + b)^2 = (a + b)(a + b)$$
$$= a(a + b) + b(a + b) = a^2 + ab + ba + b^2.$$
$$(a + b)^2 = a^2 + 2ab + b^2.$$

Since a is the first term of $a + b$ and b is the second term, our result means that

$$(a + b)^2 = \begin{pmatrix}\text{square of}\\ \text{first term}\end{pmatrix} + \begin{pmatrix}\text{twice the}\\ \text{product of}\\ \text{the terms}\end{pmatrix} + \begin{pmatrix}\text{square of}\\ \text{second term}\end{pmatrix}.$$

Similarly, $(a - b)^2 = a^2 - 2ab + b^2$. These are our first two special products.

SQUARE OF A BINOMIAL

$$(a + b)^2 = a^2 + 2ab + b^2.$$
$$(a - b)^2 = a^2 - 2ab + b^2.$$

Note that the square of a binomial has exactly *three* terms. Such expressions are called **trinomials**.

$(a + b)^2 \neq a^2 + b^2$ and $(a - b)^2 \neq a^2 - b^2$. *For example,*

$$(6 + 3)^2 \neq 6^2 + 3^2 \text{ because } 81 \neq 45.$$

EXAMPLE 1

Square of a binomial: $(a + b)^2 = a^2 + 2ab + b^2$.

a. $(x + 3)^2$. Here x plays the role of a, and 3 plays the role of b.

$$(x + 3)^2 = x^2 + 2(x)(3) + 3^2 = x^2 + 6x + 9.$$

b. $(2x + 4)^2$. Here $2x$ plays the role of a.

$$(2x + 4)^2 = (2x)^2 + 2(2x)(4) + 4^2 = 4x^2 + 16x + 16.$$

c. $(x^2 + 4y)^2$. Here x^2 is a and $4y$ is b.

$$\begin{aligned}(x^2 + 4y)^2 &= (x^2)^2 + 2(x^2)(4y) + (4y)^2 \\ &= x^4 + 8x^2y + 16y^2.\end{aligned}$$

d. $-3y(2a + b)^2$. First we square the binomial.

$$\begin{aligned}-3y(2a + b)^2 &= (-3y)[(2a)^2 + 2(2a)(b) + b^2] \\ &= (-3y)[4a^2 + 4ab + b^2] \\ &= -12a^2y - 12aby - 3b^2y.\end{aligned}$$

EXAMPLE 2

Square of a binomial: $(a - b)^2 = a^2 - 2ab + b^2$.

a. $(x - 4)^2 = x^2 - 2(x)(4) + 4^2 = x^2 - 8x + 16$.

b. $(3 - 2x)^2$. Here 3 plays the role of a, and $2x$ plays the role of b.

$$\begin{aligned}(3 - 2x)^2 &= 3^2 - 2(3)(2x) + (2x)^2 \\ &= 9 - 12x + 4x^2.\end{aligned}$$

c. $(x^3 - 2y^2)^2$. Here x^3 is a and $2y^2$ is b.

$$\begin{aligned}(x^3 - 2y^2)^2 &= (x^3)^2 - 2(x^3)(2y^2) + (2y^2)^2 \\ &= x^6 - 4x^3y^2 + 4y^4.\end{aligned}$$

d. $(-m + n)^2 = (n - m)^2 = n^2 - 2mn + m^2$.

Formulas for the cube of a binomial are given in Problems 81 and 83 of Exercise 5.1. Our next special product involves the product of the sum and difference of two terms.

PRODUCT OF THE SUM AND DIFFERENCE

$$(a+b)(a-b)=a^2-b^2.$$

That is, the product of the sum and difference of two terms is equal to the square of the first term, minus the square of the second term.

EXAMPLE 3

Product of the sum and difference: $(a+b)(a-b)=a^2-b^2$.

a. $(x+5)(x-5)=x^2-5^2=x^2-25$.

b. $(7+y)(7-y)=7^2-y^2=49-y^2$.

c. $(4x-3)(4x+3)$. Here $4x$ plays the role of a, and 3 plays the role of b.

$$(4x-3)(4x+3)=(4x)^2-3^2=16x^2-9.$$

d. $(x^2-1)(x^2+1)=(x^2)^2-1^2=x^4-1$.

e. $(2x^2+3y)(3y-2x^2)$. This will have the special form if we rewrite the first factor.

$$\begin{aligned}(3y+2x^2)(3y-2x^2)&=(3y)^2-(2x^2)^2\\&=9y^2-4x^4.\end{aligned}$$

f. $(x-\sqrt{5})(x+\sqrt{5})=x^2-(\sqrt{5})^2=x^2-5$.

Another special product involving binomials is

$$(x+a)(x+b)=x^2+(a+b)x+ab.$$

EXAMPLE 4

Product of two binomials: $(x+a)(x+b)=x^2+(a+b)x+ab$.

a. $$\begin{aligned}(x+3)(x+4)&=x^2+(3+4)x+(3\cdot 4)\\&=x^2+7x+12.\end{aligned}$$

b. $$\begin{aligned}(x+5)(x+1)&=x^2+(5+1)x+(5\cdot 1)\\&=x^2+6x+5.\end{aligned}$$

c. $(x - 2)(x + 6)$. Here -2 is a since $x - 2 = x + (-2)$.

$$(x - 2)(x + 6) = x^2 + (-2 + 6)x + (-2)(6)$$
$$= x^2 + 4x - 12.$$

d. $(y + 1)(y - 5)$. Here -5 is b.

$$(y + 1)(y - 5) = y^2 + (1 - 5)y + (1)(-5)$$
$$= y^2 - 4y - 5.$$

e. $(x - 3)(x - 2)$. Here -3 is a and -2 is b.

$$(x - 3)(x - 2) = x^2 + (-3 - 2)x + (-3)(-2)$$
$$= x^2 - 5x + 6.$$

The products in Example 4 can be found another way. We shall show you how by redoing Example 4c: $(x - 2)(x + 6)$.

Step 1. Multiply the first terms in the binomials to get the first term of the result.

$$(x - 2)(x + 6) = \underline{x^2\qquad\qquad}$$

Step 2. The middle term of the result is the product of the outer terms of the binomials, *plus* the product of the inner terms.

$$(x - 2)(x + 6) = \underline{x^2 + 4x\qquad}$$

$-2x$

$6x$

$4x$

Step 3. Multiply the last terms in the binomials to get the last term of the result.

$$(x - 2)(x + 6) = \underline{x^2 + 4x - 12.}$$

The 3-step method above can be used for multiplying many types of binomials.

EXAMPLE 5

Find $(4x + 1)(5x - 3)$ *by the* 3-*step method.*

Step 1. $(4x + 1)(5x - 3) = \underline{20x^2\qquad\qquad}$

Step 2. $(4x + 1)(5x - 3) = \underline{20x^2 - 7x}$

$$\begin{array}{r} 5x \\ -12x \\ \hline -7x \end{array}$$

Step 3. $(4x + 1)(5x - 3) = \underline{20x^2 - 7x - 3.}$

In Example 5 we spread our work over three lines so that you could follow it. But you really should do it in one line.

After some practice you should be able to find special products in your head. Memorize the formulas and the 3-step method until you know them "cold."

EXAMPLE 6

Problems involving special products.

a. $(x^2 + 1)(x - 1)(x + 1) = (x^2 + 1)[(x - 1)(x + 1)]$

$$= (x^2 + 1)[x^2 - 1] = x^4 - 1.$$

b. $(x + 2)(x - 2) + (x + 3)^2 = [x^2 - 4] + [x^2 + 6x + 9]$

$$= 2x^2 + 6x + 5.$$

c. $(x - 3)^2 - (4 - x)^2 = x^2 - 6x + 9 - (16 - 8x + x^2)$

$$= x^2 - 6x + 9 - 16 + 8x - x^2$$

$$= 2x - 7.$$

d. $(x - 1)(x + 3)^2 = (x - 1)(x^2 + 6x + 9)$

$$= x(x^2 + 6x + 9) - 1(x^2 + 6x + 9)$$

$$= x^3 + 6x^2 + 9x - x^2 - 6x - 9$$

$$= x^3 + 5x^2 + 3x - 9.$$

Exercise 5.1

In Problems **1–62**, *perform the operation by using special products. Do as many as you can in your head.*

1. $(x + 4)^2$. **2.** $(x + 6)^2$. **3.** $(y + 10)^2$.

4. $(z + 5)^2$. **5.** $(4x + 1)^2$. **6.** $(2x + 10)^2$.

7. $(x + \frac{1}{2})^2$. **8.** $(2 + x)^2$. **9.** $(x - 6)^2$.

10. $(x - 2)^2$. **11.** $(3x - 2)^2$. **12.** $(3 - x)^2$.

13. $(2x - y)^2$. **14.** $(2x - 3)^2$. **15.** $(3x + 3)^2$.

16. $(2-t)^2$.
17. $(2-4y)^2$.
18. $(3y-4x)^2$.
19. $(x+3)(x-3)$.
20. $(x+2)(x-2)$.
21. $(x-9)(x+9)$.
22. $(7-x)(7+x)$.
23. $(1+x)(1-x)$.
24. $(4+2t)(4-2t)$.
25. $(3x^2+5)(3x^2-5)$.
26. $(2-3x)(2+3x)$.
27. $(2y+3x)(2y-3x)$.
28. $(y-8)(y+8)$.
29. $(6-x)(x+6)$.
30. $(x+yz)(x-yz)$.
31. $(x+8)(x+3)$.
32. $(x-6)(x-1)$.
33. $(x+4)(x+1)$.
34. $(x+3)(x+6)$.
35. $(x-2)(x+1)$.
36. $(x+5)(x+4)$.
37. $(t+7)(t-5)$.
38. $(x-2)(x+14)$.
39. $(x-2)(x-3)$.
40. $(x-6)(x-2)$.
41. $(x-4)(x-5)$.
42. $(x-1)(x-2)$.
43. $(y-2)(y+3)$.
44. $(t-4)(t+3)$.
45. $(x+3)(2x-4)$.
46. $(2x+1)(2x+2)$.
47. $(4x-2)(2x-1)$.
48. $(x-3)(4x-3)$.
49. $(2x+1)(3x-3)$.
50. $(2x+1)(x-2)$.
51. $(5x-4)(2x-1)$.
52. $(x+\sqrt{13})(x-\sqrt{13})$.
53. $(2-5t)(1+7t)$.
54. $(4-t)(4+3t)$.
55. $(xy^2+a)^2$.
56. $(x^3-1)(x^3-1)$.
57. $(x^2-y^2)(x^2+y^2)$.
58. $(x^2-2)^2$.
59. $(x-\sqrt{7})(x+\sqrt{7})$.
60. $(ab-c)(ab+c)$.
61. $(-2x+1)^2$.
62. $(t-2s)(t+2s)$.

In Problems **63–80**, *perform the operations.*

63. $4(x+3)^2$.
64. $x(x-2)^2$.
65. $2x(y-3)(y+3)$.
66. $-2a(a+4)^2$.
67. $(a^2b-2m^2n)^2$.
68. $(x-1)(x+1)(x)$.
69. $(x+2)(x-2)(x^2+4)$.
70. $(x+1)(x+1)(x+2)$.
71. $(a-b)^2-(b-a)^2$.
72. $(x+y)^2-(y-x)^2$.
73. $(2x-3)^2+(x+1)(x+2)$.
74. $(t+3)(t-3)-(t+3)^2$.
75. $(x+2)(x-2)^2$.
76. $a(a-b)^2-b(b-a)^2$.
77. $[4x(x-3)]^2$.
78. $[2x^2(x-y)]^2$.
79. $3x(x+2)-(\sqrt{3}x-1)(\sqrt{3}x+1)$.
80. $(5x^2+1)(\sqrt{5}x+1)(\sqrt{5}x-1)$.

81. Use the formula

$$(a+b)^3=a^3+3a^2b+3ab^2+b^3$$

to find $(x+2)^3$.

82. Use the formula in Problem 81 to find $(3x+1)^3$.

83. Use the formula

$$(a-b)^3=a^3-3a^2b+3ab^2-b^3$$

to find $(2x-3)^3$.

84. Use the formula in Problem 83 to find $(3x^2 - 2y)^3$.

85. The kinetic energy (KE) of an object of mass m moving with a speed v is given by

$$\text{KE} = \frac{1}{2}mv^2.$$

If the speed of an object at time t is $v = 2t + 1$, find an expression for the kinetic energy in terms of t and m. Perform the multiplications and simplify.

86. The speed v of a model rocket at time t is given by $v = t + 2$. The mass m of the rocket decreases as fuel is burned. If $m = 12 - .1t$, find the kinetic energy of the rocket in terms of t. (See Problem 85.) Perform the multiplications and simplify.

87. In studies of battery power, the expression

$$(R + r)^2(1) - R(2)(R + r)$$

arises. Perform the indicated operations and simplify.

5.2 FACTORING

In this section we'll write an expression as a product of factors. This is called **factoring**. It will come in handy when we solve equations in the next chapter.

Let's factor $3x + 3y$. Here 3 is a factor of *each* of the terms $3x$ and $3y$. This **common factor** 3 can be pulled out of $3x + 3y$ by the distributive law:

$$3x + 3y = 3(x + y).$$

Thus $3x + 3y$ is factored. We say that a *common factor was removed* from each term.

Similarly, in the expression $4x - 6y + 2$ we see that 2 is a factor of $4x$ (since $4x = 2 \cdot 2x$), $6y$ (since $6y = 2 \cdot 3y$), and 2 (since $2 = 2 \cdot 1$). Thus we can remove the common factor 2 from each term.

$$4x - 6y + 2 = 2(2x - 3y + 1).$$

Common factors can be numbers, letters, or a combination of these.

EXAMPLE 1

Removing a common factor.

a. $5x - 30 = 5(x - 6)$.

b. $8x + 12y = 4(2x + 3y)$.

c. $6x + 2 = 2(3x + 1)$.

d. $ax - ab = a(x - b)$.

e. $xy - 3xz - x = x(y - 3z - 1)$.

*Be careful when removing common factors. You can remove only factors that are common to **all** terms. In $6x + 6y + z$, the 6 is a factor of only two terms.*

$$6x + 6y + z \neq 6(x + y + z).$$

In Example 1c, we factored $6x + 2$ as $2(3x + 1)$. You can check that another factored form for $6x + 2$ is $6(x + \frac{1}{3})$. However, when factoring an expression that has integer coefficients, we shall usually choose factors whose terms also have integer coefficients. Thus we factor $6x + 2$ as $2(3x + 1)$, not as $6(x + \frac{1}{3})$. (Later in the book we may find it convenient to lift this restriction.) Also, when factoring polynomials we shall choose factors that are themselves polynomials. For example, you can check that

$$6x + 2 = 2x^{-2}(3x^3 + x^2).$$

But $2x^{-2}$ is not a polynomial, so we do not consider it as a factor of $6x + 2$.

Let's now factor $4bx^2 + 16bx^6$. Both 4 and b are factors of both terms.

$$4bx^2 + 16bx^6 = 4b(x^2 + 4x^6).$$

But this is not completely factored. We can factor some more, since x^2 is common to x^2 and $4x^6 (= 4 \cdot x^2 \cdot x^4)$. Thus

$$4bx^2 + 16bx^6 = 4b(x^2 + 4x^6) = 4bx^2(1 + 4x^4).$$

We've now factored as much as possible. *Always factor completely.*

EXAMPLE 2

Factoring completely.

a. $7ax + 14bx = 7x(a + 2b)$.

b. $x^4 + x^3 + x^2 = x^2(x^2 + x + 1)$. *Note*: If only an x were factored out, this would give $x(x^3 + x^2 + x)$, which is not completely factored.

c. $3a^2b^5 - 4a^3b^3 = a^2b^3(3b^2 - 4a)$.

d. $9x^2y + 3xy^2 - 6xy = 3xy(3x + y - 2)$.

EXAMPLE 3

Completely factor $2(x - 1)^3(x + 2) + 3(x - 1)^2(x + 2)^2$.

Calculus students often come across an expression like this and must completely factor it. The expression consists of *two* terms:

$$2(x - 1)^3(x + 2) \quad \text{and} \quad 3(x - 1)^2(x + 2)^2.$$

The factors that are common to both terms are $(x-1)^2$ and $(x+2)$. Thus

$$\begin{aligned}&2(x-1)^3(x+2)+3(x-1)^2(x+2)^2\\&=(x-1)^2(x+2)[2(x-1)+3(x+2)]\\&=(x-1)^2(x+2)[2x-2+3x+6]\\&=(x-1)^2(x+2)(5x+4) \qquad \text{[simplifying last factor].}\end{aligned}$$

From the last section we know that $(a+b)(a-b)=a^2-b^2$. Thus the *difference of two squares* will factor into a sum and difference.

DIFFERENCE OF TWO SQUARES

$$a^2-b^2=(a+b)(a-b).$$

EXAMPLE 4

Factoring the difference of two squares.

a. $x^2-4=x^2-2^2=(x+2)(x-2)$.

b. $4x^2-1=(2x)^2-1^2=(2x+1)(2x-1)$.

c. $9x^2-16y^2=(3x+4y)(3x-4y)$.

d. $x^4-1=(x^2)^2-1^2=(x^2+1)(x^2-1)$. Although x^2+1 doesn't factor, x^2-1 does, and so we have

$$x^4-1=(x^2+1)(x+1)(x-1).$$

In general, a^2+b^2 does **not** factor.

Sometimes we can factor trinomials that have leading coefficient 1, such as $x^2-4x-12$, into a product of two binomials of the form $(x+a)$ and $(x+b)$. Suppose

$$x^2-4x-12=(x+a)(x+b).$$

Now we must determine the values of a and b that do the job.

From the last section,

$$(x+a)(x+b)=x^2+(a+b)x+ab.$$

Thus

$$x^2-4x-12=x^2+(a+b)x+ab.$$

Matching the expression on the right side with $x^2 - 4x - 12$, we must have

$$a + b = -4 \quad \text{and} \quad ab = -12.$$

There are several choices for a and b such that their product is -12:

$$\begin{array}{ll} -1 \text{ and } 12, & 1 \text{ and } -12, \\ -2 \text{ and } 6, & 2 \text{ and } -6, \\ -3 \text{ and } 4, & 3 \text{ and } -4. \end{array}$$

But since the sum of a and b must be -4, we choose $a = 2$ and $b = -6$. Thus

$$x^2 - 4x - 12 = (x + 2)(x - 6).$$

You can check this: multiply the right side and see if you get the left side.

EXAMPLE 5

Completely factor $x^2 - 5x + 4$.

We need to find two numbers whose product is $+4$ and whose sum is -5. Obviously both numbers must be negative; -4 and -1 will do the job.

$$x^2 - 5x + 4 = (x - 4)(x - 1).$$

EXAMPLE 6

Factoring trinomials.

a. $x^2 + 9x + 20 = (x + 5)(x + 4)$.

b. $y^2 + 6y + 9 = (y + 3)(y + 3) = (y + 3)^2$. Therefore $y^2 + 6y + 9$ is the square of a binomial.

c. $z^2 + 6z - 16 = (z - 2)(z + 8)$.

d. $z^2 - 6z - 16 = (z + 2)(z - 8)$. Compare the signs here with those in part c.

Let's tackle a trinomial whose leading coefficient is not 1, like $6x^2 + 5x - 4$. If it factors into a product of two binomials, then we must have something like

$$6x^2 + 5x - 4 = (\underline{\quad} + \underline{\quad})(\underline{\quad} + \underline{\quad}).$$

(last terms: -4; first terms: $6x^2$)

The product of the first terms of the binomials must be $6x^2$, and the product of the last terms must be -4. There are many combinations of binomials that will do this. One is $(x + 4)(6x - 1)$. But here the sum of the products of the outer terms and inner terms, $23x$, is not equal to the middle term of $6x^2 + 5x - 4$, which is $5x$. By trial and error we try different combinations until we hit upon one that works.

$$
\begin{array}{ll}
(x+4)(6x-1), & (2x+2)(3x-2), \\
(x-4)(6x+1), & (2x-2)(3x+2), \\
(x+1)(6x-4), & (2x+4)(3x-1), \\
(x-1)(6x+4), & (2x-4)(3x+1), \\
(x+2)(6x-2), & (2x+1)(3x-4), \\
(x-2)(6x+2), & (2x-1)(3x+4).
\end{array}
$$

The combination that works is $(2x - 1)(3x + 4)$.

$$6x^2 + 5x - 4 = (2x - 1)(3x + 4).$$

Check it. Here we factored a polynomial of degree two. This method may work for other types of trinomials, as Example 7d shows.

EXAMPLE 7

Factoring trinomials completely.

a. $2x^2 + 5x + 2 = (2x + 1)(x + 2)$.

b. $8y^2 - 6y - 9 = (2y - 3)(4y + 3)$.

c. $9y^2 - 12y + 4 = (3y - 2)(3y - 2) = (3y - 2)^2$, the square of a binomial.

d. $x^4 - 3x^2 + 2 = (x^2 - 1)(x^2 - 2) = (x + 1)(x - 1)(x^2 - 2)$. Note that we can write $x^2 - 2$ as $(x + \sqrt{2})(x - \sqrt{2})$, but not all the terms in these factors have *integer* coefficients. Thus we leave $x^2 - 2$ alone.

When factoring, it is best to first remove any common factors, as Example 8 shows.

EXAMPLE 8

Factoring completely.

a. $4x^3 - 16x = 4x(x^2 - 4) = 4x(x + 2)(x - 2)$.

b. $8ay^2 + 8ay + 2a = 2a(4y^2 + 4y + 1)$
$= 2a(2y + 1)(2y + 1) = 2a(2y + 1)^2$.

c. $7a^5 + 7a^3 = 7a^3(a^2 + 1)$.

d. $6x^5 - 4x^3 - 2x = 2x(3x^4 - 2x^2 - 1)$

$= 2x(3x^2 + 1)(x^2 - 1)$

$= 2x(3x^2 + 1)(x + 1)(x - 1).$

Exercise 5.2

Factor completely.

1. $8x + 8.$

2. $9x - 9.$

3. $10x - 5y + 25.$

4. $12x^2 + 24y - 4.$

5. $5cx + 9x.$

6. $16mx + 4m.$

7. $6xy + 3xz.$

8. $4xyz - 5yz.$

9. $2x^3 - x^2.$

10. $5x^8 - 4x^7.$

11. $2x^3y^3 + x^5y^5.$

12. $m^2y - y^2m.$

13. $4m^2x^3 - 8mx^4.$

14. $25a^5x^9 - 15a^4x^{10}.$

15. $9a^4y^3 + 3a^2y^5 - 6a^3y^4z.$

16. $by^5 - 2b^3y^4 - 8b^2y^2.$

17. $x^2 - 1.$

18. $x^2 - 49.$

19. $x^2 + 4x + 3.$

20. $x^2 + 5x + 6.$

21. $x^2 - 9x + 20.$

22. $x^2 + 3x - 10.$

23. $y^2 + 2y - 24.$

24. $y^2 - 10y + 9.$

25. $y^2 - 36.$

26. $4 - y^2.$

27. $x^2 + 12x + 36.$

28. $x^2 - 3x - 28.$

29. $x^2 - 4x - 32.$

30. $x^2 - 8x + 12.$

31. $y^2 - 10y + 25.$

32. $y^2 + 8y + 16.$

33. $3x^2 + 7x + 2.$

34. $5x^2 - 12x + 4.$

35. $2y^2 - 7y + 3.$

36. $7y^2 + 9y + 2.$

37. $16x^2 + 8x + 1.$

38. $4x^2 - 4x + 1.$

39. $9 - 4x^2y^2.$

40. $a^2b^2 - c^2d^2.$

41. $4y^2 + 7y - 2.$

42. $8y^2 + 2y - 3.$

43. $6x^2 - 11x - 10.$

44. $5x^2 + 14x - 3.$

45. $2x^2 + 4x - 6.$

46. $a^2x^2 + a^2x - 20a^2.$

47. $3x^3 + 18x^2 + 27x.$

48. $3x^4 - 15x^3 + 18x^2.$

49. $16s^2t^3 - 4s^2t.$

50. $a^2b^2 - a^4b^4.$

51. $4y^2 - 6y - 18.$

52. $30y^2 + 55y + 15.$

53. $(x + 3)^3(x - 1) + (x + 3)^2(x - 1)^2.$

54. $(x + 5)^2(x + 1)^3 + (x + 5)^3(x + 1)^2.$

55. $(x + 4)(2x + 1) + (x + 4).$

56. $(x - 3)(2x + 3) - (2x + 3)(x + 5).$

57. $x^4 - 16.$

58. $81x^4 - y^4.$

59. $y^8 - 1.$

60. $t^4 - 4.$

61. $x^4 + x^2 - 2.$

62. $x^4 - 5x^2 + 4.$

63. $x^5 - 2x^3 + x.$

64. $4x^3 - 6x^2 - 4x.$

5.3 FACTORING, CONTINUED

Sometimes we can factor an expression by first grouping its terms in a suitable way. The next example shows how this may be done.

EXAMPLE 1

Factor $x^2 + xy - 5x - 5y$ *by grouping.*

Grouping the first two terms and the last two terms, we have

$$\begin{aligned} & x^2 + xy - 5x - 5y \\ &= (x^2 + xy) - (5x + 5y) && \text{[note the signs for the second grouping]} \\ &= x(x + y) - 5(x + y) && \text{[factoring each group]} \\ &= (x + y)(x - 5) && \text{[removing the common factor } (x + y) \text{ from both terms].} \end{aligned}$$

Another way to factor the expression is by grouping the first and third terms and the second and last terms:

$$\begin{aligned} & x^2 + xy - 5x - 5y \\ &= (x^2 - 5x) + (xy - 5y) \\ &= x(x - 5) + y(x - 5) && \text{[factoring each group]} \\ &= (x - 5)(x + y) && \text{[removing the common factor } (x - 5) \text{ from both terms].} \end{aligned}$$

There are formulas for factoring a sum or difference of cubes:

SUM OR DIFFERENCE OF CUBES

$$a^3 + b^3 = (a + b)(a^2 - ab + b^2).$$

$$a^3 - b^3 = (a - b)(a^2 + ab + b^2).$$

EXAMPLE 2

Factoring the sum or difference of cubes.

a. $x^3 + 8 = x^3 + 2^3 = (x + 2)[x^2 - (x)(2) + 2^2]$

$$= (x + 2)(x^2 - 2x + 4).$$

b. $2y^6 - 54 = 2(y^6 - 27)$ [removing common factor]

$$= 2[(y^2)^3 - 3^3]$$

$$= 2(y^2 - 3)[(y^2)^2 + (y^2)(3) + 3^2]$$

$$= 2(y^2 - 3)(y^4 + 3y^2 + 9).$$

Factoring can be a useful tool for solving equations, as Example 3 shows.

EXAMPLE 3

Solve $x = ax + b$ for x.

$$x = ax + b,$$

$$x - ax = b \quad \text{[getting terms involving } x \text{ on one side]},$$

$$x(1 - a) = b \quad \text{[factoring]},$$

$$\frac{x(1-a)}{1-a} = \frac{b}{1-a} \quad \text{[dividing both sides by } 1 - a],$$

$$\boxed{x = \frac{b}{1-a}}.$$

Exercise 5.3

In Problems **1–10**, *factor completely by grouping.*

1. $x + xy + 4 + 4y$.

2. $2ax - ay + 6bx - 3by$.

3. $xy - 2x - 4y + 8$.

4. $x^3 + x^2 + x + 1$.

5. $2x^3 - 3x^2 - 8x + 12$.

6. $x^4 - x^3 + x^2 - 1$.

7. $ax^2 - ay^2 + bx^2 - by^2$.

8. $x^2 + x - y^2 - y$.

9. $x^2 + 4x + 4 - a^2$.

10. $a^2 - b^2 + 2bc - c^2$.

In Problems **11–18**, *use the formulas for the sum or difference of cubes to factor completely.*

11. $x^3 + 27$.

12. $x^3 - 1$.

13. $8y^3 - 27$.

14. $64y^3 + 8$.

15. $24x^3 + 3y^6$.

16. $a^2 - b^2 + a^3 - b^3$.

17. $(a + 1)^3 - (b + 1)^3$.

18. $(1 - y^2)^3 - (1 - x^2)^3$.

In Problems **19–26**, *solve each equation for the given letter.*

19. $S = P + Prt$; P.

20. $ax + cx = 2ax + a^2$; x.

21. $ax + b = cx + d$; x.

22. $2at - t = 4bt + 6$; t.

23. $mgh + \frac{1}{2}mv^2 = c$; m.

24. $Q = mc(t_2 - t_1) + mL$; m.

25. $a(s - t) + b(t + s) = 7$; s.

26. $a(x - y) + b(x + y) = cxy$; x.

27. The total area of a certain closed cylinder of radius r and altitude h is given by $2\pi r^2 + 2\pi rh$. Completely factor this expression.

28. If a ball of mass m is tied to a string and whirled around in a vertical circle, at the ball's lowest point we have

$$T - mg = m\frac{v^2}{r},$$

where T is the tension in the string, g is the acceleration due to gravity, v is the speed of the ball, and r is the radius of the circle. Solve for m.

29. To study a predator-prey relationship, an experiment* was conducted in which a blindfolded subject, the "predator," stood in front of a 3-foot square table on which uniform sandpaper discs, the "prey," were placed. For one minute the "predator" searched for the discs by tapping with a finger. Whenever a disc was found, it was removed and searching resumed. The experiment was repeated for various disc densities (number of discs per 9 ft^2). If y is the number of discs picked up in one minute when x discs are on the table, it was estimated that

$$y = a(1 - by)x,$$

where a and b are constants. Solve this equation for y.

5.4 BINOMIAL THEOREM†

Let's look at some powers of the binomial $a + b$. We know the special product

$$(a + b)^2 = a^2 + 2ab + b^2. \qquad (1)$$

Now we'll find $(a + b)^3$.

$$\begin{aligned}(a + b)^3 &= (a + b)(a + b)^2 \\ &= (a + b)(a^2 + 2ab + b^2).\end{aligned}$$

---

* C. S. Holling, "Some Characteristics of Simple Types of Predation and Parasitism," *The Canadian Entomologist*, XCI, no. 7 (1959), 385–398.

† This section is not needed for the rest of the book.

By the distributive law,

$$\begin{aligned}(a+b)(a^2+2ab+b^2) &= a(a^2+2ab+b^2)+b(a^2+2ab+b^2)\\ &= a^3+2a^2b+ab^2+a^2b+2ab^2+b^3\\ &= a^3+3a^2b+3ab^2+b^3.\end{aligned}$$

Thus

$$(a+b)^3 = a^3+3a^2b+3ab^2+b^3. \qquad (2)$$

There are similarities in the *expansions* of $(a+b)^2$ and $(a+b)^3$ given in Eqs. (1) and (2). In both expansions the number of terms is one more than the power to which $a+b$ is raised:

$$(a+b)^2 \text{ has three terms;}$$

$$(a+b)^3 \text{ has four terms.}$$

The first and last terms of $(a+b)^2$ are the *squares* of a and b; the first and last terms of $(a+b)^3$ are the *cubes* of a and b. In both expansions the powers of a *decrease* from left to right, whereas from the second term on, the powers of b *increase*. Also, for the terms involving both a and b, the sum of the exponents of a and b equals the power to which $a+b$ is raised. For example, the second term in $(a+b)^2$ is $2ab = 2a^1b^1$ and $1+1=2$. Similarly, the second term in $(a+b)^3$ is $3a^2b = 3a^2b^1$ and $2+1=3$.

We could determine the expansions of $(a+b)^4$, $(a+b)^5$, etc., by successive multiplication. However, there is a formula, called the *binomial theorem*, that gives the expansion of $(a+b)^n$ where n is any positive integer. The patterns that we observed in $(a+b)^2$ and $(a+b)^3$ carry over to $(a+b)^n$.

BINOMIAL THEOREM

If n is a positive integer, then

$$(a+b)^n = a^n + \frac{n}{1}a^{n-1}b + \frac{n(n-1)}{1\cdot 2}a^{n-2}b^2 + \frac{n(n-1)(n-2)}{1\cdot 2\cdot 3}a^{n-3}b^3 + \cdots + \frac{n(n-1)(n-2)\cdots 2}{1\cdot 2\cdots (n-1)}ab^{n-1} + b^n.$$

Notice that in the next-to-last term above, the numerator of the coefficient is $n(n-1)(n-2)\cdots 2$, which is the product of the integers from n to 2, and the denominator is $1\cdot 2\cdots (n-1)$, which is the product of the integers from 1 to $n-1$.

In the formula, the first and last terms of the expansion of $(a+b)^n$ are a^n and b^n, and there are $n+1$ terms. Also, the powers of a *decrease* from left to right (from n to 1). From the second term on, the powers of b *increase* from 1 to n. For the terms involving both a and b, the sum of the exponents of a and b is n.

EXAMPLE 1

Using the binomial theorem for $(a + b)^n$.

a. $(x + 3)^4$.

Here $n = 4$, x plays the role of a, and 3 plays the role of b. The expansion will have five terms.

$$(a + b)^4 = a^4 + \frac{4}{1}a^3 b + \frac{4 \cdot 3}{1 \cdot 2}a^2 b^2 + \frac{4 \cdot 3 \cdot 2}{1 \cdot 2 \cdot 3}ab^3 + b^4.$$

$$\begin{aligned}(x + 3)^4 &= x^4 + \frac{4}{1}x^3(3) + \frac{4 \cdot 3}{1 \cdot 2}x^2(3)^2 + \frac{4 \cdot 3 \cdot 2}{1 \cdot 2 \cdot 3}x(3)^3 + 3^4 \\ &= x^4 + 4x^3(3) + 6x^2(9) + 4x(27) + 81 \\ &= x^4 + 12x^3 + 54x^2 + 108x + 81.\end{aligned}$$

b. $(y - 2)^5 = [y + (-2)]^5$.

Here $n = 5$, y plays the role of a, and -2 plays the role of b. The expansion will have six terms.

$$(a + b)^5 = a^5 + \frac{5}{1}a^4 b + \frac{5 \cdot 4}{1 \cdot 2}a^3 b^2 + \frac{5 \cdot 4 \cdot 3}{1 \cdot 2 \cdot 3}a^2 b^3 + \frac{5 \cdot 4 \cdot 3 \cdot 2}{1 \cdot 2 \cdot 3 \cdot 4}ab^4 + b^5.$$

$$\begin{aligned}(y - 2)^5 &= y^5 + \frac{5}{1}y^4(-2) + \frac{5 \cdot 4}{1 \cdot 2}y^3(-2)^2 + \frac{5 \cdot 4 \cdot 3}{1 \cdot 2 \cdot 3}y^2(-2)^3 \\ &\quad + \frac{5 \cdot 4 \cdot 3 \cdot 2}{1 \cdot 2 \cdot 3 \cdot 4}y(-2)^4 + (-2)^5 \\ &= y^5 + 5y^4(-2) + 10y^3(4) + 10y^2(-8) + 5y(16) + (-32) \\ &= y^5 - 10y^4 + 40y^3 - 80y^2 + 80y - 32.\end{aligned}$$

c. $(3x - 1)^4 = [3x + (-1)]^4$.

Here $n = 4$, $3x$ plays the role of a, and -1 plays the role of b.

$$\begin{aligned}(3x - 1)^4 &= (3x)^4 + \frac{4}{1}(3x)^3(-1) + \frac{4 \cdot 3}{1 \cdot 2}(3x)^2(-1)^2 + \frac{4 \cdot 3 \cdot 2}{1 \cdot 2 \cdot 3}(3x)(-1)^3 + (-1)^4 \\ &= 3^4 x^4 + 4(3)^3 x^3(-1) + 6(3)^2 x^2(1) + 4(3)x(-1) + 1 \\ &= 81x^4 - 108x^3 + 54x^2 - 12x + 1.\end{aligned}$$

EXAMPLE 2

Use the binomial theorem to find the first three terms in the expansion of $(x^2 + 4)^8$.

Here $n = 8$, x^2 plays the role of a, and 4 plays the role of b.

first term: $(x^2)^8 = x^{16}$.

$$\text{second term: } \frac{8}{1}(x^2)^7(4) = 32x^{14}.$$

$$\text{third term: } \frac{8 \cdot 7}{1 \cdot 2}(x^2)^6(4)^2 = 28x^{12}(16) = 448x^{12}.$$

Exercise 5.4

In Problems **1–8**, *use the binomial theorem to expand each expression.*

1. $(x+4)^3$.

2. $(x-3)^3$.

3. $(y-2)^4$.

4. $(y+3)^5$.

5. $(3x+1)^5$.

6. $(x+h)^6$.

7. $(2z-y)^4$.

8. $(z^2-2)^3$.

In Problems **9–16**, *find the first three terms in the binomial expansion of each expression.*

9. $(x+1)^{100}$.

10. $(x-1)^{45}$.

11. $(2x-3y)^7$.

12. $(x+2)^8$.

13. $(y^2-5)^{10}$.

14. $(y^3-6)^{12}$.

15. $(3z^2+1)^5$.

16. $(2z^2+2)^5$.

5.5 REVIEW

IMPORTANT TERMS

binomial *(5.1)*

trinomial *(5.1)*

common factor *(5.2)*

square of a binomial *(5.1)*

factoring *(5.2)*

binomial theorem *(5.4)*

REVIEW PROBLEMS

In Problems **1–20**, *find the special products.*

1. $(x+6)^2$.

2. $(x-7)^2$.

3. $(x-5)^2$.

4. $(x+3y)^2$.

5. $(2x+4y)^2$.

6. $(3x-6)^2$.

7. $(x-8)(x+8)$.

8. $(9-2x)(9+2x)$.

9. $(3x+2)(3x-2)$.

10. $(1-3x)(3x+1)$.

11. $(2x-4y)(2x+4y)$.

12. $(a-2b)(a+2b)$.

13. $(x-6)(x+4)$.
14. $(x+3)(x-2)$.
15. $(x-6)(x-7)$.

16. $(x+5)(x+8)$.
17. $(2x-3)(2x-4)$.
18. $(3x-2)(2x+3)$.

19. $(y^2+4)(y^2-4)$.
20. $(y^3+1)(y^3-1)$.

In Problems **21–26**, *perform the indicated operations.*

21. $2x^2(x-3)(x+4)$.
22. $-3x(x+3)^2$.

23. $(y-\sqrt{2})(y+\sqrt{2})-(y-4)^2$.
24. $5y(1+2y)^2$.

25. $(4x+3)(4x-3)(x+2)$.
26. $3(3x-1)^2+5x(x+2)^2$.

In Problems **27–46**, *factor completely.*

27. $6x^3y^4+4xy^6$.
28. $10abc^8-15ab^2c$.
29. $x^2-11x+30$.

30. x^2+6x+8.
31. $16-y^2$.
32. $2x^3-5x^2+6x-15$.

33. x^3-x^2-56x.
34. $2x^2+4x-96$.
35. $3x^2+10x-8$.

36. $2x^2+9x-35$.
37. $8x^2-50$.
38. $15y^2+2y-8$.

39. $2z^3-128$.
40. $8y^2+6y+1$.
41. x^4-2x^2-8.

42. x^4+10x^2+25.
43. $x^3(x-6)^2+x^4(x-6)$.

44. $(x+4)^3(x+6)^4+(x+4)^4(x+6)^3$.

45. $2x^3+8x^2-3x-12$.
46. $8x^6+27$.

In Problems **47** *and* **48**, *solve for x.*

47. $a(x+2)=b(x+3)$.
48. $ax+b=c-\dfrac{bx}{2}$.

In Problems **49–52**, *use the binomial theorem to expand each expression.*

49. $(x-4)^4$.
50. $(x+y)^5$.

51. $(2x+1)^5$.
52. $(x^3+2y)^3$.

6 Quadratic Equations

6.1 SOLUTION BY FACTORING AND SQUARE ROOT

In Chapter 4 we solved equations where the variable was raised to the first power only. Now we want to solve an equation that can be expressed in the form $ax^2 + bx + c = 0$.* This is called a **second-degree**, or **quadratic**, **equation** in x. Here the greatest power of the variable is the second power.

> **QUADRATIC EQUATION**
> $$ax^2 + bx + c = 0.$$

EXAMPLE 1

Quadratic equations: $ax^2 + bx + c = 0$.

a. $3x^2 - 5x + 2 = 0$. Here $a = 3$, $b = -5$, and $c = 2$.

b. $y^2 - 4 = 0$. Here the variable is y, and $a = 1$, $b = 0$, and $c = -4$.

c. $x^2 = x$. By rewriting this as $x^2 - x = 0$, we have $a = 1$, $b = -1$, and $c = 0$.

---

* We assume $a \neq 0$.

Factoring is useful in solving many quadratic equations. We'll use it to solve

$$x^2 + 5x + 6 = 0.$$

Factoring the left side gives

$$(x + 3)(x + 2) = 0.$$

Now we make use of an important fact:

> If the product of two or more factors is *zero*, then at least one of the factors must be zero.

Applying this to the two factors of our equation gives

$$\text{either } x + 3 = 0 \quad \textit{or} \quad x + 2 = 0.$$

So

$$\begin{aligned} &\text{if } x + 3 = 0, \text{ then } x = -3, \\ \text{and} \quad &\text{if } x + 2 = 0, \text{ then } x = -2. \end{aligned}$$

Thus there are *two* solutions:

$$\boxed{x = -3, -2.}$$

Let's check both values in the original equation.

$\underline{x = -3}$	$\underline{x = -2}$
$x^2 + 5x + 6 = 0,$	$x^2 + 5x + 6 = 0,$
$(-3)^2 + 5(-3) + 6 = 0,$	$(-2)^2 + 5(-2) + 6 = 0,$
$9 - 15 + 6 = 0,$	$4 - 10 + 6 = 0,$
$0 = 0.$ ✓	$0 = 0.$ ✓

Both check. Thus we see that *a quadratic equation can have two different solutions.*

EXAMPLE 2

Solve $2x^2 - 4x = 0$ *by factoring.*

We factor the left side and set each factor equal to 0.

$$\begin{aligned} 2x^2 - 4x &= 0, \\ 2x(x - 2) &= 0. \end{aligned}$$

There are three factors on the left: 2, x, and $x - 2$. Thus either

$$2 = 0 \quad \text{or} \quad x = 0 \quad \text{or} \quad x - 2 = 0.$$

Certainly $2 \neq 0$.

$x = 0$ is one solution.

If $x - 2 = 0$, then $x = 2$.

$$\boxed{x = 0, 2.}$$

Let's look again at the equation $2x^2 - 4x = 0$ in Example 2. Since 2 is a factor of each term, we can divide both sides by 2 to make the coefficients easier to work with.

$$\frac{2x^2 - 4x}{2} = \frac{0}{2},$$

$$\frac{2x^2}{2} - \frac{4x}{2} = 0,$$

$$x^2 - 2x = 0,$$

$$x(x - 2) = 0 \qquad \text{[factoring]}.$$

Setting each factor equal to 0 gives the solutions 0 and 2 as before. You may have noticed that x is *also* a factor of each term of the equation $2x^2 - 4x = 0$. But we can get into trouble if we divide both sides by x! Watch.

$$\frac{2x^2 - 4x}{x} = \frac{0}{x},$$

$$\frac{2x^2}{x} - \frac{4x}{x} = 0,$$

$$2x - 4 = 0,$$

$$2x = 4,$$

$$x = 2.$$

Here the only solution is 2. The last equation ($x = 2$) is *not* equivalent to the given equation, which has solutions 0 and 2. We *lost* a solution. This occurred because when we divided both sides by x, we had to assume that x was not 0 (we cannot divide by 0). Thus 0 had no chance of being a solution to the equations that followed. In general, *if you divide both sides of an equation by an expression involving the variable, you may lose solutions. Therefore this operation should be avoided!*

EXAMPLE 3

Solve $x - 2 = \frac{x^2}{2} - x$ *by factoring.*

First, to clear of fractions we multiply both sides by 2:

$$2(x-2) = 2\left(\frac{x^2}{2} - x\right),$$
$$2x - 4 = x^2 - 2x.$$

Next, we rewrite this equation so that one side is 0. Then we factor.

$$0 = x^2 - 4x + 4,$$
$$0 = (x-2)(x-2).$$

$$x - 2 = 0, \qquad x - 2 = 0,$$
$$x = 2. \qquad x = 2.$$

$$\boxed{x = 2.}$$

The given quadratic equation has only one solution. Since more than one factor gave rise to the same solution, 2, we say that 2 is a **repeated root**.

EXAMPLE 4

Solve $(3x-4)(x+1) = -2$.

The left side is factored, but the right side is -2. We want one side to be 0. Thus we first multiply and combine terms.

$$(3x-4)(x+1) = -2,$$
$$3x^2 - x - 4 = -2,$$
$$3x^2 - x - 2 = 0,$$
$$(3x+2)(x-1) = 0.$$

$$3x + 2 = 0, \qquad x - 1 = 0,$$
$$3x = -2, \qquad x = 1.$$
$$x = -\frac{2}{3}.$$

$$\boxed{x = -\frac{2}{3}, 1.}$$

*If a product of factors is -2, this does not mean that one of the factors must be -2. In the equation of Example 4, $(3x-4)(x+1) = -2$, you should **not** set each factor equal to -2. That is, the equations $3x - 4 = -2$ and $x + 1 = -2$ do not provide solutions to the given equation.*

EXAMPLE 5

Solve $x^2 = 3$ by factoring.

$$x^2 = 3,$$

$$x^2 - 3 = 0. \tag{1}$$

Since $3 = (\sqrt{3})^2$, the left side of Eq. 1 is a difference of squares and thus we can factor it.

$$(x + \sqrt{3})(x - \sqrt{3}) = 0.$$

(In this chapter we allow factors involving coefficients that are not integers.) Thus

$$x + \sqrt{3} = 0, \qquad x - \sqrt{3} = 0,$$

$$x = -\sqrt{3}. \qquad x = \sqrt{3}.$$

$$\boxed{x = \sqrt{3}, -\sqrt{3}.}$$

Sometimes the pair of numbers $\sqrt{3}$ and $-\sqrt{3}$ is written as $\pm\sqrt{3}$, which is read "plus or minus $\sqrt{3}$." Thus we could write the answer as $x = \pm\sqrt{3}$.

A more general form of the equation $x^2 = 3$ in Example 5 is $u^2 = k$. In the same manner as in Example 5, we can show the following:

$$\boxed{\text{If } u^2 = k, \text{ then } u = \pm\sqrt{k}.}$$

The formula $u = \pm\sqrt{k}$ is called the **square-root method** of solving $u^2 = k$. It is a good method of solving a quadratic equation having no x-term. For example, the solution of $x^2 = 4$ is $x = \pm\sqrt{4} = \pm 2$. *Don't forget the "±" sign!*

EXAMPLE 6

Solving by the square-root method.

a. *Solve* $4x^2 - 3 = 0$.

$$4x^2 - 3 = 0,$$

$$4x^2 = 3,$$

$$x^2 = \frac{3}{4},$$

$$x = \pm\sqrt{\frac{3}{4}} = \pm\frac{\sqrt{3}}{\sqrt{4}} = \boxed{\pm\frac{\sqrt{3}}{2}.}$$

b. *Solve* $(x+2)^2 = 7$.

This equation has the special form $u^2 = k$, where u is $x+2$ and k is 7. Thus

$$(x+2)^2 = 7,$$
$$x + 2 = \pm\sqrt{7},$$
$$\boxed{x = -2 \pm \sqrt{7}.}$$

Here we have two solutions: one when we use the "+" sign, $-2+\sqrt{7}$, and another when we use the "−" sign, $-2-\sqrt{7}$.

Sometimes factoring can be used to solve equations that are not quadratic, as Example 7 shows.

EXAMPLE 7

Solve $x^4 - 4x^2 = 0$.

This is called a fourth-degree equation. Can you tell why?

$$x^4 - 4x^2 = 0,$$
$$x^2(x^2 - 4) = 0,$$
$$x^2(x+2)(x-2) = 0,$$
$$x \cdot x(x+2)(x-2) = 0.$$

Setting all *four* factors equal to 0, we have

$x = 0.$	$x = 0.$	$x + 2 = 0,$	$x - 2 = 0,$
		$x = -2.$	$x = 2.$

$$\boxed{x = 0, \pm 2.}$$

Here 0 is a repeated root.

Exercise 6.1

In Problems **1–50**, *solve by the methods of this section.*

1. $x^2 + 3x + 2 = 0.$ **2.** $(x+2)(x-5) = 0.$ **3.** $(x-3)(2x+1) = 0.$

4. $x^2 + 8x + 15 = 0.$ **5.** $t^2 - 7t + 12 = 0.$ **6.** $t^2 - 4t + 4 = 0.$

7. $z^2 + 2z - 3 = 0.$ **8.** $z^2 + z - 12 = 0.$ **9.** $x^2 - 12x + 36 = 0.$

10. $x^2 - 1 = 0.$ **11.** $x^2 - 8x = 0.$ **12.** $t^2 = 16.$

13. $2x^2 + 10x = 0.$

14. $0 = 3x - 4x^2.$

15. $0 = 3t^2 - 7t.$

16. $x^2 = 32.$

17. $4 - x^2 = 0.$

18. $10x^2 - x - 3 = 0.$

19. $x^2 = 25.$

20. $x^2 = 8.$

21. $x^2 = 6.$

22. $9x^2 = 36.$

23. $\frac{x^2}{3} = 4.$

24. $\frac{x^2}{7} = 1.$

25. $9z^2 = 81.$

26. $3x^2 - 12x + 12 = 0.$

27. $6x^2 + 7x - 3 = 0.$

28. $3 = z^2.$

29. $\frac{4}{3}t^2 = 5.$

30. $7t^2 = 7.$

31. $2x^2 - 14 = 0.$

32. $x^2 + 2 = 18.$

33. $2x^2 + 7x = 4.$

34. $x^2 = 2x + 3.$

35. $-x^2 + 3x + 10 = 0.$

36. $2x^2 + 3x - 2 = 0.$

37. $4x^2 + 4x = -1.$

38. $9x^2 - 1 = 0.$

39. $6(x^2 + 2x) + 6 = 0.$

40. $4x^2 - 12x + 9 = 0.$

41. $t(t + 4) = 5.$

42. $2y^2 = 4y.$

43. $3x^2 + 5(2 - x) = 2x^2 + 4.$

44. $2(x^2 - 5) - 3(x^2 - 7) = 3.$

45. $4 - x^2 = (x + 1)^2 + 3.$

46. $(2x - 5)(x + 5) = -22.$

47. $\frac{x^2}{2} - x - 4 = 0.$

48. $x^2 + \frac{7}{2}x - 2 = 0.$

49. $6y^2 + \frac{5}{2}y + \frac{1}{4} = 0.$

50. $\frac{x^2}{2} + \frac{10}{3}x + 2 = 0.$

In Problems **51–56**, *solve by the square-root method.*

51. $(x - 3)^2 = 16.$

52. $(x + 5)^2 = 6.$

53. $(x + 4)^2 = 8.$

54. $(w - 6)^2 = 1.$

55. $(y + \frac{1}{2})^2 = 1.$

56. $(x - 4)^2 = \frac{9}{4}.$

In Problems **57–68**, *solve.*

57. $x(x - 1)(x + 2) = 0.$

58. $x^2(x - 4) = 0.$

59. $(x - 2)^2(x + 1)^2 = 0.$

60. $x(x - 1)(x + 1) = 0.$

61. $7x^2(x - 2)^2(x + 3)(x - 4) = 0.$

62. $x(x^2 - 1)(x^2 - 4) = 0.$

63. $x(x^2 - 1)(x^2 - 1) = 0.$

64. $x^3 - x = 0.$

65. $x^3 - 64x = 0.$

66. $x^3 - 4x^2 - 5x = 0.$

67. $3y^3 + 18y^2 + 24y = 0.$

68. $3x^4 + 11x^3 - 4x^2 = 0.$

To solve $x^4 - 5x^2 + 4 = 0$, *we can factor the left side.*

$$(x^2 - 1)(x^2 - 4) = 0,$$

$$(x - 1)(x + 1)(x - 2)(x + 2) = 0.$$

Thus

$$\boxed{x = \pm 1, \pm 2.}$$

In a similar way, solve Problems **69–74** *by factoring.*

69. $x^4 - 10x^2 + 9 = 0.$

70. $x^4 - 29x^2 + 100 = 0.$

71. $x^2(x^4 - 2x^2 + 1) = 0.$

72. $x^5 - 6x^3 + 8x = 0.$

73. $x^4 - 13x^2 + 36 = 0.$

74. $x^5 - 4x^3 + 4x = 0.$

6.2 COMPLETING THE SQUARE

In the last section you solved quadratic equations by factoring. This is a good method, but sooner or later you'll run into a quadratic equation for which factoring is difficult. For example, it's hard to factor the left side of $3x^2 - 6x - 4 = 0$.

You may still solve this equation, but other methods are used. One way is the method of **completing the square.** It involves rewriting the equation until one side factors as a square of a binomial.

We'll use this method to solve $3x^2 - 6x - 4 = 0$. The "rewriting" procedure perhaps will seem strange to you, but the reasons for each step will be obvious later.

Method of completing the square for $3x^2 - 6x - 4 = 0$.

We first add 4 to both sides so that only terms involving x remain on the left side.

$$3x^2 - 6x = 4.$$

Next, we get the coefficient of the x^2-term to be 1. Thus we divide both sides by 3.

$$\frac{3x^2 - 6x}{3} = \frac{4}{3},$$

$$\frac{3x^2}{3} - \frac{6x}{3} = \frac{4}{3},$$

$$x^2 - 2x = \frac{4}{3}.$$

We take half of the coefficient of the x-term and square the result.

$$\text{Half of } -2\text{:} \quad \frac{-2}{2} = -1. \qquad \text{Squaring:} \quad (-1)^2 = 1.$$

We add this square, 1, to both sides.

$$x^2 - 2x + 1 = \frac{4}{3} + 1,$$

$$x^2 - 2x + 1 = \frac{7}{3}.$$

Look at the left side of the equation. It's the square of a binomial.

$$(x-1)^2 = \frac{7}{3}.$$

This was the reason for doing all the operations. The equation now has the form $u^2 = k$. Thus, by the square-root method,

$$x - 1 = \pm\sqrt{\frac{7}{3}}.$$

Before we find x, we'll simplify the radical.

$$\sqrt{\frac{7}{3}} = \sqrt{\frac{7\cdot 3}{3\cdot 3}} = \sqrt{\frac{21}{3^2}} = \frac{\sqrt{21}}{\sqrt{3^2}} = \frac{\sqrt{21}}{3}.$$

Thus

$$x - 1 = \pm\frac{\sqrt{21}}{3},$$

so

$$\boxed{x = 1 \pm \frac{\sqrt{21}}{3}.}$$

EXAMPLE 1

Solve $4x^2 + 12x + 3 = 0$ *by completing the square.*

$$4x^2 + 12x + 3 = 0,$$

$$4x^2 + 12x = -3 \qquad \text{[getting constant on right]},$$

$$\frac{4x^2 + 12x}{4} = \frac{-3}{4} \qquad \text{[dividing both sides by 4]},$$

$$x^2 + 3x = -\frac{3}{4}.$$

Half of 3: $\frac{3}{2}$. Squaring: $\left(\frac{3}{2}\right)^2 = \frac{9}{4}$.

$$x^2 + 3x + \frac{9}{4} = -\frac{3}{4} + \frac{9}{4} \qquad \text{[adding } \tfrac{9}{4} \text{ to both sides]},$$

$$\left(x + \frac{3}{2}\right)^2 = \frac{6}{4} \qquad \text{[factoring left side]},$$

$$x + \frac{3}{2} = \pm\sqrt{\frac{6}{4}}.$$

Since $\sqrt{\frac{6}{4}} = \frac{\sqrt{6}}{\sqrt{4}} = \frac{\sqrt{6}}{2}$,

$$x = -\frac{3}{2} \pm \frac{\sqrt{6}}{2}.$$

Exercise 6.2

Solve by completing the square.

1. $x^2 + 6x - 1 = 0$.
2. $2x^2 + 4x - 8 = 0$.
3. $5x^2 - 20x + 5 = 0$.
4. $x^2 - 12x + 8 = 0$.
5. $y^2 - 3y - 1 = 0$.
6. $x^2 + 5x + 5 = 0$.
7. $x^2 + x - 4 = 0$.
8. $4w^2 - 8w - 3 = 0$.
9. $2x^2 - 14x + 1 = 0$.
10. $9x^2 + 9x - 2 = 0$.
11. $x^2 + \frac{1}{2}x - 1 = 0$.
12. $x^2 + 2x - \frac{1}{4} = 0$.
13. $7x + 3(x^2 - 5) = x - 3$.
14. $2x(4x - 1) = 4 + 2x$.

6.3 THE QUADRATIC FORMULA

Another way to solve quadratic equations is to use a special formula called the *quadratic formula.**

QUADRATIC FORMULA

If $ax^2 + bx + c = 0$, then

$$x = \frac{-b \pm \sqrt{b^2 - 4ac}}{2a}.$$

EXAMPLE 1

Solve $2x^2 + 5x - 3 = 0$ *by the quadratic formula.*

Here $a = 2$, $b = 5$, and $c = -3$.

$$x = \frac{-b \pm \sqrt{b^2 - 4ac}}{2a} = \frac{-5 \pm \sqrt{(5)^2 - 4(2)(-3)}}{2(2)}$$

$$= \frac{-5 \pm \sqrt{25 + 24}}{4} = \frac{-5 \pm \sqrt{49}}{4} = \frac{-5 \pm 7}{4}.$$

---

* The way we obtain the quadratic formula is explained in Section 6.6.

Thus the solutions are

$$x = \frac{-5+7}{4} = \frac{2}{4} = \frac{1}{2} \quad \text{and} \quad x = \frac{-5-7}{4} = \frac{-12}{4} = -3.$$

$$\boxed{x = \frac{1}{2}, -3.}$$

Make sure you write the quadratic formula correctly.

$$x \neq -b \pm \frac{\sqrt{b^2 - 4ac}}{2a}.$$

EXAMPLE 2

Solve $4y^2 + 9 = 12y$ *by the quadratic formula.*

First we put the equation in the form $ay^2 + by + c = 0$.

$$4y^2 - 12y + 9 = 0.$$

Here $a = 4$, $b = -12$, and $c = 9$.

$$y = \frac{-b \pm \sqrt{b^2 - 4ac}}{2a} = \frac{-(-12) \pm \sqrt{(-12)^2 - 4(4)(9)}}{2(4)}$$

$$= \frac{12 \pm \sqrt{144 - 144}}{8} = \frac{12 \pm 0}{8}.$$

Notice that in $y = \frac{12 \pm 0}{8}$ we get the same value whether we use the "+" sign or the "−" sign. Thus the only solution is

$$y = \frac{12}{8} = \boxed{\frac{3}{2}.}$$

We consider $\frac{3}{2}$ to be a repeated root.

EXAMPLE 3

Solve $2x(x - 3) = x^2 - 1$.

$$2x^2 - 6x = x^2 - 1 \quad \text{[removing parentheses]},$$

$$x^2 - 6x + 1 = 0 \quad \text{[rewriting so one side is 0]}.$$

This equation is quadratic. Factoring does not work here, so we shall use the quadratic formula.

$$a = 1, \qquad b = -6, \qquad c = 1.$$

$$x = \frac{-b \pm \sqrt{b^2 - 4ac}}{2a} = \frac{-(-6) \pm \sqrt{(-6)^2 - 4(1)(1)}}{2(1)}$$

$$= \frac{6 \pm \sqrt{36 - 4}}{2} = \frac{6 \pm \sqrt{32}}{2}.$$

But $\sqrt{32} = \sqrt{16 \cdot 2} = \sqrt{16} \cdot \sqrt{2} = 4\sqrt{2}$. Thus

$$x = \frac{6 \pm 4\sqrt{2}}{2} = \frac{6}{2} \pm \frac{4\sqrt{2}}{2} = \boxed{3 \pm 2\sqrt{2}.}$$

The expression $b^2 - 4ac$ in the quadratic formula is called the **discriminant** of the quadratic equation $ax^2 + bx + c = 0$. If you look back at Examples 1 and 3 in this section, you will see that the discriminants 49 and 32 are positive and there are two different solutions to each equation. But in Example 2, the discriminant is 0 and the equation has only one solution (repeated root). In general,

a positive discriminant gives two different solutions, but a zero discriminant gives only one solution.

Negative discriminants will be considered in the next section.

What is the best way of solving a quadratic equation $ax^2 + bx + c = 0$? The quadratic formula can be used for *any* quadratic equation. However, if there is no x-term involved, use the square-root method. For example, let's solve $5x^2 - 2 = 0$. We have

$$5x^2 - 2 = 0,$$

$$5x^2 = 2,$$

$$x^2 = \frac{2}{5},$$

$$x = \pm\sqrt{\frac{2}{5}} = \pm\sqrt{\frac{2 \cdot 5}{5 \cdot 5}} = \pm\frac{\sqrt{10}}{\sqrt{5^2}} = \pm\frac{\sqrt{10}}{5}.$$

If there are x^2- and x-terms but no constant term ($c = 0$), use factoring. For example, to solve $4x^2 + 3x = 0$, we have

$$4x^2 + 3x = 0,$$

$$x(4x + 3) = 0.$$

$$x = 0. \quad \Bigg| \quad \begin{aligned} 4x + 3 &= 0, \\ 4x &= -3, \\ x &= -\frac{3}{4}. \end{aligned}$$

For the equation $ax^2 + bx + c = 0$, where a, b, and c are all different from 0, try factoring first. If that doesn't work, use the quadratic formula.

Exercise 6.3

all

In Problems **1–12**, *solve by the quadratic formula.*

1. $x^2 + 3x + 1 = 0$.

2. $x^2 - 4x + 2 = 0$.

3. $x^2 - 6x + 9 = 0$.

4. $x^2 + x - 3 = 0$.

5. $2y^2 + 3y - 4 = 0$.

6. $3x^2 + 6x - 2 = 0$.

7. $4x^2 = -20x - 25$.

8. $9z(z - 1) = 3z - 4$.

9. $5x(x + 2) + 6 = 3$.

10. $(4x - 1)(2x + 3) = 18x - 4$.

11. $2 - 2x - 3x^2 = 0$.

12. $1 + 8x - 4x^2 = 0$.

In Problems **13–48**, *solve by any method. The choice is yours for each problem.*

13. $x^2 - 36 = 0$.

14. $x^2 + 4x - 1 = 0$.

15. $x^2 + 6x - 7 = 0$.

16. $x^2 - x - 20 = 0$.

17. $z^2 = 2z$.

18. $4z^2 - 25 = 0$.

19. $x^2 + 4x + 2 = 0$.

20. $x^2 - 5x + 3 = 0$.

21. $y^2 + 16y + 64 = 0$.

22. $3y^2 - y - 4 = 0$.

23. $3x^2 - 4x - 2 = 0$.

24. $3 + x - 4x^2 = 0$.

25. $5x^2 - 21x + 4 = 0$.

26. $2x^2 - 6x + 2 = 0$.

27. $(x + 2)^2 = 5$.

28. $2x^2 - 11 = 0$.

29. $5x^2 + 4x = 0$.

30. $y(y - 5) = 0$.

31. $4 + 4y - y^2 = 0$.

32. $4y^2 + 12y + 9 = 0$.

33. $1 - 6y + 9y^2 = 0$.

34. $7 - y - y^2 = 0$.

35. $x^2 + 9 = 7x$.

36. $x^2 - 16 = 8 - 2x$.

37. $x^2 + 3x = 12 - 2x - x^2$.

38. $3x(2x - 5) = -4x - 3$.

39. $y(y + 4) = 5$.

40. $(y + 1)^2 = 2y^2$.

41. $z^2 = 2(z - 1)(z + 2)$.

42. $z^4 - 6z^2 + 7 = z^2(z^2 + 1)$.

43. $(2x + 1)^2 = 8x$.

44. $2x(x + 1) = (x - 1)(x + 1)$.

45. $\dfrac{2x^2 - 5x}{3} = x - 1$.

46. $\dfrac{x^2}{3} + 2x = x + 1$.

47. $\dfrac{x^2}{3} = \dfrac{11}{6}x + 1$.

48. $5x^2 - \dfrac{7}{2}x = \dfrac{x + 2}{2}$.

6.4 COMPLEX NUMBERS

When you use the quadratic formula, sometimes the square root of a negative number occurs. For example, solving

$$x^2 + x + 1 = 0$$

gives

$$x = \frac{-1 \pm \sqrt{1^2 - 4(1)(1)}}{2(1)} = \frac{-1 \pm \sqrt{1-4}}{2}$$
$$= \frac{-1 \pm \sqrt{-3}}{2}.$$

Now, $\sqrt{-3}$ is a number whose square is -3. This number cannot be real, since the square of any real number cannot be negative. To handle $\sqrt{-3}$ and other square roots of negative numbers, we work with $\sqrt{-1}$.

To represent $\sqrt{-1}$, we use the symbol i, called the **imaginary unit**. It has the property that $i^2 = -1$. Thus i is not a real number. We can handle square roots of all other negative numbers by using the following.

> If a is a positive number, then
>
> $$\sqrt{-a} = i\sqrt{a}.$$

EXAMPLE 1

Square roots of negative numbers.

a. $\sqrt{-4} = i\sqrt{4} = i(2) = 2i$.

b. $\sqrt{-16} = i\sqrt{16} = 4i$.

c. $\sqrt{-8} = i\sqrt{8} = i(2\sqrt{2}) = 2i\sqrt{2}$.

d. $-\sqrt{-3} = -(i\sqrt{3}) = -i\sqrt{3}$.

As in Example 1, it will be our custom to express square roots of negative numbers in terms of i.

$\sqrt[3]{-8} = -2$. ***Don't write*** $\sqrt[3]{-8} = i\sqrt[3]{8} = 2i$. *The symbol i involves a square root, not a cube root.*

Any number of the form bi, where b is a real number, is called a **pure imaginary number**. Thus $2i$ and $-3i$ are pure imaginary. From Example 1 you can see that square roots of negative numbers are pure imaginary. Since $0 = 0i$, then 0 is not only a real number but also pure imaginary.

Combining a real number and a pure imaginary number by addition gives a so-called *complex number*, such as $4 + 2i$.

> A **complex number** is one of the form $a + bi$, where a and b are real numbers and i is the imaginary unit.

EXAMPLE 2

Complex numbers: $a + bi$.

a. $2 + 3i$. Here $a = 2$ and $b = 3$.

b. $-3 - 5i$. Here $a = -3$ and $b = -5$, since $-3 - 5i = -3 + (-5)i$.

c. 6. Here $a = 6$ and $b = 0$, since $6 = 6 + 0i$. Every real number is also a complex number.

d. $7i$. Here $a = 0$ and $b = 7$, since $7i = 0 + 7i$.

From Example 2c you can see that every real number is also a complex number. Complex numbers that are not real are called **imaginary numbers**. Thus $2 + 3i$ is an imaginary number. Also, $7i$ is imaginary as well as pure imaginary.

As you saw at the beginning of this section, some quadratic equations have imaginary numbers for solutions.

EXAMPLE 3

Solve $x^2 + 2x + 6 = 0$.

$$x = \frac{-2 \pm \sqrt{2^2 - 4(1)(6)}}{2(1)}$$

$$= \frac{-2 \pm \sqrt{-20}}{2} = \frac{-2 \pm i\sqrt{20}}{2}$$

$$= \frac{-2 \pm 2i\sqrt{5}}{2} = -\frac{2}{2} \pm \frac{2i\sqrt{5}}{2}$$

$$= \boxed{-1 \pm i\sqrt{5}.}$$

Thus the solutions are $x = -1 + i\sqrt{5}$ and $x = -1 - i\sqrt{5}$, which are imaginary numbers.

In Example 3, notice that the discriminant is negative, -20. Whenever this happens, the solutions of a quadratic equation will be two different imaginary numbers.

Here is a summary of the possibilities of the discriminant and the corresponding type of solutions of a quadratic equation.

$b^2 - 4ac$	Type of Solutions
positive	two different real solutions
zero	one real solution (repeated root)
negative	two different imaginary solutions

The discriminant is useful because from it you can determine something about the solutions of a quadratic equation without actually having to find them.

EXAMPLE 4

Describe the solutions and solve.

a. $2x^2 - 4x + 5 = 0$.

First, we find the discriminant.

$$b^2 - 4ac = (-4)^2 - 4(2)(5) = 16 - 40 = -24.$$

Since the discriminant is negative, there are two different imaginary solutions.

$$x = \frac{-(-4) \pm \sqrt{-24}}{2(2)} = \frac{4 \pm 2i\sqrt{6}}{4}$$

$$= \frac{4}{4} \pm \frac{2i\sqrt{6}}{4} = \boxed{1 \pm \frac{i\sqrt{6}}{2}.}$$

b. $x^2 - 6x + 6 = 0$.

The discriminant is

$$b^2 - 4ac = (-6)^2 - 4(1)(6) = 36 - 24 = 12.$$

A positive discriminant means two different real solutions.

$$x = \frac{-(-6) \pm \sqrt{12}}{2(1)} = \frac{6 \pm 2\sqrt{3}}{2} = \boxed{3 \pm \sqrt{3}.}$$

c. $x^2 - 10x + 25 = 0$.

The discriminant is

$$b^2 - 4ac = (-10)^2 - 4(1)(25) = 100 - 100 = 0.$$

A discriminant of 0 means one real solution.

$$x = \frac{-(-10) \pm \sqrt{0}}{2(1)} = \frac{10}{2} = \boxed{5.}$$

Whenever a discriminant is 0 we get a repeated root.

EXAMPLE 5

Solve $x^2 + 16 = 0$.

We certainly don't need to use the quadratic formula on so easy a problem.

$$x^2 + 16 = 0,$$

$$x^2 = -16 \qquad [\text{form } u^2 = k],$$

$$x = \pm\sqrt{-16} = \boxed{\pm 4i} \qquad [\text{square-root method}].$$

In Chapter 8 you will learn how to add, subtract, multiply, and divide complex numbers.

Exercise 6.4

In Problems **1–12**, *write each number in terms of i.*

1. $\sqrt{-81}$. **2.** $\sqrt{-36}$. **3.** $\sqrt{-\frac{1}{16}}$.

4. $\sqrt{-.04}$. **5.** $\sqrt{-12}$. **6.** $\sqrt{-50}$.

7. $\sqrt{-32}$. **8.** $\sqrt{-28}$. **9.** $\sqrt{-2}$.

10. $-\sqrt{-9}$. **11.** $-\sqrt{-25}$. **12.** $\sqrt{-27}$.

In Problems **13–28**, *solve the equation.*

13. $x^2 - 4x + 5 = 0$. **14.** $x^2 - 2x + 2 = 0$. **15.** $x^2 + 2x + 3 = 0$.

16. $2x^2 - 4x + 5 = 0$. **17.** $x^2 - 2x + 4 = 0$. **18.** $t^2 - 3t + 1 = 0$.

19. $x^2 + 4 = 0$. **20.** $x^2 + 8 = 0$. **21.** $6r^2 + 8r + 3 = 0$.

22. $x^2 = x - 1$. **23.** $-3r^2 + 5r = 4$. **24.** $x(x + 5) = 5(x - 5)$.

25. $1 + x^2 = 0$. **26.** $(x + 1)^2 = -4$. **27.** $x(4x - 3) = x^2 - 2$.

28. $\frac{x^2}{5} + 2 = x$.

In Problems **29–36**, *first find the discriminant. Next, use just the discriminant to determine the type of solutions. Then solve.*

29. $3x^2 - 5x - 2 = 0$. **30.** $x^2 + 4x + 5 = 0$. **31.** $9t^2 + 12t + 4 = 0$.

32. $4t^2 + 12t + 9 = 0$. **33.** $3x^2 - 4x + 3 = 0$. **34.** $4x^2 - 17x + 15 = 0$.

35. $(3x - 1)(x + 2) = 2x$. **36.** $x(x - 4) - 3(2 - x) + 7 = 0$.

6.5 APPLICATIONS OF QUADRATIC EQUATIONS

Here are some word problems that lead to quadratic equations.

EXAMPLE 1

A rectangular observation deck overlooking a scenic valley is to be built. See Fig. 6-1(a). *It is to have dimensions* 6 m *by* 12 m. *A rectangular shelter of area* 40 m^2 *is to be centered over the deck. The uncovered part of the deck is to be a walkway of uniform width. How wide should this walkway be?*

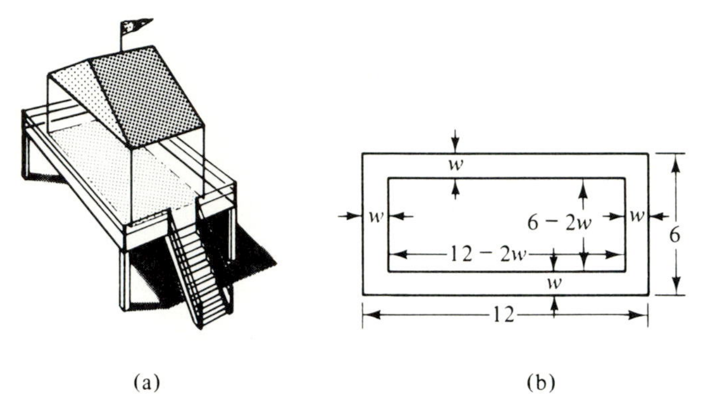

FIGURE 6-1

A diagram of the deck is shown in Fig. 6-1(b). Let w = width (in meters) of the walkway. Then the part of the deck for the shelter has dimensions $12 - 2w$ by $6 - 2w$. Since this area must be 40 m^2, where area = (length)(width), we have

$$(12 - 2w)(6 - 2w) = 40,$$

$$72 - 36w + 4w^2 = 40 \quad \text{[multiplying]},$$

$$4w^2 - 36w + 32 = 0,$$

$$w^2 - 9w + 8 = 0 \quad \text{[dividing both sides by 4]},$$

$$(w - 8)(w - 1) = 0,$$

$$w = 8, 1.$$

Although 8 is a solution to the equation, it is *not* a solution to our problem, because one of the dimensions of the deck itself is only 6 m. Thus the only possible solution is 1 m.

EXAMPLE 2

A subscription TV cable service estimates that 1000 *subscribers will pay* \$3.00 *each to watch an off-Broadway show in their own homes. They also believe that the number of subscribers will increase by* 50 *for each* \$.10 *decrease in the charge. If the firm must receive* \$3125 *to meet its costs and to earn a profit, how much should they charge each subscriber?*

Method 1.

For each \$.10 decrease in the charge, the number of subscribers increases by 50. If there are n decreases, the charge decreases by $.10n$ and the number of subscribers increases by $50n$. Thus there will be $1000 + 50n$ subscribers, each of which pays $3.00 - .10n$. Now,

$$\text{(number of subscribers)(charge per subscriber)} = \text{total income}.$$

Consequently,

$$(1000 + 50n)(3.00 - .10n) = 3125,$$

$$3000 + 50n - 5n^2 = 3125,$$

$$5n^2 - 50n + 125 = 0 \quad \text{[multiplying both sides by } -1 \text{ and simplifying]},$$

$$n^2 - 10n + 25 = 0 \quad \text{[dividing both sides by 5]},$$

$$(n - 5)^2 = 0,$$

$$n = 5.$$

Thus there are five \$0.10 decreases (for a total decrease of \$.50), and the charge should be 3.00 − .50, or \$2.50.

Method 2.

Another way of handling the problem is as follows. Suppose that c is the charge (in dollars) to each subscriber. Then the total decrease from the \$3 charge is $3 - c$. Thus the number of \$0.10 decreases is $\frac{3-c}{.10}$. Since each \$0.10 decrease gives 50 more subscribers, the total increase in subscribers will be $50\left(\frac{3-c}{.10}\right)$. Thus the total number of subscribers will be $1000 + 50\left(\frac{3-c}{.10}\right)$. Now,

$$\text{total income} = \text{(charge per subscriber)(number of subscribers)}.$$

Thus

$$\begin{aligned} 3125 &= c\left[1000 + 50\left(\frac{3-c}{.10}\right)\right] \\ &= c[1000 + 500(3 - c)] \\ &= c[1000 + 1500 - 500c] \\ &= c[2500 - 500c]. \end{aligned}$$

$$3125 = 2500c - 500c^2.$$

Consequently,

$$500c^2 - 2500c + 3125 = 0.$$

Dividing both sides by 125, we obtain

$$4c^2 - 20c + 25 = 0.$$

Factoring gives $(2c - 5)(2c - 5) = 0$, from which $c = 5/2 = 2.50$. The charge should be \$2.50.

Discussion.

Both methods give the same result. However, the nice thing about Method 2 is that the unknown is the charge you're looking for. Unfortunately, the equation that you get is very messy. In Method 1, the unknown is not the charge, but the equation is much simpler.

EXAMPLE 3

The board of directors of a corporation agrees to redeem some of its stock in two years. At that time \$1,123,600 *will be needed. Suppose that the board presently sets aside* \$1,000,000. *At what annual rate of interest, compounded annually, will the money have to be invested so that its future value will be enough to redeem the stock?*

Let r be the annual rate of interest (in decimal form). At the end of the first year, the accumulated amount will be \$1,000,000 plus the interest on this, which is $1{,}000{,}000r$, for a total of

$$1{,}000{,}000 + 1{,}000{,}000r$$

or, after factoring,

$$1{,}000{,}000(1 + r).$$

At the end of the second year, the accumulated amount will be $1{,}000{,}000(1 + r)$ plus the interest on this, which is $[1{,}000{,}000(1 + r)]r$, for a total of

$$1{,}000{,}000(1 + r) + 1{,}000{,}000(1 + r)r.$$

This must equal \$1,123,600.

$$1{,}000{,}000(1 + r) + 1{,}000{,}000(1 + r)r = 1{,}123{,}600.$$

Removing the common factor $1{,}000{,}000(1 + r)$ on the left side, we have

$$1{,}000{,}000(1 + r)[1 + r] = 1{,}123{,}600,$$

$$1{,}000{,}000(1 + r)^2 = 1{,}123{,}600,$$

$$(1 + r)^2 = \frac{1{,}123{,}600}{1{,}000{,}000},$$

$$(1 + r)^2 = \frac{11{,}236}{10{,}000},$$

$$(1 + r)^2 = \frac{2809}{2500}.$$

Thus

$$1 + r = \pm\sqrt{\frac{2809}{2500}} \qquad \text{[square-root method]}$$

$$= \pm\frac{\sqrt{2809}}{\sqrt{2500}} = \pm\frac{53}{50}.$$

Either $r = -1 + \frac{53}{50} = \frac{3}{50} = .06$, or $r = -1 - \frac{53}{50} = -2.06$. We throw out -2.06, since we don't want r to be negative. Thus $r = .06$, so the rate that we want is 6%.

Exercise 6.5

1. A ball is thrown up from the ground at a speed of 56 ft/s. Its height s (in feet) from the ground after t seconds is given by

$$s = 56t - 16t^2.$$

When the ball hits the ground, s will be 0. Find how many seconds it takes to hit the ground.

2. A ball is dropped from a cliff 144 ft above the ground. The distance s (in feet) it falls in t seconds is given by

$$s = 16t^2.$$

How many seconds does it take for the ball to hit the ground?

3. Given the formula of motion,

$$s = v_0 t + \frac{1}{2}at^2,$$

find t (in seconds) to one decimal place if $s = 15$ m, $v_0 = 18$ m/s, and $a = -9.8$ m/s^2. Here s, v_0, and a are displacement, initial velocity, and acceleration, respectively.

4. The monthly revenue R of a certain company is given by $R = 800p - 7p^2$, where p is the price in dollars of each unit of the product that they manufacture. At what price will the revenue be $10,000 if the price must be greater than $50?

5. An economics instructor told his class that the demand equation for a certain product is $p = 400 - q^2$ and its supply equation is $p = 20q + 100$. If the $400 - q^2$ is set equal to the $20q + 100$, then the *positive* solution to the resulting equation gives the *equilibrium quantity*. The instructor asked his class to find this quantity. What answer should the class give?

6. A sociologist is hired by a city to study different programs that aid the education of preschool-age children. The sociologist estimates that n years after the beginning of a particular program, p thousand preschoolers will be enrolled, where

$$p = \frac{5}{4}n(12 - n).$$

How many years after the start of the program will 25,000 preschoolers *first* be enrolled?

7. A group of biologists studied the nutritional effects on rats that were fed a diet containing 10% protein.* The protein was made up of yeast and corn flour. By changing the percentage P (expressed as a decimal) of yeast in the protein mix, the group estimated that the average weight gain g (in grams) of a rat over a period of time was given by

$$g = -200P^2 + 200P + 20.$$

What percentage of yeast gave an average weight gain of 70 grams?

8. The product of two consecutive integers is 42. Find the integers.

9. The sum of a number and its square is 12. Find the number.

10. A lumber company owns a forest that is of rectangular shape, 1 mi by 2 mi. If the company cuts a uniform strip of trees along the outer edges of this forest, how wide should the strip be if $\frac{3}{4}$ sq mi of forest is to remain?

11. A rectangular plot, 4 m by 8 m, is to be used for a garden. It is decided to put a pavement inside the entire border so that 12 m^2 of the plot is left for flowers. How wide should the pavement be?

12. A company parking lot is 120 ft long and 80 ft wide. Because of an increase in personnel, it is decided to double the area of the lot by adding strips of equal width to one end and one side. Find the width of one such strip.

13. Suppose that consumers will buy q units of a product when the price is $(80 - q)/4$ dollars *each*. How many units must be sold in order that sales revenue be $400?

14. *Calculator problem.* The diameter of a circular ventilating duct is 140 mm. It is joined to a square duct system as shown in Fig. 6-2. To ensure smooth air flow, the areas of the circle and square sections must be equal. To the nearest millimeter, what should the length x of a side of the square section be?

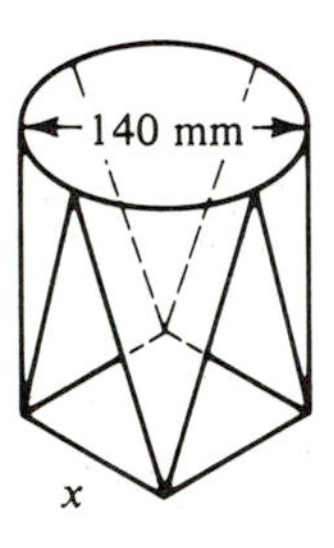

FIGURE 6-2

15. A real estate firm owns a garden apartment development, which consists of 70 apartments. At $250 per month each apartment can be rented. However, for each $10

* Adapted from R. Bressani, "The Use of Yeast in Human Foods," ed. Mateles and Tannenbaum, *Single-Cell Protein*, (Cambridge, Mass.: M. I. T. Press, 1968).

per month increase there will be two vacancies with no possibility of filling them. If the firm wants to receive $17,980 per month from rents, what rent should be charged for each apartment?

16. Imperial Educational Services (I.E.S.) wants to offer a workshop in pollution control to key personnel at Acme Corporation. I.E.S. will offer the course to thirty persons at a charge of $50 each. Moreover, I.E.S. will agree to reduce the charge for *everybody* by $1.00 for each person over the thirty who attends, up to a total group size of fifty. It has been determined that the greatest revenue that I.E.S. can receive is $1600. What group size will give this revenue?

17. For security reasons a company will enclose a rectangular area of 11,200 ft^2 in the rear of its plant. One side will be bounded by the building and the other three sides by fencing. See Fig. 6-3. If 300 ft of fencing will be used, what will the dimensions of the rectangular area be?

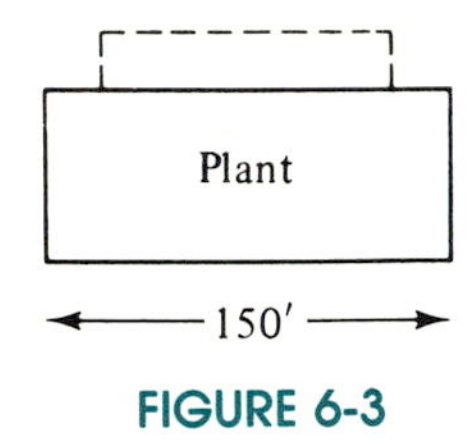

FIGURE 6-3

18. An open box is to be made from a square piece of tin by cutting out a 3-in. square from each corner and folding up the sides. See Fig. 6-4. The box is to contain 75 cubic inches. Find the dimensions of the square piece of tin that must be used.

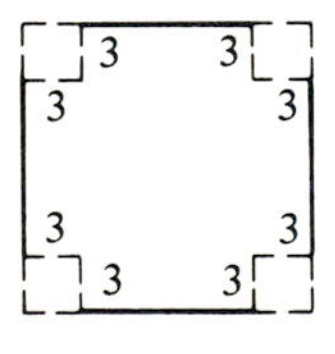

FIGURE 6-4

19. The Dandy Candy Company makes the popular Dandy Bar. The rectangular-shaped bar is 10 cm long, 5 cm wide, and 2 cm thick. See Fig. 6-5. Because of increasing costs, the company has decided to cut the volume of the bar by a drastic 28%. The thickness will be the same, but the length and width will be reduced by equal amounts. What will the length and width of the new bar be?

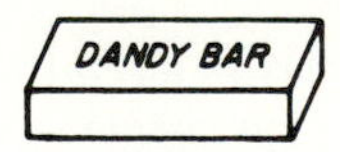

FIGURE 6-5

20. The Dandy Candy Company, besides making the Dandy Bar, also makes a washer-shaped candy (a candy with a hole in it), called Life Rings. See Fig. 6-6. Because of increasing costs, the company will cut the volume of candy in each piece by 20%. To

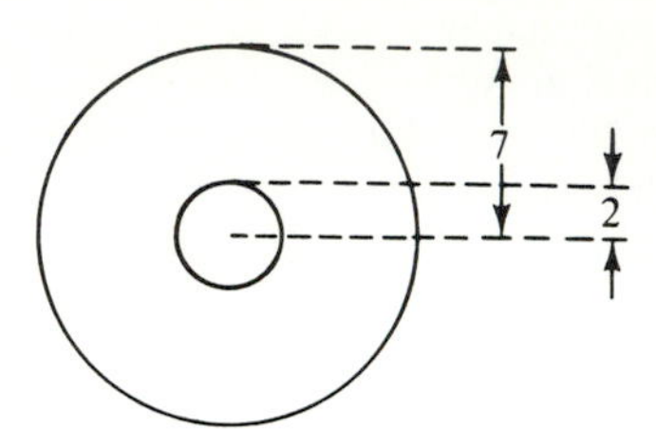

FIGURE 6-6

do this they will keep the same thickness and outer radius, but will make the inner radius larger. At present the thickness is 2 mm, the inner radius is 2 mm, and the outer radius is 7 mm. Find the inner radius of the new-style candy. *Hint*: The volume V of a solid disc is $\pi r^2 h$, where r is the radius and h is the thickness.

21. An oil dealer decides to tear down two cylindrical storage tanks and replace them by one new tank. The old tanks are each 16 ft high. One has a radius of 15 ft and the other a radius of 20 ft. The new tank will also be 16 ft high. Find its radius if it is to hold the same volume of oil as the old tanks combined. *Hint*: The volume V of a cylindrical tank is $V = \pi r^2 h$, where r is the radius and h is the height.

22. A person deposits \$50 in a bank and in two years it increases to \$56.18. If the bank compounds interest annually, what annual rate of interest does it pay?

23. In two years a company will begin an expansion program. It has decided to invest \$1,000,000 now so that in two years the total value of the investment will be \$1,102,500, the amount required for the expansion. What is the annual rate of interest, compounded annually, that the company must receive to achieve its purpose?

24. Researchers at a university need a 10,000-m^2 rectangular plot on which to grow three hybrids of corn. The plot is to be enclosed by fencing. Fencing will also be used to separate the different hybrids. See Fig. 6-7. If 600 m of fence are to be used for the project, what are the dimensions of this plot?

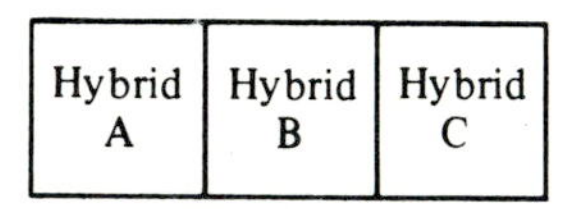

FIGURE 6-7

25. A machine company has an incentive plan for its salespeople. For each machine sold by a salesperson, the commission is \$20. The commission for *every* machine sold will increase by \$0.02 for each machine sold over 600. For example, the commission on each of 602 machines sold is \$20.04. How many machines must a salesperson sell in order to earn \$15,400?

26. A chemist took some acid from a full beaker containing 81 cm^3 of pure acid and then filled up the beaker with pure water. He then took the same amount from the acid-water mixture and found that 64 cm^3 of pure acid remained in the beaker. How many cubic centimeters did he take each time?

6.6 SUPPLEMENT ON QUADRATIC FORMULA*

In Section 6.3 we stated the quadratic formula. This formula can be derived using the method of completing the square.

Suppose that $ax^2 + bx + c = 0$ is a quadratic equation. Then

$$ax^2 + bx = -c,$$

$$\frac{ax^2 + bx}{a} = \frac{-c}{a},$$

$$x^2 + \frac{b}{a}x = -\frac{c}{a}.$$

$$\text{Half of } \frac{b}{a}\text{:} \quad \frac{\frac{b}{a}}{2} = \frac{b}{2a}. \qquad \text{Squaring:} \quad \left(\frac{b}{2a}\right)^2 = \frac{b^2}{4a^2}.$$

$$x^2 + \frac{b}{a}x + \frac{b^2}{4a^2} = \frac{b^2}{4a^2} - \frac{c}{a}. \tag{1}$$

The left side is $\left(x + \frac{b}{2a}\right)^2$ and the right side is

$$\frac{b^2}{4a^2} - \frac{c}{a} = \frac{b^2}{4a^2} - \frac{4ac}{4a^2} = \frac{b^2 - 4ac}{4a^2}.$$

Thus Eq. 1 becomes

$$\left(x + \frac{b}{2a}\right)^2 = \frac{b^2 - 4ac}{4a^2},$$

$$x + \frac{b}{2a} = \pm\sqrt{\frac{b^2 - 4ac}{4a^2}},$$

$$x = -\frac{b}{2a} \pm \frac{\sqrt{b^2 - 4ac}}{2a},$$

$$x = \frac{-b \pm \sqrt{b^2 - 4ac}}{2a},$$

which is the quadratic formula.

---

* This topic is not needed in the rest of the book.

6.7 REVIEW

IMPORTANT TERMS AND SYMBOLS

second-degree equation *(6.1)*
quadratic equation *(6.1)*
repeated root *(6.1)*
square-root method *(6.1)*
completing the square *(6.2)*
quadratic formula *(6.3)*
discriminant *(6.3)*
$\sqrt{-1}, i$ *(6.4)*
imaginary unit *(6.4)*
pure imaginary number *(6.4)*
complex number *(6.4)*
imaginary number *(6.4)*

REVIEW PROBLEMS

In Problems **1–10**, *solve by factoring.*

1. $x^2 - 10x + 25 = 0.$
2. $x^2 - 2x - 8 = 0.$
3. $x^2 - 2x - 24 = 0.$
4. $x^2 + 6x + 9 = 0.$
5. $12x^2 - 20x + 3 = 0.$
6. $4x^2 - 5x - 6 = 0.$
7. $x^2 - 12 = 0.$
8. $x^2 - 28 = 0.$
9. $2x^3 - x^2 = 0.$
10. $x^4 - 9x^2 = 0.$

In Problems **11–14**, *solve by completing the square.*

11. $x^2 - 10x + 1 = 0.$
12. $x^2 + 8x - 5 = 0.$
13. $4x^2 + 12x - 2 = 0.$
14. $2x^2 - 2x - 3 = 0.$

In Problems **15–20**, *solve by the quadratic formula.*

15. $x^2 - 6x + 7 = 0.$
16. $x^2 + 3x - 5 = 0.$
17. $4x^2 + 4x + 1 = 0.$
18. $3x^2 - 2x + 6 = 0.$
19. $2x - 5 - 2x^2 = 0.$
20. $25 - 20x + 4x^2 = 0.$

In Problems **21–32**, *solve.*

21. $16x^2 - 9 = 0.$
22. $25x^2 - 3 = 1.$
23. $y^2 + 2y - 24 = 0.$
24. $y^2 + 4y - 21 = 0.$
25. $(z + 4)^2 = 36.$
26. $4z(z + 2) = -8 - 4z.$
27. $3(t - 1) = 2t^2.$
28. $100t^2 = 100t.$
29. $(x + 1)(x + 2) = 4.$
30. $x^2 + 5 = -(1 - x).$
31. $4x^2 + 10x = -\frac{25}{4}.$
32. $\frac{3}{4}(x^2 - 2) = x.$

33. The formula $S = 2\pi r^2 + 2\pi rh$ gives the total surface area S of a cylinder of radius r and height h. Find r so that a cylinder of height 2 in. will have an area of 48π in.^{2}.

34. An open box is to be made from a square piece of tin by cutting out a 6-in. square from each corner and turning up the sides. The box will contain 150 in.^{3}. Find the *area* of the original square.

35. Imperial Educational Services (I.E.S.) is offering a workshop in data processing to key personnel at Zeta Corporation. The price per person is \$50, and currently fifty persons will attend. Suppose I.E.S. offers to reduce the charge for *everybody* by \$.50 for each person over the fifty who attends. How many people, over the fifty, can attend so that the cost to Zeta Corporation is the same as that for fifty people?

36. A square plot, 12 yd by 12 yd, is to be used for a garden. It is decided to put a pavement of uniform width inside the plot bordering three of the sides so that 80 yd^2 of the plot is left for flowers. How wide should the pavement be?

37. You are the chief financial advisor to a corporation that owns an office building consisting of 50 offices. At \$400 per month, every office can be rented. However, for each \$20 per month increase, there will be two vacancies with no possibility of filling them. The corporation wants to receive a total of \$20,240 per month from rents in the building. You are asked to determine the rent that should be charged for each office. What is your reply?

38. A certain projectile is fired. Its height s (in feet) above ground level after t seconds is given by

$$s = 152t - 16t^2.$$

The projectile will reach a maximum height and fall down. At what times will the projectile be 280 ft above ground level?

7 Fractions

7.1 REDUCTION, MULTIPLICATION, DIVISION

With the fraction $\frac{-2}{3}$, which can be thought of as

$$+\frac{-2}{+3},$$

we can associate three signs:

the sign in front of the fraction (+),
the sign of the numerator (−), and
the sign of the denominator (+).

Furthermore, from the rules for signed numbers, we know that

$$\frac{-2}{3} = -\frac{2}{3}.$$

Notice that $-\frac{2}{3}$ can be obtained from $\frac{-2}{3}$ by changing *two* signs of $\frac{-2}{3}$: the sign in front of the fraction and the sign of the numerator. We also know that

$$\frac{2}{-3} = -\frac{2}{3}.$$

Again, by changing *two* signs of $\frac{2}{-3}$, namely, the sign in front of the fraction and the sign of the denominator, we can obtain $-\frac{2}{3}$.

In general, **by changing any two of the three signs of a fraction, we obtain a fraction equal** (equivalent) **to the original one**. Thus $-\frac{-6}{7}$ can be written in the following ways:

$$+\frac{+6}{+7}, \qquad -\frac{+6}{-7}, \qquad +\frac{-6}{-7}.$$

EXAMPLE 1

Changing signs of a fraction.

a. $\frac{-x}{y} = -\frac{x}{y}$.

b. $\frac{-x}{-y} = \frac{x}{y}$.

c. $-\frac{x(-y)}{z} = -\frac{-(xy)}{z} = \frac{xy}{z}$.

d. $\frac{w(-x)}{(-y)(-z)} = \frac{-(wx)}{yz} = -\frac{wx}{yz}$.

e. $\frac{-a-b}{-c} = \frac{-(a+b)}{-c} = \frac{a+b}{c}$.

f. $\frac{y}{(-1)ab} = \frac{y}{-ab} = -\frac{y}{ab}$.

If the numerator and denominator of a fraction have a *common factor*, then we can *reduce* the fraction. This is the same as canceling common factors.

$$\frac{ab}{ac} = \frac{b}{c} \quad \text{or} \quad \frac{\not{a}b}{\not{a}c} = \frac{b}{c}.$$

A fraction is *reduced* when its numerator and denominator have no common factors except 1 or -1.

To **reduce** a fraction, first completely factor both the numerator and denominator. Then cancel all common factors.

EXAMPLE 2

Reducing fractions.

a. $\frac{3x-9}{x-3} = \frac{3(\cancel{x-3})}{\cancel{x-3}} = 3$.

b. $\frac{4x^2}{8x^2-4x^3} = \frac{\cancel{4x^2}}{\cancel{4x^2}(2-x)} = \frac{1}{2-x}$.

c. $\frac{2x+6}{6x}$. Here 2 is a common factor of the numerator and denominator.

$$\frac{2x+6}{6x} = \frac{\cancel{2}(x+3)}{\underset{3}{\cancel{6}}x} = \frac{x+3}{3x}.$$

EXAMPLE 3

Reducing fractions.

a. $$\frac{x^2-1}{x^2+2x+1} = \frac{(x-1)\cancel{(x+1)}}{\cancel{(x+1)^2}_{x+1}} = \frac{x-1}{x+1}.$$

b. $$\frac{2x^2-2x-12}{4x^2-8x-12} = \frac{\cancel{2}(x^2-x-6)}{\cancel{4}_2(x^2-2x-3)} = \frac{(x+2)\cancel{(x-3)}}{2(x+1)\cancel{(x-3)}} = \frac{x+2}{2(x+1)}.$$

c. $$\frac{(y-2)^3(y-4)^2}{(y-2)^2(y-4)^4} = \frac{y-2}{(y-4)^2}.$$

Remember, you can cancel only ***factors,*** *and these must be factors of both the* ***entire*** *numerator and* ***entire*** *denominator. Note that,*

$$\frac{4+3x}{x} \neq \frac{4+3\cancel{x}}{\cancel{x}},$$

because x is a factor of a term in the numerator; it is ***not*** *a factor of the entire numerator. Also,* ***don't do things like***

$$\frac{\cancel{x+5}}{x+5\cancel{(x+5)}}$$ [*x* + 5 *is* ***not*** *a factor of the denominator*].

Sometimes, to reduce a fraction we may have to rewrite an expression of the form $a - b$. For example, the expressions $2 - 1$ and $1 - 2$ may look alike, but they have different values: $2 - 1 = 1$ and $1 - 2 = -1$. One value is the opposite of the other: $2 - 1 = -(1 - 2)$. In general,

$$\boxed{a - b = -(b - a) = (-1)(b - a).}$$

This fact can sometimes be used to reduce fractions in a sneaky way.

EXAMPLE 4

Sneaky reductions.

a. $$\frac{3-x}{x-3} = \frac{-(x-3)}{x-3} = \frac{(-1)\cancel{(x-3)}}{\cancel{x-3}} = -1.$$

Another way to reduce the fraction is by rewriting the denominator:

$$\frac{3-x}{x-3} = \frac{\cancel{3-x}}{(-1)\cancel{(3-x)}} = \frac{1}{-1} = -1.$$

b. $\dfrac{x^2+3x-4}{1-x^2} = \dfrac{(x-1)(x+4)}{(1-x)(1+x)} = \dfrac{\cancel{(x-1)}(x+4)}{(-1)\cancel{(x-1)}(1+x)} = -\dfrac{x+4}{1+x}.$

In arithmetic we multiply two fractions by using the next rule.

MULTIPLICATION OF FRACTIONS

$$\frac{a}{b} \cdot \frac{c}{d} = \frac{ac}{bd} \quad \begin{matrix} \leftarrow \textit{product of numerators} \\ \leftarrow \textit{product of denominators} \end{matrix}$$

This rule works not only when a, b, c, and d are numbers, but also when they are expressions.

EXAMPLE 5

Multiplication.

a. $\dfrac{x}{x-1} \cdot \dfrac{x+1}{x-2} = \dfrac{x(x+1)}{(x-1)(x-2)}$. We usually leave our result in factored form.

b. $\dfrac{3}{-x} \cdot \dfrac{x^2+4}{x} \cdot \dfrac{x+5}{x} = \dfrac{3(x^2+4)(x+5)}{-x^3} = -\dfrac{3(x^2+4)(x+5)}{x^3}.$

c. $z^3 \cdot \dfrac{z+2}{z+1} = \dfrac{z^3}{1} \cdot \dfrac{z+2}{z+1} = \dfrac{z^3(z+2)}{z+1}.$

Here we wrote z^3 as $\dfrac{z^3}{1}$ so that we could multiply fractions. More simply, we can directly multiply the numerator of $\dfrac{z+2}{z+1}$ by z^3:

$$z^3 \cdot \frac{z+2}{z+1} = \frac{z^3(z+2)}{z+1}.$$

Multiplying $\dfrac{x}{x+1}$ by $\dfrac{x+1}{x-5}$, we have

$$\frac{x}{x+1} \cdot \frac{x+1}{x-5} = \frac{x\cancel{(x+1)}}{\cancel{(x+1)}(x-5)} = \frac{x}{x-5}.$$

Here we multiplied and then canceled. Instead, we can do the cancellation first.

$$\frac{x}{\cancel{x+1}} \cdot \frac{\cancel{x+1}}{x-5} = \frac{x}{x-5}.$$

EXAMPLE 6

Multiplication.

a. $\cancel{(x+2)} \cdot \dfrac{x-4}{\cancel{x+2}} = x - 4.$

b. $\dfrac{3x-3}{x} \cdot \dfrac{5}{x^2-1} = \dfrac{3(x-1)}{x} \cdot \dfrac{5}{(x-1)(x+1)}$ [factoring]

$$= \frac{3\cancel{(x-1)}}{x} \cdot \frac{5}{\cancel{(x-1)}(x+1)} = \frac{15}{x(x+1)}.$$

c. $\dfrac{x^3}{4x^2-9} \cdot \dfrac{2x+3}{x^2+x} = \dfrac{\overset{x^2}{\cancel{x^3}}}{(2x-3)\cancel{(2x+3)}} \cdot \dfrac{\cancel{2x+3}}{\cancel{x}(x+1)} = \dfrac{x^2}{(2x-3)(x+1)}.$

EXAMPLE 7

Multiplication.

a. $\dfrac{x^2+5x+4}{x^2+5x+6} \cdot \dfrac{x^2+3x+2}{x^2+2x-8} = \dfrac{(x+1)\cancel{(x+4)}}{(x+3)\cancel{(x+2)}} \cdot \dfrac{(x+1)\cancel{(x+2)}}{\cancel{(x+4)}(x-2)}$

$$= \frac{(x+1)^2}{(x+3)(x-2)}.$$

b. $\dfrac{6}{x^2-3x+2} \cdot \dfrac{2-x}{x+3} = \dfrac{6}{(x-1)\cancel{(x-2)}} \cdot \dfrac{(-1)\cancel{(x-2)}}{x+3}$

$$= \frac{-6}{(x-1)(x+3)} = -\frac{6}{(x-1)(x+3)}.$$

After factoring $x^2 - 3x + 2$, we wrote $2 - x$ as $(-1)(x-2)$ so that we could cancel.

To divide two fractions, we "invert" the divisor (which is a fraction) and then multiply. Then we reduce if possible.

DIVISION OF FRACTIONS

$$\frac{\frac{a}{b}}{\frac{c}{d}} = \frac{a}{b} \cdot \frac{d}{c} = \frac{ad}{bc}.$$

Notice the switch from division to multiplication. We multiply $\frac{a}{b}$ by the *reciprocal* of the divisor $\frac{c}{d}$. Sometimes we use the division symbol, $\div$, to denote the original division:

$$\frac{\frac{a}{b}}{\frac{c}{d}} = \frac{a}{b} \div \frac{c}{d} = \frac{a}{b} \cdot \frac{d}{c} = \frac{ad}{bc}.$$

EXAMPLE 8

Division.

a. $$\frac{\frac{4y}{3z}}{\frac{z}{9}} = \frac{4y}{3z} \div \frac{z}{9} = \frac{4y}{3z} \cdot \frac{9}{z} = \frac{4y}{\cancel{3}z} \cdot \frac{\overset{3}{\cancel{9}}}{z} = \frac{12y}{z^2}.$$

b. $$\frac{x}{x+2} \div \frac{x+3}{x-5} = \frac{x}{x+2} \cdot \frac{x-5}{x+3} = \frac{x(x-5)}{(x+2)(x+3)}.$$

c. $$\frac{\frac{4x}{x^2-1}}{\frac{2x^2+8x}{x-1}} = \frac{4x}{x^2-1} \cdot \frac{x-1}{2x^2+8x} = \frac{\overset{2}{\cancel{4x}}}{\cancel{(x-1)}(x+1)} \cdot \frac{\cancel{x-1}}{\cancel{2x}(x+4)}$$

$$= \frac{2}{(x+1)(x+4)}.$$

EXAMPLE 9

Division.

a. $$\frac{\frac{x}{x-3}}{2x} = \frac{\frac{x}{x-3}}{\frac{2x}{1}} = \frac{\cancel{x}}{x-3} \cdot \frac{1}{\cancel{2x}} = \frac{1}{2(x-3)}.$$

b. $$\frac{4x^2-4x+1}{\frac{x}{2x-1}} = (4x^2-4x+1) \cdot \frac{2x-1}{x}$$

$$= (2x-1)^2 \cdot \frac{2x-1}{x} = \frac{(2x-1)^3}{x}.$$

The properties in this section are useful in solving some equations, as Example 10 shows.

EXAMPLE 10

Solve.

a. $-\dfrac{x+1}{-3} = 2.$

$$-\frac{x+1}{-3} = 2,$$

$$\frac{x+1}{3} = 2 \qquad \text{[two sign changes]},$$

$$x + 1 = 6 \qquad \text{[multiplying both sides by 3]},$$

$$\boxed{x = 5.}$$

b. $\dfrac{x}{\frac{1}{4}} = \dfrac{3}{5}.$

$$\frac{x}{\frac{1}{4}} = \frac{3}{5},$$

$$4x = \frac{3}{5} \qquad \left[\text{since } \frac{x}{\frac{1}{4}} = x \cdot \frac{4}{1} = 4x\right],$$

$$x = \frac{\frac{3}{5}}{4} \qquad \text{[dividing both sides by 4]},$$

$$x = \frac{3}{5} \cdot \frac{1}{4} = \boxed{\frac{3}{20}.}$$

Exercise 7.1

In Problems **1–6**, *write each fraction so that minus signs do not appear in the numerator and in the denominator. For example, write* $\dfrac{ab}{-c}$ *as* $-\dfrac{ab}{c}$.

1. $\dfrac{-x}{-y}$.

2. $-\dfrac{-wx}{-z}$.

3. $-\dfrac{a}{b(-c)}$.

4. $-\dfrac{ab}{(-c)(-d)}$.

5. $\dfrac{x}{-y-z}$.

6. $\dfrac{-x(y+z)}{(-a)(b)(-c)}$.

In Problems **7–20**, *reduce.*

7. $\dfrac{2x+4}{6}$.

8. $\dfrac{10x-15}{20}$.

9. $\dfrac{12x^2+6x}{3x^2-9x}$.

10. $\dfrac{6x+2x^2}{8x-8x^2}$.

11. $\dfrac{y^4+y^2}{y^6-4y^5}$.

12. $\dfrac{x^4-x^2}{x^2+x}$.

13. $\dfrac{x+1}{x^2+7x+6}$.

14. $\dfrac{z^2-9}{z^2-6z+9}$.

15. $\dfrac{x^2-3x-10}{x^2-4x-5}$.

16. $\dfrac{2x^2+11x+12}{2x^2+8x}$.

17. $\dfrac{3z^2+z-2}{6z^2-z-2}$.

18. $\dfrac{a^2-b^2}{b^2-a^2}$.

19. $\dfrac{25 - x^2}{x^2 - 2x - 15}$.

20. $\dfrac{x^2 + 3x - 4}{2 - x - x^2}$.

In Problems **21–36**, *do the multiplication and reduce.*

21. $\dfrac{2x}{x - 1} \cdot \dfrac{3x}{x + 2}$.

22. $\dfrac{x - 3}{x^2} \cdot \dfrac{x - 3}{x^4}$.

23. $\dfrac{-x}{x + 2} \cdot \dfrac{x + 2}{x^2}$.

24. $\dfrac{3}{x + 1} \cdot \dfrac{(x + 1)^2}{-(x + 3)}$.

25. $\dfrac{5x - 10}{5(x + 3)} \cdot \dfrac{x}{2x - 4}$.

26. $\dfrac{4}{6x^2} \cdot \dfrac{3x + 9}{x^2 + 3x}$.

27. $(x^2 - 9) \cdot \dfrac{x - 3}{4x + 12}$.

28. $\dfrac{5}{(x + 1)^2} \cdot (x^2 + 2x + 1)$.

29. $\dfrac{(x - 1)^2}{x + 3} \cdot \dfrac{x^3}{x - 1} \cdot \dfrac{(x + 3)^2}{x - 1}$.

30. $\dfrac{2x - 1}{x^2 + 4x} \cdot \dfrac{x}{4x^2 - 1} \cdot \dfrac{2x + 1}{6}$.

31. $\dfrac{x^2 + 6x + 8}{x^2 - 5x + 6} \cdot \dfrac{x - 2}{x^2 + 5x + 4}$.

32. $\dfrac{2x^2 + 3x + 1}{x^2 - 4x - 5} \cdot \dfrac{x^2 - 25}{4x^2 - 1}$.

33. $\dfrac{8x^2 + 32}{x^2 + 2x} \cdot \dfrac{x^2 + 4x + 4}{8x^2 - 32}$.

34. $\dfrac{x^3 + 8x^2 + 15x}{x^2 + 4x + 3} \cdot \dfrac{x - 1}{x^2 + 5x}$.

35. $\dfrac{x^2 + 3x + 2}{(1 - x)^2}\left(\dfrac{x - 1}{x + 2}\right)^2$.

36. $\dfrac{8 - 4x}{6x^3} \cdot \dfrac{4 - x}{x^2 - 6x + 8}$.

In Problems **37–54**, *do the division and reduce.*

37. $\dfrac{\frac{x^2}{6}}{\frac{x}{3}}$.

38. $\dfrac{\frac{4x^3}{9x}}{\frac{x}{18}}$.

39. $\dfrac{\frac{2m}{n^3}}{\frac{4m}{n^2}}$.

40. $\dfrac{c + d}{c} \div \dfrac{c - d}{2c}$.

41. $\dfrac{4x}{3} \div 2x$.

42. $(-9x^3) \div \dfrac{x}{3}$.

43. $\dfrac{\frac{x - 5}{x^2 - 7x + 10}}{x - 2}$.

44. $\dfrac{\frac{x^2 + 6x + 9}{x}}{x + 3}$.

45. $\dfrac{\frac{2x - 4}{-6x}}{\frac{x - 2}{3x^2}}$.

46. $\dfrac{\frac{10x^3}{x^2 - 1}}{\frac{5x}{x + 1}}$.

47. $\dfrac{\frac{x^2 - 4}{x^2 + 2x - 3}}{\frac{x^2 - x - 6}{x^2 - 9}}$.

48. $\dfrac{\frac{x^2 + 7x + 10}{x^2 - 2x - 8}}{\frac{x^2 + 6x + 5}{x^2 - 3x - 4}}$.

49. $\dfrac{\frac{2x^2 + 5x + 3}{6x^2 + 4x}}{\frac{x^2 - 1}{3x^2 - x - 2}}$.

50. $\dfrac{\frac{2x^2 + 5x - 3}{4x^2 - 1}}{\frac{x^2 + 4x + 3}{6x^2 + x - 1}}$.

51. $\dfrac{\frac{(x + 2)^2}{3x - 2}}{\frac{9x + 18}{4 - 9x^2}}$.

52. $\dfrac{\frac{3 - 2x - x^2}{-20x^3}}{\frac{x^2 - 1}{x^4 + x^3}}$.

53. $\dfrac{\frac{3}{5}(x^2 + 4x + 4)}{\frac{9}{8}(x^2 - 4)}$.

54. $\dfrac{3x}{x - 1} \cdot \dfrac{x}{x + 1} \div \dfrac{6x}{x + 1}$.

In Problems **55–66**, *solve.*

55. $-\frac{-x}{5} = 2.$

56. $-\frac{4x}{-5} = 8.$

57. $-\frac{y-4}{-2} = y.$

58. $\frac{-y}{3} - \frac{y}{-4} = 1.$

59. $\frac{-(x+1)}{2} - \frac{-7x}{-4} = 4.$

60. $\frac{x}{-2} - \frac{-x}{3} = \frac{x}{-4}.$

61. $2x = \frac{3}{7}.$

62. $5x = \frac{6}{5}.$

63. $\frac{y}{\frac{2}{3}} = 5.$

64. $\frac{y}{\frac{1}{3}} = 9.$

65. $\frac{x-1}{\frac{4}{5}} = 16.$

66. $\frac{x}{\frac{3}{4}} = \frac{\frac{3}{2}}{6}.$

67. When a certain yo-yo is allowed to fall, the tension in the supporting string is

$$\frac{\frac{1}{2}\left(\frac{W}{g}\right)\left(\frac{1}{6}\right)^2(6a)}{\frac{1}{6}},$$

where W is weight and g and a are accelerations. Simplify this expression.

68. The power P dissipated in a resistor is given by $P = i^2R$, where i is current and R is resistance. In terms of time t, the current and resistance are given by

$$i = \frac{2t+1}{3t^2} \quad \text{and} \quad R = \frac{3t}{2t+1}.$$

In $P = i^2R$, replace i and R by these expressions and simplify the result.

69. In a scale drawing, the scale is $\frac{1}{4}$ in. = 2 ft. The scale length (in inches) of a 56-ft object is given by l, where

$$\frac{l}{\frac{1}{4}} = \frac{56}{2}.$$

Find l.

70. In 1983 a homeowner installed an energy-saving heating furnace that reduced the 1982 heating bill (H) by 20%. In 1984 the homeowner installed storm windows that reduced the 1983 heating bill by 10%. Compared to 1982, what is the total annual percentage of savings due to these energy-conservation efforts?

7.2 ADDITION AND SUBTRACTION OF FRACTIONS

To add or subtract fractions with a *common denominator*, we use the next rules.

$$\frac{a}{c} + \frac{b}{c} = \frac{a+b}{c}.$$

$$\frac{a}{c} - \frac{b}{c} = \frac{a-b}{c}.$$

That is, the sum (or difference) is a fraction whose denominator is the common denominator and whose numerator is the sum (or difference) of the numerators of the fractions.

There are similar rules for handling sums and differences involving any number of fractions having a common denominator. For example,

$$\frac{a}{d}-\frac{b}{d}+\frac{c}{d}=\frac{a-b+c}{d}.$$

After adding or subtracting fractions, reduce the result if possible.

EXAMPLE 1

a. $$\frac{3}{x-4}-\frac{2+3x}{x-4}=\frac{3-(2+3x)}{x-4} \quad \text{[Note use of parentheses.]}$$

$$=\frac{3-2-3x}{x-4}=\frac{1-3x}{x-4}.$$

b. $$\frac{x^2-5}{x-2}+\frac{2x-3}{x-2}=\frac{(x^2-5)+(2x-3)}{x-2}$$

$$=\frac{x^2+2x-8}{x-2} \quad \text{[simplifying numerator]}$$

$$=\frac{\cancel{(x-2)}(x+4)}{\cancel{x-2}} \quad \text{[factoring and canceling]}$$

$$=x+4.$$

To add or subtract fractions with different denominators, we first rewrite the fractions as equivalent fractions that *do* have the same denominator. Then we add or subtract as we did before.

The key to rewriting fractions is the *fundamental principle of fractions*:

FUNDAMENTAL PRINCIPLE OF FRACTIONS

$$\frac{a}{b}=\frac{ac}{bc}.$$

That is, *multiplying both numerator and denominator of a fraction by the same number*, *except* 0, *gives an equivalent fraction.*

Let's apply this to adding fractions with different denominators. We shall find the sum

$$\frac{2}{x}+\frac{3}{x+4}.$$

Since the denominators are x and $x + 4$, we shall rewrite each fraction as an equivalent fraction with denominator $x(x + 4)$. The first fraction becomes

$$\frac{2(x+4)}{x(x+4)} \quad \text{[multiplying numerator and denominator by } x + 4\text{];}$$

the second fraction becomes

$$\frac{3x}{(x+4)x} \quad \text{[multiplying numerator and denominator by } x\text{].}$$

Since these fractions have a common denominator, we can combine them.

$$\begin{aligned}\frac{2}{x} + \frac{3}{x+4} &= \frac{2(x+4)}{x(x+4)} + \frac{3x}{(x+4)x} \\ &= \frac{2(x+4)+3x}{x(x+4)} \\ &= \frac{2x+8+3x}{x(x+4)} \\ &= \frac{5x+8}{x(x+4)}.\end{aligned}$$

We could have rewritten the original fractions with other common denominators. But we chose to rewrite them as fractions with the denominator $x(x + 4)$, which is called the **least common denominator** (L.C.D.) of the fractions $\frac{2}{x}$ and $\frac{3}{x+4}$.

Here's a general rule for finding an L.C.D.

FINDING THE L.C.D.

To find the L.C.D. of two or more fractions, first factor all of the denominators. Then, from the different factors that occur, form a product in which each factor is raised to the greatest power to which that factor occurs in any single denominator.

EXAMPLE 2

Find the L.C.D. *of the fractions*

$$\frac{2x}{(x+1)(x-2)}, \quad \frac{x-3}{x^2(x+1)}, \quad \textit{and} \quad \frac{x^2+x}{x(x-2)^2}.$$

There are three different factors in the denominators:

$$x + 1, \quad x - 2, \quad \text{and} \quad x.$$

The factor $x + 1$ occurs at most one time in any denominator. The factor $x - 2$ occurs at most two times (in the third fraction). The factor x occurs at most two times (in the second fraction). The L.C.D. is the product of these factors, each raised to the greatest power to which it occurs in any denominator. Thus the L.C.D. is

$$(x+1)(x-2)^2x^2.$$

EXAMPLE 3

Find the L.C.D. *of the fractions*

$$\frac{2}{x^2}, \quad \frac{x}{x+1}, \quad \frac{3x}{x^3(x-1)}, \quad \text{and} \quad \frac{5}{\underbrace{x^2+2x+1}_{(x+1)^2}}.$$

Note that we wrote the denominator x^2+2x+1 in factored form: $(x+1)^2$. In the denominators, the factor x occurs at most three times, the factor $x+1$ occurs at most twice, and the factor $x-1$ occurs at most once. Thus the L.C.D. is

$$x^3(x+1)^2(x-1).$$

To add or subtract fractions with different denominators, here's what to do.

> **ADDING OR SUBTRACTING FRACTIONS**
>
> For each fraction, multiply both its numerator and denominator by a quantity that makes its denominator equal to the L.C.D. of the fractions. Then combine and, if possible, simplify.

EXAMPLE 4

Find $\frac{3}{x+2}+\frac{x-1}{x-6}$.

The L.C.D. is $(x+2)(x-6)$. To get the denominator of the first fraction equal to the L.C.D., we multiply the *numerator and denominator* by $x-6$. In the second fraction, we multiply *numerator and denominator* by $x+2$.

$$\frac{3}{x+2}+\frac{x-1}{x-6}$$

$$=\frac{3(x-6)}{(x+2)(x-6)}+\frac{(x-1)(x+2)}{(x-6)(x+2)} \quad \text{[putting the L.C.D. in each fraction]}$$

$$=\frac{3(x-6)+(x-1)(x+2)}{(x+2)(x-6)} \quad \text{[combining fractions with common denominators]}$$

$$=\frac{3x-18+x^2+x-2}{(x+2)(x-6)}$$

$$=\frac{x^2+4x-20}{(x+2)(x-6)}.$$

In Example 4, note that we cannot add the fractions by adding their numerators and adding their denominators:

$$\frac{3}{x+2}+\frac{x-1}{x-6}\neq\frac{3+(x-1)}{(x+2)+(x-6)}.$$

EXAMPLE 5

a. *Find* $\frac{2}{x}-\frac{3}{xy}+\frac{4}{xz^2}$.

The L.C.D. is xyz^2.

$$\frac{2}{x}-\frac{3}{xy}+\frac{4}{xz^2}=\frac{2(yz^2)}{xyz^2}-\frac{3(z^2)}{xyz^2}+\frac{4(y)}{xz^2y}$$

$$=\frac{2yz^2-3z^2+4y}{xyz^2}.$$

b. $(x^{-1}-y^{-1})^2=\left(\frac{1}{x}-\frac{1}{y}\right)^2=\left(\frac{y}{xy}-\frac{x}{yx}\right)^2=\left(\frac{y-x}{xy}\right)^2$

$$=\frac{(y-x)^2}{(xy)^2}=\frac{y^2-2xy+x^2}{x^2y^2}.$$

EXAMPLE 6

Find $3x-4+\frac{2}{x-1}$.

Since $3x-4=\frac{3x-4}{1}$, the L.C.D. of $3x-4$ and $\frac{2}{x-1}$ is $x-1$.

$$3x-4+\frac{2}{x-1}=\frac{(3x-4)(x-1)}{x-1}+\frac{2}{x-1}$$

$$=\frac{(3x-4)(x-1)+2}{x-1}$$

$$=\frac{3x^2-7x+4+2}{x-1}$$

$$=\frac{3x^2-7x+6}{x-1}.$$

EXAMPLE 7

Find $\frac{6x-17}{x^2-5x+6}-\frac{1}{x-3}+3$.

The first denominator factors into $(x-3)(x-2)$. Thus the denominators of the three terms are $(x-3)(x-2)$, $x-3$, and 1. The L.C.D. is $(x-3)(x-2)$.

$$\frac{6x-17}{\underbrace{x^2-5x-6}_{(x-3)(x-2)}}-\frac{1}{x-3}+3$$

$$=\frac{6x-17}{(x-3)(x-2)}-\frac{x-2}{(x-3)(x-2)}+\frac{3(x-3)(x-2)}{(x-3)(x-2)}$$

$$=\frac{6x-17-(x-2)+3\overbrace{(x-3)(x-2)}^{x^2-5x+6}}{(x-3)(x-2)}$$

$$=\frac{6x-17-x+2+3x^2-15x+18}{(x-3)(x-2)}$$

$$=\frac{3x^2-10x+3}{(x-3)(x-2)}=\frac{(3x-1)\cancel{(x-3)}}{\cancel{(x-3)}(x-2)}$$

$$=\frac{3x-1}{x-2}.$$

EXAMPLE 8

Find $\dfrac{3}{x-2}+\dfrac{2}{2-x}$.

You might be tempted to say that the L.C.D. is $(x-2)(2-x)$. However, we can rewrite the second fraction so that its denominator is $x-2$.

$$\frac{3}{x-2}+\frac{2}{2-x}=\frac{3}{x-2}+\frac{2}{-(x-2)}$$

$$=\frac{3}{x-2}-\frac{2}{x-2}$$

$$=\frac{3-2}{x-2}=\frac{1}{x-2}.$$

EXAMPLE 9

Find $\dfrac{x-2}{x^2+6x+9}-\dfrac{x+2}{2(x^2-9)}$.

Since $x^2+6x+9=(x+3)^2$ and $2(x^2-9)=2(x+3)(x-3)$, the L.C.D. is $2(x+3)^2(x-3)$.

$$\frac{x-2}{(x+3)^2}-\frac{x+2}{2(x+3)(x-3)}$$
$$=\frac{(x-2)(2)(x-3)}{(x+3)^2(2)(x-3)}-\frac{(x+2)(x+3)}{2(x+3)(x-3)(x+3)}$$
$$=\frac{(x-2)(2)(x-3)-(x+2)(x+3)}{2(x+3)^2(x-3)}$$
$$=\frac{2(x^2-5x+6)-[x^2+5x+6]}{2(x+3)^2(x-3)}$$
$$=\frac{2x^2-10x+12-x^2-5x-6}{2(x+3)^2(x-3)}$$
$$=\frac{x^2-15x+6}{2(x+3)^2(x-3)}.$$

Exercise 7.2

In Problems **1–8**, *find the* L.C.D.

1. $\frac{6}{(x-4)^2}, \quad \frac{7}{(x-4)^5}.$

2. $\frac{x}{x+1}, \quad \frac{2}{x-3}.$

3. $\frac{4}{x^2y}, \quad \frac{5}{xy^3}.$

4. $\frac{x+1}{x^2y^3z}, \quad \frac{y-1}{xyz^4}.$

5. $\frac{3x}{x^2+6x+9}, \quad \frac{x^2}{x^2-9}.$

6. $\frac{2}{x^2+3x-4}, \quad \frac{1}{x-1}.$

7. $\frac{1}{2x+2}, \quad \frac{x}{x^2+x}, \quad \frac{2}{x+1}.$

8. $\frac{x}{4x+2}, \quad \frac{4}{4x^2-1}, \quad \frac{x}{3}.$

In Problems **9–48**, *perform the indicated operations and simplify.*

9. $\frac{x+1}{x-3}+\frac{4}{x-3}.$

10. $\frac{x^2}{x-2}+\frac{x-6}{x-2}.$

11. $\frac{3x}{x+1}+\frac{4}{x+1}-\frac{x+2}{x+1}.$

12. $\frac{2x}{x^2-1}-\frac{2}{x^2-1}.$

13. $\frac{3x^2+6x}{x^2+x-2}-\frac{x^2+2x+1}{x^2-1}.$

14. $\frac{3x+4}{x+2}+\frac{x^2-9}{x^2+5x+6}.$

15. $\frac{2}{x}+\frac{3}{y}.$

16. $3+\frac{x}{y}.$

17. $\frac{x-4}{6}-\frac{x-2}{9}.$

18. $\frac{x-2}{3}+1.$

19. $\frac{3}{2x}-\frac{2}{xy}.$

20. $\frac{4}{x}+\frac{2}{y}-\frac{x+1}{xy}.$

21. $\frac{5}{x-2}+\frac{3}{x-3}$.

22. $\frac{y}{y-2}-\frac{3}{y}$.

23. $\frac{5y}{x^2}-\frac{2}{xy}+\frac{3}{y}$.

24. $\frac{x}{a^2}+\frac{y}{ab}$.

25. $\frac{x+3}{x-1}+4$.

26. $\frac{a}{b}+\frac{c}{d}$.

27. $\frac{x}{x-y}+\frac{y}{x+y}$.

28. $\frac{2}{x+1}-\frac{3}{x-1}$.

29. $\frac{x+3}{x-3}-\frac{x-3}{2(x+3)}$.

30. $\frac{4}{2x-1}+\frac{x}{x+2}$.

31. $\frac{6x+12}{x^2+5x+4}+\frac{x}{x+4}$.

32. $\frac{5}{x^2+3x-4}+\frac{1}{x+4}$.

33. $\frac{1}{x^2-1}-\frac{1}{x-1}+\frac{1}{x+1}$.

34. $\frac{x}{x+1}-\frac{2x}{x^2+3x+2}$.

35. $\frac{x-1}{x^2+6x+9}+\frac{2}{x^2-9}$.

36. $x^2+2-\frac{x^4}{x^2-2}$.

37. $\frac{x+1}{x^2+7x+10}-\frac{2x}{x^2+6x+5}$.

38. $\frac{y}{3y^2-5y-2}-\frac{2}{3y^2-7y+2}$.

39. $\frac{2x-6}{x^2-5x+6}+\frac{x^2+8x+16}{x^2+6x+8}$.

40. $\frac{2}{x^3(x-3)}+\frac{3}{x(x-3)^2}$.

41. $2x+3+\frac{2}{x-1}$.

42. $\frac{3}{x-1}-\frac{4}{1-x}$.

43. $\frac{x-2}{x^2+x}+\frac{3}{x^3+2x^2}-\frac{2x-3}{x^2+3x+2}$.

44. $\frac{2}{x^2-5x+6}-\frac{1}{x^2-3x+2}+\frac{4}{x^2-4x+3}$.

45. $\frac{y}{2y^2+7y+3}-\frac{2}{4y^2+4y+1}$.

46. $\frac{3}{x+3}+\frac{1}{x-3}-\frac{4}{x+2}$.

47. $\frac{y}{x^2+2xy+y^2}+\frac{3x}{x^2-y^2}-\frac{2}{x+y}$.

48. $1-\frac{2y^2}{x^2-y^2}+\frac{2xy}{x^2+y^2}$.

In Problems **49–52**, *perform the operations and simplify as single fractions. Give all answers with positive exponents only.*

49. $(x+y^{-1})^2$.

50. $x^{-2}-y^{-2}$.

51. $[x^{-1}(x+y^{-1})]^{-1}$.

52. $(x^{-1}+y^{-1})^{-1}$.

53. When two springs are connected in series as shown in Fig. 7-1, the reciprocal of the effective spring constant k is given by

$$\frac{1}{k}=\frac{1}{k_1}+\frac{1}{k_2},$$

where k_1 and k_2 are the spring constants of the two springs. Express $\frac{1}{k}$ as a single fraction.

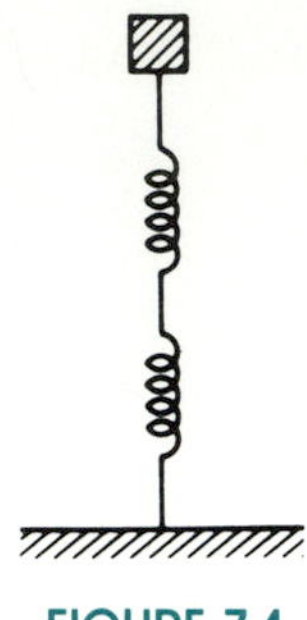

FIGURE 7-1

54. Repeat Problem 53, given that three springs are in series and

$$\frac{1}{k} = \frac{1}{k_1} + \frac{1}{k_2} + \frac{1}{k_3}.$$

55. The formula for the stress factor f of a particular coil spring is given by

$$f = \frac{4k - 1}{4k - 4} + \frac{.5}{k}.$$

Express f as a single fraction and simplify.

7.3 COMBINED OPERATIONS

We are now going to simplify a fraction whose numerator or denominator involves addition or subtraction of fractions. There are two methods that are used.

Method 1. Perform the operations indicated in the numerator and in the denominator. Then divide the numerator by the denominator.

Method 2. Multiply the numerator and denominator of the given fraction by the L.C.D. of the fractions that appear in the numerator and denominator. By the fundamental principle of fractions, this gives an equivalent fraction.

In Example 1 both methods will be shown.

EXAMPLE 1

Simplify $\dfrac{2 - \dfrac{3}{x}}{4x}$.

Method 1.

This is a fraction with numerator $2-\frac{3}{x}$ and denominator $4x$. To simplify, we'll combine $2-\frac{3}{x}$ into one fraction and then divide that result by $4x$.

$$\frac{2-\frac{3}{x}}{4x}=\frac{\frac{2x}{x}-\frac{3}{x}}{4x}=\frac{\frac{2x-3}{x}}{4x}$$

$$=\frac{\frac{2x-3}{x}}{\frac{4x}{1}}=\frac{2x-3}{x}\cdot\frac{1}{4x}=\frac{2x-3}{4x^2}.$$

Method 2.

We multiply the numerator and denominator by the L.C.D. of the fractions that appear in the numerator or denominator. The only fraction is $\frac{1}{x}$, so the L.C.D. is x.

$$\frac{2-\frac{3}{x}}{4x}=\frac{x\left(2-\frac{3}{x}\right)}{x(4x)}$$

$$=\frac{x(2)-x\left(\frac{3}{x}\right)}{4x^2} \qquad \text{[distributive law]}$$

$$=\frac{2x-3}{4x^2}.$$

EXAMPLE 2

Simplify $\dfrac{\frac{x}{x-1}}{\frac{1}{x-1}+\frac{1}{x+1}}$.

We shall use Method 2 and multiply the numerator and denominator by the L.C.D. of the fractions that appear. The L.C.D. is $(x-1)(x+1)$.

$$\frac{\frac{x}{x-1}}{\frac{1}{x-1}+\frac{1}{x+1}}=\frac{\cancel{(x-1)}(x+1)\left[\frac{x}{\cancel{x-1}}\right]}{(x-1)(x+1)\left[\frac{1}{x-1}+\frac{1}{x+1}\right]}$$

$$=\frac{(x+1)(x)}{\cancel{(x-1)}(x+1)\cdot\frac{1}{\cancel{x-1}}+(x-1)\cancel{(x+1)}\cdot\frac{1}{\cancel{x+1}}}$$

$$=\frac{(x+1)(x)}{(x+1)+(x-1)}$$

$$=\frac{(x+1)\cancel{(x)}}{2\cancel{x}}=\frac{x+1}{2}.$$

EXAMPLE 3

Simplify $\dfrac{3-\dfrac{1}{2x}}{6x+\dfrac{11x}{x-2}}$.

We shall use Method 1.

$$\frac{3-\dfrac{1}{2x}}{6x+\dfrac{11x}{x-2}}=\frac{\dfrac{3(2x)}{2x}-\dfrac{1}{2x}}{\dfrac{6x(x-2)}{x-2}+\dfrac{11x}{x-2}}$$

$$=\frac{\dfrac{6x-1}{2x}}{\dfrac{6x^2-12x+11x}{x-2}}$$

$$=\frac{6x-1}{2x}\cdot\frac{x-2}{6x^2-x}$$

$$=\frac{\cancel{(6x-1)}(x-2)}{2x(x)\cancel{(6x-1)}}$$

$$=\frac{x-2}{2x^2}.$$

Exercise 7.3

In Problems **1–18**, *simplify.*

1. $\dfrac{\dfrac{1}{x}+\dfrac{3}{x}}{4}$.

2. $\dfrac{\dfrac{x}{x-1}-\dfrac{1}{x-1}}{x-1}$.

3. $\dfrac{4-\dfrac{6}{x}}{2}$.

4. $\dfrac{x-1}{1-\dfrac{1}{x}}$.

5. $\dfrac{7}{3x-\dfrac{1}{2}}$.

6. $\dfrac{x-\dfrac{1}{x}}{x+1}$.

7. $\dfrac{3-\dfrac{1}{y}}{2+\dfrac{1}{x}}$.

8. $\dfrac{\dfrac{a}{b}+2}{\dfrac{b}{a}-2}$.

9. $\dfrac{\dfrac{1}{x}+\dfrac{x}{2x-3}}{\dfrac{x-1}{x}}$.

10. $\dfrac{\dfrac{x^2-9}{4}}{\dfrac{1}{x}-\dfrac{1}{3}}$.

11. $\dfrac{x+1+\dfrac{1}{x+3}}{\dfrac{x+2}{3}}$.

12. $\dfrac{\dfrac{x}{x^2+4x+3}}{\dfrac{1}{x^2-1}+1}$.

13. $\dfrac{3x+\dfrac{x}{x-3}}{3x-\dfrac{x}{x-3}}$.

14. $\dfrac{\dfrac{1}{x^4}+\dfrac{1}{x^2}+1}{\dfrac{1}{x^3}+\dfrac{1}{x}+x}$.

15. $\dfrac{2+x^{-1}}{x}$.

16. $\dfrac{x + x^{-1}}{x^{-1} - x}$.

17. $\left(\dfrac{x^{-1} + y^{-1}}{x^{-2} - y^{-2}}\right)^{-2}$.

18. $\left(\dfrac{1 + y^{-2}}{y + y^{-1}}\right)^{-2}$.

19. In a model of traffic flow on a lane of a freeway,* the number N of cars the lane can carry per unit time is given by

$$N = \frac{-2a}{-2at_r + v - \dfrac{2al}{v}},$$

where a is acceleration of a car when stopping, t_r is reaction time to begin braking, v is average speed of the cars, and l is length of a car. Simplify the right side of the equation.

20. When the system of masses m_1 and m_2 shown in Fig. 7-2 is released from rest, the acceleration a of the masses is given by

$$a = \frac{g}{1 + \dfrac{m_1}{m_2}},$$

where g is a constant. Simplify the right side of the equation.

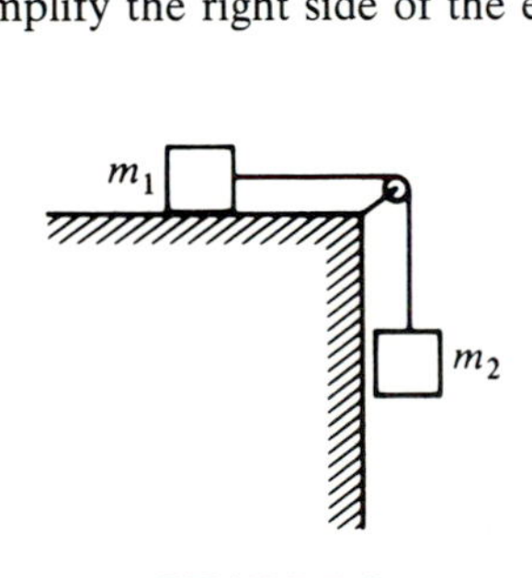

FIGURE 7-2

7.4 FRACTIONAL EQUATIONS

An equation with an unknown in a denominator is called a **fractional equation.** We usually solve such equations by clearing of fractions.

For example, let's solve

$$\frac{6}{x-3} = 5. \tag{1}$$

We clear of fractions by multiplying both sides by $x - 3$.

---

*J. I. Shonle, *Environmental Applications of General Physics* (Reading, Mass.: Addison-Wesley, 1975.)

$$(x-3)\cdot\frac{6}{x-3}=(x-3)\cdot 5,$$

$$\cancel{(x-3)}\cdot\frac{6}{\cancel{x-3}}=5x-15,$$

$$6=5x-15,$$

$$21=5x,$$

$$\frac{21}{5}=x. \tag{2}$$

We're not done yet! Recall that if we multiply both sides of an equation by a number different from zero, then the resulting equation is equivalent to the original. Here we multiplied both sides of Eq. 1 by the polynomial $x-3$, which is 0 when $x=3$. Thus Eqs. 1 and 2 are equivalent provided $x \neq 3$. When x is $\frac{21}{5}$ (which is not 3), we *do* have equivalence. Thus $\frac{21}{5}$ must be a solution of the original equation.

You may wonder whether 3 itself is also a solution. It can be shown that when both sides of an equation (with variable x) are multiplied by the same polynomial (in x), then the resulting equation has *all* the solutions of the original and perhaps *more*. Thus if 3 is a solution to the original equation, it has to be a solution of Eq. 2, which it is not. Hence the only solution of the original equation is $\frac{21}{5}$.

In summary, when you multiply both sides of a given equation by a polynomial, any solution of the resulting equation will be a solution of the given equation as long as it does not give the polynomial a value of zero. If the polynomial does have a value of zero, then the solution of the resulting equation must be checked by substitution in the *given* equation to see if it is a solution there.

EXAMPLE 1

Solve $\frac{1}{x+2}+\frac{1}{x-2}=\frac{4}{x^2-4}$. (3)

We clear of fractions by multiplying both sides by the L.C.D. of all fractions involved. Since $x^2-4=(x+2)(x-2)$, clearly the L.C.D. is $(x+2)(x-2)$.

$$\frac{1}{x+2}+\frac{1}{x-2}=\frac{4}{(x+2)(x-2)},$$

$$(x+2)(x-2)\left[\frac{1}{x+2}+\frac{1}{x-2}\right]=(x+2)(x-2)\left[\frac{4}{(x+2)(x-2)}\right],$$

$$\frac{\cancel{(x+2)}(x-2)}{\cancel{x+2}}+\frac{(x+2)\cancel{(x-2)}}{\cancel{x-2}}=\frac{\cancel{(x+2)}\cancel{(x-2)}\cdot 4}{\cancel{(x+2)}\cancel{(x-2)}} \quad \text{[distributive law and cancellation]},$$

$$(x-2)+(x+2)=4, \tag{4}$$

$$2x=4,$$

$$x=2.$$

Since we multiplied both sides by the polynomial $(x + 2)(x - 2)$, we must check its value when x is 2.

$$\text{If } x = 2, \quad \text{then} \quad (x + 2)(x - 2) = (2 + 2)(2 - 2) = 0.$$

Since the value is 0, *we must see if* 2 *is a solution of the original equation*. When $x = 2$ in that equation, a denominator is 0. But we cannot divide by 0. Thus we conclude that **there is no solution.*** Although 2 is a solution of Eq. 4, it is not a solution of the original equation and is sometimes called an **extraneous solution** of Eq. 3. In general, *multiplying both sides of an equation by a polynomial may lead to extraneous solutions.*

As Example 1 showed, fractional equations can be very tricky. You may think you have a solution when you really don't. Thus you must be very careful when solving such equations.

EXAMPLE 2

Solve $\dfrac{x}{x+4} = \dfrac{2}{x}$.

Multiplying both sides by the L.C.D., $x(x + 4)$, and simplifying, we get

$$x\cancel{(x+4)} \cdot \frac{x}{\cancel{x+4}} = \cancel{x}(x+4) \cdot \frac{2}{\cancel{x}},$$

$$x^2 = 2(x + 4),$$

$$x^2 = 2x + 8,$$

$$x^2 - 2x - 8 = 0 \qquad \text{[quadratic equation]},$$

$$(x + 2)(x - 4) = 0 \qquad \text{[factoring]},$$

$$x + 2 = 0, \quad \Big| \quad x - 4 = 0,$$

$$x = -2. \quad \Big| \quad x = 4.$$

We now check the value of $x(x + 4)$ when $x = -2$ and when $x = 4$.

$$\text{If } x = -2, \quad \text{then} \quad x(x + 4) = -2(-2 + 4) \neq 0.$$

$$\text{If } x = 4, \quad \text{then} \quad x(x + 4) = 4(4 + 4) \neq 0.$$

Because $x(x + 4) \neq 0$ when $x = -2$ and $x = 4$, the solution is

$$\boxed{x = -2, 4.}$$

---

*The solution set is a set with no elements in it, { }. It is called the *empty set* or *null set*. We can also write the empty set with the symbol $\varnothing$.

EXAMPLE 3

Solve $\dfrac{y+1}{y+3}+\dfrac{y+5}{y-2}=\dfrac{7(2y+1)}{y^2+y-6}$.

Since $y^2+y-6=(y+3)(y-2)$, we see that the L.C.D. is $(y+3)(y-2)$. Multiplying both sides by the L.C.D., we have

$$(y+3)(y-2)\left(\frac{y+1}{y+3}+\frac{y+5}{y-2}\right)=(y+3)(y-2)\cdot\frac{7(2y+1)}{(y+3)(y-2)},$$

$$\frac{\cancel{(y+3)}(y-2)(y+1)}{\cancel{y+3}}+\frac{(y+3)\cancel{(y-2)}(y+5)}{\cancel{y-2}}=\frac{\cancel{(y+3)}\cancel{(y-2)}\cdot 7(2y+1)}{\cancel{(y+3)}\cancel{(y-2)}},$$

$$(y-2)(y+1)+(y+3)(y+5)=7(2y+1),$$

$$y^2-y-2+y^2+8y+15=14y+7,$$

$$2y^2+7y+13=14y+7,$$

$$2y^2-7y+6=0 \qquad \text{[quadratic equation]},$$

$$(2y-3)(y-2)=0 \qquad \text{[factoring]}.$$

$$2y-3=0, \quad 2y=3, \quad y=\frac{3}{2}. \qquad \Big| \qquad y-2=0, \quad y=2.$$

We now check the value of $(y+3)(y-2)$ when $y=\frac{3}{2}$ and when $y=2$.

$$\text{If } y=\frac{3}{2}, \quad \text{then} \quad (y+3)(y-2)=\left(\frac{3}{2}+3\right)\left(\frac{3}{2}-2\right)\neq 0.$$

$$\text{If } y=2, \quad \text{then} \quad (y+3)(y-2)=(2+3)(2-2)=0.$$

Because $(y+3)(y-2)\neq 0$ when $y=\frac{3}{2}$, this value of y must be a solution of the original equation. Because $(y+3)(y-2)=0$ when $y=2$, we must check this value in the original equation. Substituting 2 for y in that equation gives a denominator of 0. Thus the only solution is

$$\boxed{y=\frac{3}{2}.}$$

EXAMPLE 4

A construction firm has a government contract to build a swimming pool for the use of certain public officials. According to the contract, it must be completed within the next 21 days. The supervisor of the job knows that his regular crew would take 45 days to build it. To meet the deadline, the supervisor decides to use a second crew, who can build the pool by themselves in 30 days. How long will it take both crews to construct the pool if they work together?

In one day the regular crew does $\frac{1}{45}$ of the construction work and the second crew does $\frac{1}{30}$. When both crews work, in one day a total of $\frac{1}{45} + \frac{1}{30}$ of the construction work is done. Let n = the number of days it takes to build the pool when both crews work. Then in one day, $\frac{1}{n}$ of the work is done. Thus $\frac{1}{45} + \frac{1}{30}$ is $\frac{1}{n}$ of the work.

$$\frac{1}{45} + \frac{1}{30} = \frac{1}{n},$$

$$90n\left[\frac{1}{45} + \frac{1}{30}\right] = 90n\left(\frac{1}{n}\right) \qquad \text{[multiplying both sides by } 90n\text{, the L.C.D.]},$$

$$2n + 3n = 90,$$

$$5n = 90,$$

$$n = 18.$$

If $n = 18$, then $90n \neq 0$. Thus it will take 18 days to build the pool.

EXAMPLE 5

The rate of the current in a stream is 3 mi/h. *A man rowed upstream for* 3 mi *and then returned to his starting point. The round trip took a total of* 1 h, 20 min. *How fast could the man row in still water?*

Let r be the rate (in miles per hour) at which the man can row in still water. Since the rate of the current is 3 mi/h, the man's rate upstream was $r - 3$, and downstream it was $r + 3$. Since time = distance/rate and distance = 3 for each rate, we have

$$\left(\begin{matrix}\text{time}\\ \text{upstream}\end{matrix}\right) + \left(\begin{matrix}\text{time}\\ \text{downstream}\end{matrix}\right) = \text{total time},$$

$$\frac{3}{r-3} + \frac{3}{r+3} = \frac{4}{3}. \qquad \left[1 \text{ h } 20 \text{ min} = \frac{4}{3} \text{ h}\right]$$

Multiplying both sides by $3(r-3)(r+3)$ and simplifying, we obtain

$$3(3)(r+3) + 3(3)(r-3) = 4(r-3)(r+3),$$

$$9r + 27 + 9r - 27 = 4[r^2 - 9],$$

$$18r = 4r^2 - 36,$$

$$0 = 4r^2 - 18r - 36,$$

$$0 = 2r^2 - 9r - 18 \qquad \text{[dividing both sides by 2]},$$

$$0 = (2r+3)(r-6).$$

$$2r + 3 = 0, \quad r = -\frac{3}{2}. \qquad \bigg| \qquad r - 6 = 0, \quad r = 6.$$

The values $-\frac{3}{2}$ and 6 do not make $3(r-3)(r+3)$ equal 0. Thus they are solutions of the original equation. But r is a rate, a positive number, and so we choose $r = 6$ mi/h.

Exercise 7.4

In Problems **1–34**, *solve the fractional equations.*

1. $\frac{3}{x} = 12.$

2. $\frac{1}{x} + \frac{1}{5} = \frac{4}{5}.$

3. $\frac{x}{3x - 4} = 3.$

4. $\frac{4}{x - 1} = 2.$

5. $\frac{10}{3r} - \frac{9r + 2}{6r} = 3.$

6. $\frac{4x}{7 - x} = 1.$

7. $\frac{x}{3} = \frac{6}{x} - 1.$

8. $\frac{2x - 3}{4x - 5} = 6.$

9. $1 + \frac{2}{x} = \frac{2(x + 1)}{x}.$

10. $\frac{1}{2x} - \frac{2x - 3}{4x} = \frac{3x}{4}.$

11. $\frac{3}{4y} - \frac{5}{6y} = \frac{1}{6}.$

12. $\frac{1}{x} - \frac{2}{3x} = 3x.$

13. $\frac{1}{x^2} + \frac{6}{x} + 8 = 0.$

14. $\frac{1}{x^2} + \frac{1}{x} - 12 = 0.$

15. $\frac{1}{x - 1} = \frac{2}{x - 2}.$

16. $\frac{x}{x - 1} = \frac{4}{x}.$

17. $\frac{3}{x - 4} + \frac{x - 3}{x} = 2.$

18. $\frac{4 - x}{x} + \frac{8}{4 + x} = 1.$

19. $\frac{2}{x - 2} = \frac{x + 1}{x + 4}.$

20. $\frac{4}{x - 3} = \frac{3}{x - 4}.$

21. $\frac{x^2}{x - 1} + 1 = \frac{1}{x - 1}.$

22. $\frac{9}{x - 3} = \frac{3x}{x - 3}.$

23. $\frac{3x - 2}{2x + 3} = \frac{3x - 1}{2x + 1}.$

24. $\frac{y - 3}{y + 3} = \frac{y - 3}{y + 2}.$

25. $\frac{y - 6}{y} - \frac{6}{y} = \frac{y + 6}{y - 6}.$

26. $\frac{x + 1}{x} - \frac{6}{x + 5} = \frac{3}{x}.$

27. $\frac{2}{x - 1} - \frac{6}{2x + 1} = 5.$

28. $x + \frac{2x}{x - 2} = \frac{4}{x - 2}.$

29. $\frac{3}{x} - \frac{4}{x + 2} = \frac{5}{3}.$

30. $\frac{x}{x + 3} - \frac{x}{x - 3} = \frac{3x - 4}{x^2 - 9}.$

31. $\frac{3x + 4}{x + 2} - \frac{3x - 5}{x - 4} = \frac{12}{x^2 - 2x - 8}.$

32. $\frac{4}{x - 4} - \frac{3}{x - 3} = \frac{1}{x - 5}.$

33. $\frac{3}{x + 1} + \frac{4}{x} = \frac{12}{x + 2}.$

34. $\frac{2}{x^2 - 1} - \frac{1}{x(x - 1)} = \frac{2}{x^2}.$

35. An important equation for lenses is

$$\frac{1}{p} + \frac{1}{q} = \frac{1}{f},$$

where f is focal length, p is object distance, and q is image distance, and f, p, and q

are in centimeters. Suppose that for a converging lens the focal length is 12 cm and the object distance is 24 cm. Find the image distance.

36. *Calculator problem.* An object is 120 in. from a wall. In order to focus the image of the object on the wall, a converging lens with a focal length of 24 in. is used. The lens is placed between the object and the wall at a distance of p inches from the object, where

$$\frac{1}{p}+\frac{1}{120-p}=\frac{1}{24}.$$

Find p to one decimal place.

37. In a certain wildlife preserve, the number (y) of prey consumed by an individual predator over a given period of time is given by

$$y=\frac{10x}{1+.1x},$$

where x is *prey density* (the number of prey per unit of area). What prey density would allow a predator to survive if it needs to consume 50 prey over the given time period?

38. If a sound source moves towards you, the pitch of the sound seems to change. This is known as the Doppler effect. For example, if you were standing near a railroad track, you would hear an increase of pitch in the train's whistle as the train approaches. If the actual frequency of a sound source is f, and the sound source has speed v, and V is the velocity of sound, then the apparent frequency heard, f' (read "f prime"), is given by

$$f'=\left(\frac{V}{V-v}\right)f.$$

Suppose a train whistle has a frequency of 480 vibrations/s and approaches you at 44 ft/s. Also, suppose sound travels at 1100 ft/s. (a) Find the apparent frequency you would hear. (b) At what speed would the train have to approach you so that you would hear an apparent frequency of 507 vibrations/s?

39. The combined resistance R of two resistors R_1 and R_2 connected in parallel is given by

$$R=\frac{1}{\dfrac{1}{R_1}+\dfrac{1}{R_2}}.$$

If one resistor has a resistance of 60 ohms (Ω) and the combined resistance is 10 Ω, find the resistance of the other resistor. (Here, Ω is the Greek letter omega.)

40. Solve the equation

$$\frac{P_1V_1}{T_1}=\frac{P_2V_2}{T_2}$$

for T_1. This equation relates to gases.

41. Entering into a storage tank are three pipes: A, B, and C. Pipe A can fill the tank in 2 h, pipe B in 3 h, and pipe C in 4 h. How long will it take to fill the tank if all three pipes are used?

42. Water is flowing into a tank by means of pipes A and B. Pipe A can fill the tank in 2 h, and pipe B can fill it in 5 h. However, water is also flowing out of the tank into another tank by a pipe C by which the original tank can be completely emptied in 4 h. How long would it take to fill the original tank if it were initially empty and pipes A, B, and C were all opened?

43. A boat traveled 36 mi upstream on a river where the rate of the current was 3 mi/h. It then returned. The round trip took 5 h. Find the speed of the boat in still water.

44. A person rowed downstream for 10 mi and then rowed upstream for the same period of time. However, going back, only 5 mi was covered. If the rate of the stream was $1\frac{1}{4}$ mi/h, find how fast the person can row in still water.

45. A land investment company purchased a parcel of land for $7200. After having sold all but 20 acres at a profit of $30 per acre, the entire cost of the parcel had been regained. How many acres were sold?

46. The *margin of profit* of a company is the net income divided by total revenue. A company's margin of profit increased by .02 from last year. In that year the company sold its product at $3.00 each and had a net income of $4500. This year the company increased the price of its product by $0.50 each, sold 2000 more, and had a net income of $7140. The company has never had a margin of profit greater than .15. How many of its product were sold last year and how many were sold this year?

47. There are several rules for determining doses of medicine for children when the adult dose has been specified. Such rules may be based on weight, height, and so on. If A = age of child, d = adult dose, and c = child's dose, then here are two rules.

$$\text{Young's rule:} \quad c = \frac{A}{A+12}d.$$

$$\text{Cowling's rule:} \quad c = \frac{A+1}{24}d.$$

At what age are the children's doses the same under both rules? Round your answer to the nearest year.

48. A company manufactures products A and B. The cost of producing each unit of A is $2 more than that of B. The total costs of production for all units of A and B are $1500 and $1000, respectively, and 25 more units of A are produced than of B. How many of each are produced?

7.5 REVIEW

IMPORTANT TERMS

fundamental principle of fractions *(7.2)* L.C.D. *(7.2)*
fractional equation *(7.4)* extraneous solution *(7.4)*

REVIEW PROBLEMS

In Problems **1–30**, *perform the indicated operations and simplify your answers.*

1. $\frac{2}{x}+\frac{4}{x-6}$.

2. $\frac{x-7}{x+4}-\frac{6}{2x+8}$.

3. $\frac{x^2-64}{x^3}\cdot\frac{x^2}{2x+16}$.

4. $\frac{x-2}{4}\cdot\frac{8x+4}{x^2+2x-8}$.

5. $\frac{x+2}{\frac{2x+4}{3}}$.

6. $\frac{\frac{x^3}{x-1}}{x^5}$.

7. $\frac{3}{x-2}-\frac{x+2}{x-3}$.

8. $\frac{2}{3x}+\frac{3}{x}-\frac{4}{5x}$.

9. $\frac{9x}{x^2+2x+1}\cdot\frac{(x+1)^3}{-3}$.

10. $\frac{-(x+3)}{x^2+x}\cdot\frac{x}{-(x^2-9)}$.

11. $2+\frac{x}{x-1}-\frac{x-1}{x^2-1}$.

12. $\frac{2}{x-y}+\frac{2}{y-x}$.

13. $\frac{x^2+5x+6}{x^2-2x-8}\cdot\frac{x^2-16}{x^2+7x+12}$.

14. $\frac{4x^2+4x+1}{x^2-2x-3}\cdot\frac{x^2+2x+1}{2x^2+3x+1}$.

15. $\frac{\frac{8-4x}{2x}}{\frac{x^2-4x+4}{x^2-2x}}$.

16. $\frac{\frac{4x^2-9}{(x+1)^3}}{\frac{4x+6}{(x+1)^2}}$.

17. $\frac{x+2}{x^2+4x+4}+\frac{x-3}{x+2}$.

18. $\frac{6}{y+3}-\frac{2y}{y-3}+\frac{3}{y^2-9}$.

19. $\frac{2x+2}{x^3-x}\div\frac{x-1}{x^2}$.

20. $\frac{3x^2-12x-15}{x^2+5x+4}\div\frac{30-6x}{x+4}$.

21. $\frac{x+1}{x^2+x-12}\cdot\frac{9-x^2}{x^2+3x+2}$.

22. $\frac{4-x^2}{25-x^2}\cdot\frac{x-5}{x-2}$.

23. $\frac{x+2}{\frac{x}{x+1}+\frac{4}{x}}$.

24. $\frac{\frac{1}{x+2}-\frac{1}{x-2}}{\frac{2}{x-2}}$.

25. $\frac{1-\frac{7}{x^2-9}}{\frac{x-4}{3-x}}$.

26. $\frac{\frac{2}{x+3}+1}{1-\frac{2}{x+4}}$.

27. $(x^{-1}+x)^{-1}$.

28. $\frac{x}{y^{-1}}+\left(\frac{y}{x^{-1}}\right)^{-2}$.

29. $\frac{x^{-2}-4}{x^{-1}+2}$.

30. $\left(\frac{x+y^{-1}}{x^{-1}+y}\right)^{-3}$.

In Problems **31–38**, *solve the equations.*

31. $\frac{2}{x+5} = \frac{4}{x-5}$.

32. $\frac{2x}{x-3} - \frac{x+1}{x+2} = 1$.

33. $\frac{x}{x-1} - \frac{9}{x+3} = 0$.

34. $\frac{1}{x^2} - \frac{9}{x} + 8 = 0$.

35. $\frac{x+1}{x} + \frac{2x}{x-2} = \frac{5x+1}{x}$.

36. $\frac{6x+7}{2x+1} - \frac{6x+1}{2x} = 1$.

37. $\frac{x+2}{x-5} = \frac{7}{x-5}$.

38. $\frac{x}{x-1} - \frac{2}{x} + \frac{x-2}{x^2-x} = 0$.

39. In studies of photosynthesis, the formula

$$P = \frac{bE}{1 + aE}$$

occurs. Solve for a.

40. The total capacitance C_T of an electric circuit containing two capacitors C_1 and C_2 in series is given by

$$\frac{1}{C_T} = \frac{1}{C_1} + \frac{1}{C_2}.$$

Find C_2 if $C_1 = 2$ (microfarads) and $C_T = \frac{2}{3}$ (microfarads).

41. A chemical company can fill a tank car with an industrial solvent with their regular pump in 20 min. Another pump, one that the company keeps in reserve, can fill the tank car in 30 min. How many minutes would it take to fill the tank car if both pumps were used together?

42. A trucker on a 130-mi run decreased his speed by 10 mi/h on the last 80-mi stretch because of a rainstorm. What was the trucker's original speed if the entire run took 3 h?

8 Exponents, Radicals, and Complex Numbers

8.1 FRACTIONAL EXPONENTS

In Chapter 2 the basic rules for exponents were given. However, up to now we have only dealt with exponents that are integers. In this section we will give a meaning to a fractional exponent, such as in $a^{1/n}$ and $a^{m/n}$. We want to define $a^{1/n}$, where n is a positive integer, so that the rules of exponents hold. For instance, by the rule $(a^m)^n = a^{mn}$ we must have

$$(a^{1/n})^n = a^{n/n} = a^1 = a.$$

But recall that

$$(\sqrt[n]{a})^n = a.$$

Thus we want the nth power of $a^{1/n}$ to be the same as the nth power of $\sqrt[n]{a}$. For this reason we define $a^{1/n}$ to be the principal nth root of a:*

Rule 1. $a^{1/n} = \sqrt[n]{a}.$

For example,

$$13^{1/5} = \sqrt[5]{13} \quad \text{and} \quad \sqrt{2 + 7x} = (2 + 7x)^{1/2}.$$

---

* We want $a^{1/n}$ to be a real number. For this, we must restrict the values of a at times. For example, $(-4)^{1/2} = \sqrt{-4}$ is not a real number. In Rule 2 we must also restrict a.

With our definition of $a^{1/n}$, the rules of exponents are true whether the exponents are integers or fractions.

EXAMPLE 1

Write each of the following in radical form and find the value.

a. $4^{1/2} = \sqrt[2]{4} = \sqrt{4} = 2.$

b. $27^{1/3} = \sqrt[3]{27} = 3.$

c. $(-32)^{1/5} = \sqrt[5]{-32} = -2.$

d. $16^{-(1/2)} = \frac{1}{16^{1/2}} = \frac{1}{\sqrt{16}} = \frac{1}{4}.$

EXAMPLE 2

Write each of the following with fractional exponents.

a. $\sqrt{7} = 7^{1/2}.$

b. $\sqrt[4]{3x^2 + 4y} = (3x^2 + 4y)^{1/4}.$

c. $x\sqrt{y} + \sqrt{xy} = xy^{1/2} + (xy)^{1/2}.$

Rule 1 can be made more general. We define $a^{m/n}$, where m and n are integers and n is positive, as follows.

Rule 2. $a^{m/n} = (\sqrt[n]{a})^m = \sqrt[n]{a^m}.$

That is, $a^{m/n} = (a^{1/n})^m = (a^m)^{1/n}$. By Rule 2 we may look at $x^{4/3}$ in two ways: first, as the fourth power of the cube root of x, $(\sqrt[3]{x})^4$; and second, as the cube root of the fourth power of x, $\sqrt[3]{x^4}$. To compute $a^{m/n}$ it is often easier first to take the nth root of a and then raise the result to the mth power.

EXAMPLE 3

Find the value of each of the following.

a. $8^{2/3} = (\sqrt[3]{8})^2 = 2^2 = 4$. We could also write $8^{2/3} = \sqrt[3]{8^2} = \sqrt[3]{64} = 4.$

b. $(-27)^{4/3} = (\sqrt[3]{-27})^4 = (-3)^4 = 81.$

c. $27^{-2/3} = \frac{1}{27^{2/3}} = \frac{1}{(\sqrt[3]{27})^2} = \frac{1}{(3)^2} = \frac{1}{9}.$

d. $\left(\frac{1}{4}\right)^{3/2} = \left(\sqrt{\frac{1}{4}}\right)^3 = \left(\frac{1}{2}\right)^3 = \frac{1}{8}.$

In Rule 2 we must at times restrict a because by that rule, $\sqrt{x^2} = x^{2/2} = x$. But this is not always true. If $x = -6$, then $\sqrt{x^2} = \sqrt{36} = 6 = -(-6) = -x$. But if $x = 6$, then $\sqrt{x^2} = \sqrt{36} = 6 = x$. Thus $\sqrt{x^2}$ is x if x is positive; it is $-x$ if x is negative. More simply, this means that $\sqrt{x^2} = |x|$. **Unless we say otherwise, we shall assume that all literal numbers appearing in radicals or bases are positive. As a result, we are free to write $\sqrt{x^2} = x$.**

EXAMPLE 4

Write each of the following with fractional exponents and simplify if possible.

a. $\sqrt[5]{x^4} = x^{4/5}$.

b. $x\sqrt[6]{y^5} = xy^{5/6}$.

c. $(\sqrt{3})^4 = (3^{1/2})^4 = 3^{4/2} = 3^2 = 9$.

d. $\sqrt[4]{x^8} = x^{8/4} = x^2$.

e. $\sqrt[4]{a^4 b^8} = (a^4 b^8)^{1/4} = (a^4)^{1/4}(b^8)^{1/4} = a^{4/4} b^{8/4} = ab^2$.

Table 8-1 gives a summary of the basic rules for exponents and radicals. Keep in mind that *the rules for exponents are true for all types of exponents*. These rules will be applied to fractional exponents in Examples 5 and 6.

TABLE 8-1

$a^0 = 1$.	$a^{-n} = \frac{1}{a^n}$.
$a^m a^n = a^{m+n}$.	$(a^m)^n = a^{mn}$.
$\frac{a^m}{a^n} = a^{m-n}$.	$(ab)^n = a^n b^n$.
$\left(\frac{a}{b}\right)^n = \frac{a^n}{b^n}$.	$(\sqrt[n]{a})^n = \sqrt[n]{a^n} = a$.
$\sqrt[n]{ab} = \sqrt[n]{a}\sqrt[n]{b}$.	$\sqrt[n]{\frac{a}{b}} = \frac{\sqrt[n]{a}}{\sqrt[n]{b}}$.
$a^{1/n} = \sqrt[n]{a}$.	$a^{m/n} = (\sqrt[n]{a})^m = \sqrt[n]{a^m}$.

EXAMPLE 5

Perform the indicated operations and give the answers with positive exponents only.

a. $x^{1/2}x^{1/3} = x^{(1/2)+(1/3)} = x^{5/6}$.

b. $\frac{2x}{x^{1/4}} = \frac{2x^1}{x^{1/4}} = 2x^{1-(1/4)} = 2x^{3/4}$.

c. $(8a^3)^{2/3} = 8^{2/3}(a^3)^{2/3} = (\sqrt[3]{8})^2 a^2 = 2^2 a^2 = 4a^2.$

d. $(15^{-1/2})^4 = 15^{-4/2} = 15^{-2} = \dfrac{1}{15^2} = \dfrac{1}{225}.$

e. $(x^{2/9}y^{-4/3})^{1/2} = (x^{2/9})^{1/2}(y^{-4/3})^{1/2} = x^{1/9}y^{-2/3} = \dfrac{x^{1/9}}{y^{2/3}}.$

f. $\left(\dfrac{x^{12}}{y^6}\right)^{-1/3} = \dfrac{(x^{12})^{-1/3}}{(y^6)^{-1/3}} = \dfrac{x^{-4}}{y^{-2}} = \dfrac{y^2}{x^4}.$

g. $\left(\dfrac{x^{1/5}y^{6/5}}{z^{2/5}}\right)^{15} = \dfrac{(x^{1/5}y^{6/5})^{15}}{(z^{2/5})^{15}} = \dfrac{x^3y^{18}}{z^6}.$

EXAMPLE 6

Perform the indicated operations.

a. $x^{1/2}(2 - x^{3/2}) = 2x^{1/2} - x^{1/2}x^{3/2} = 2x^{1/2} - x^2.$

b. $2x^{1/2} + (x^{1/6})^3 = 2x^{1/2} + x^{3/6} = 2x^{1/2} + x^{1/2} = 3x^{1/2}.$

c. $$\begin{aligned}(27x)^{1/3} - (3x^{1/3} - 5x^{1/3}) &= \sqrt[3]{27}x^{1/3} - (-2x^{1/3})\\ &= 3x^{1/3} + 2x^{1/3}\\ &= 5x^{1/3}.\end{aligned}$$

d. $\dfrac{3x^4 - 2x^2}{x^{1/3}} = \dfrac{3x^4}{x^{1/3}} - \dfrac{2x^2}{x^{1/3}} = 3x^{4-(1/3)} - 2x^{2-(1/3)} = 3x^{11/3} - 2x^{5/3}.$

Exercise 8.1

In Problems **1–15**, *find the value of each expression.*

1. $25^{1/2}$. **2.** $125^{1/3}$. **3.** $81^{-1/2}$.

4. $4^{-1/2}$. **5.** $27^{2/3}$. **6.** $(\frac{1}{9})^{3/2}$.

7. $27^{-2/3}$. **8.** $9^{3/2}$. **9.** $(-8)^{4/3}$.

10. $(-8)^{-1/3}$. **11.** $16^{3/4}$. **12.** $(64^{1/2})^{1/3}$.

13. $(-\frac{1}{32})^{-1/5}$. **14.** $(\frac{1}{27})^{-2/3}$. **15.** $(4^3)^{2/3}$.

In Problems **16–30**, *rewrite each expression by using positive fractional exponents and simplify where possible.*

16. $\sqrt[3]{y}$. **17.** $\sqrt{x}$. **18.** $\sqrt[4]{x^9}$.

19. $\sqrt[3]{x^2}$.

20. $\dfrac{\sqrt{y}}{\sqrt[4]{x}}$.

21. $\sqrt[4]{x^3y^5}$.

22. $\sqrt[4]{x}\sqrt[5]{y}$.

23. $\sqrt[6]{x^5y^{12}}$.

24. $\dfrac{1}{\sqrt{2}}$.

25. $x^2\sqrt[4]{x}$.

26. $\dfrac{2}{\sqrt[3]{x-5}}$.

27. $\dfrac{3}{\sqrt{x}}$.

28. $\sqrt[3]{7-x}$.

29. $\sqrt[3]{(x^2-5x)^2}$.

30. $\dfrac{1}{\sqrt[6]{(x^2-2x)^5}}$.

In Problems **31–76**, *perform the operations and simplify. Write your answers with positive exponents only.*

31. $x^{1/2}x^{3/2}$.

32. $x^{2/3}x^{4/3}$.

33. $x^{4/3}x^{-1/3}$.

34. $x^{-1/2}x^{3/2}$.

35. $x^{-2}x^{7/2}x^{5/2}$.

36. $x^3x^{-3/8}x^{-5/8}$.

37. $x^{1/2}x^{1/4}$.

38. $x^{3/5}x^{1/3}$.

39. $x^{1/2}(3x^{1/2}y)$.

40. $(xy^{-1/2})y^{5/2}$.

41. $(x^{1/2})^3$.

42. $(y^{-1/3})^6$.

43. $(y^6)^{1/3}$.

44. $(t^{10})^{4/5}$.

45. $2(x^{-2/3})^6$.

46. $(-y^{1/3})^6$.

47. $(2x^{-2}y^{1/3})^3$.

48. $(3x^3y^{1/2})^4$.

49. $(ab^2c^3)^{3/4}(a^{1/4}b^{3/4})^5$.

50. $(x^{1/2}y^{2/3})(x^2y)^{1/3}$.

51. $x^{1/3}(x^{2/3}+3)$.

52. $x^{3/2}(1-x^{1/2})$.

53. $(27x^{15})^{-1/3}$.

54. $(8x^{-6})^{1/3}$.

55. $(-8x^{-6})^{1/3}$.

56. $(27^{-1}x^{15})^{-1/3}$.

57. $\dfrac{x^{4/9}y^{2/5}}{x^{1/9}y^{7/5}}$.

58. $\dfrac{x^{4/7}y^{3/20}}{x^{3/7}y^{17/20}}$.

59. $\dfrac{x^{2/3}y^{-9/4}}{x^{-4/3}y^{-1/4}}$.

60. $\dfrac{xy^{2/5}}{x^{1/2}y^{2/5}z^{-1/3}}$.

61. $\left(\dfrac{x^{-4/3}}{y^{-2/3}}\right)^{-3}$.

62. $\left(\dfrac{3^{1/4}x^{3/4}}{y^{1/2}}\right)^4$.

63. $\left(\dfrac{x^{-1}}{x^{1/3}}\right)^2$.

64. $\left(\dfrac{y^{2/3}}{x^{3/4}}\right)^{2/3}$.

65. $\left(\dfrac{2^{1/3}x^{2/3}}{x^{1/3}}\right)^6$.

66. $\left(\dfrac{x^{3/2}}{x^{-1}}\right)^{-4}$.

67. $2x^{1/2}-5x^{1/2}$.

68. $7x^{3/2}+2x^{3/2}-5x^{3/2}$.

69. $4x^{-1/3} - 2(x^2)^{-1/6}$.

70. $3x^{1/4} - (x^{-1/2})^{-1/2}$.

71. $\dfrac{x^{4/3} + x^2 - x^4}{x}$.

72. $\dfrac{x^{-1/2}y^{-1/3}}{2}$.

73. $\dfrac{x^{1/2} - 3x^{1/3}}{x^{1/4}}$.

74. $\dfrac{2x^{1/3} - y^3 - y^{1/4}}{y^{1/4}}$.

75. $\dfrac{(2x^{1/2}y)^3(4x)^{-1/2}}{x}$.

76. $\dfrac{(x^{1/3}y^{1/6})^3(2x^{-2})}{xy^{1/2}}$.

77. If you were to divide the value of imports by the gross national product of many individual countries, the result R would be related to the country's population P (in millions) by the formula $R = .40P^{-1/3}$.* Determine the value of R for a country with a population of 8,000,000. (*Note*: $P \neq 8{,}000{,}000$.)

78. For various athletic events, such as running, swimming, and cycling, the formula $t = ax^b$ describes the time t (in minutes) for a participant to travel a total distance x (in kilometers).† Here a and b are fixed numbers and depend on the specific event. The velocity v of a participant is given by $v = x/t$. Show by substitution that

$$v = \frac{x^{1-b}}{a}.$$

8.2 CHANGING THE FORM OF A RADICAL

At times you may replace a radical by an expression that does not have a radical. For example, just as the rule $\sqrt[n]{a^n} = a$ allows us to write $\sqrt[3]{x^3} = x$, we have

$$\sqrt[3]{x^6y^9} = \sqrt[3]{(x^2y^3)^3} = x^2y^3.$$

This shows that you may "remove" a radical of index n when the radicand is an nth power of an expression.

EXAMPLE 1

Removing radicals: $\sqrt[n]{a^n} = a$.

a. $\sqrt{x^6y^8} = \sqrt{(x^3y^4)^2} = x^3y^4$.

b. $\sqrt[4]{\dfrac{x^{16}}{y^8}} = \sqrt[4]{\left(\dfrac{x^4}{y^2}\right)^4} = \dfrac{x^4}{y^2}$.

---

* R. Taagepera, "Why the Trade/GNP Ratio Decreases with Country Size," *Social Science Research*, (1976), 385–404.

† P. S. Riegel, "Athletic Records of Human Endurance," *American Scientist*, 69 (1981), 285–290.

c. $\sqrt[5]{32x^5y^{15}} = \sqrt[5]{(2xy^3)^5} = 2xy^3$.

One way to change the form of a radical with index n is to "remove" from the radicand all factors whose nth roots can easily be found. We use the rule $\sqrt[n]{ab} = \sqrt[n]{a}\sqrt[n]{b}$. For example,

$$\sqrt{50} = \sqrt{25 \cdot 2} = \sqrt{25}\sqrt{2} = 5\sqrt{2}.$$

EXAMPLE 2

Remove as many factors as possible from the radicand.

a. $\sqrt[3]{x^3y} = \sqrt[3]{x^3}\sqrt[3]{y} = x\sqrt[3]{y}$.

b. $\sqrt{25x^7} = \sqrt{25x^6x} = \sqrt{(5x^3)^2x} = \sqrt{(5x^3)^2}\sqrt{x} = 5x^3\sqrt{x}$.

c. $\sqrt[5]{64x^6y^{14}z^2} = \sqrt[5]{32x^5y^{10}(2xy^4z^2)} = 2xy^2\sqrt[5]{2xy^4z^2}$, since $\sqrt[5]{32x^5y^{10}}$ is $2xy^2$.

When a radicand is a fraction, you may use the rule $\sqrt[n]{\frac{a}{b}} = \frac{\sqrt[n]{a}}{\sqrt[n]{b}}$ to get an equivalent expression in which the radicand is not a fraction. For example,

$$\sqrt{\frac{7}{16}} = \frac{\sqrt{7}}{\sqrt{16}} = \frac{\sqrt{7}}{4}.$$

Note that in the final result, no radical appears in the denominator, because 16 is a perfect square ($16 = 4^2$) and $\sqrt{16} = 4$. It is customary to rewrite radicals so that no fraction appears in a radicand and no radical appears in a denominator. This is called **rationalizing the denominator**. Sometimes you have to rewrite a radicand before using the above rule. This is the case in

$$\sqrt{\frac{3}{7}}.$$

Since the index is 2, we want the denominator to be a perfect square. We can make the denominator be 7^2 if we multiply numerator and denominator by 7. Thus

$$\sqrt{\frac{3}{7}} = \sqrt{\frac{3 \cdot 7}{7 \cdot 7}} = \sqrt{\frac{21}{7^2}} = \frac{\sqrt{21}}{\sqrt{7^2}} = \frac{\sqrt{21}}{7}.$$

The denominator is now rationalized. Actually, multiplying numerator and denominator by 7 is equivalent to multiplying $\frac{3}{7}$ by $\frac{7}{7}$. That is,

$$\sqrt{\frac{3}{7}} = \sqrt{\frac{3}{7} \cdot \frac{7}{7}} = \sqrt{\frac{21}{7^2}} = \frac{\sqrt{21}}{\sqrt{7^2}} = \frac{\sqrt{21}}{7}.$$

EXAMPLE 3

Rationalizing the denominator.

a. $\sqrt[4]{\dfrac{y}{x^8}} = \dfrac{\sqrt[4]{y}}{\sqrt[4]{x^8}} = \dfrac{\sqrt[4]{y}}{x^2}.$

b. $\sqrt{\dfrac{21}{x}} = \sqrt{\dfrac{21}{x} \cdot \dfrac{x}{x}} = \dfrac{\sqrt{21x}}{\sqrt{x^2}} = \dfrac{\sqrt{21x}}{x}.$

c. $\sqrt[5]{\dfrac{x}{y^2}}.$

Since the index is 5, we want the denominator to be a fifth power. We'll multiply numerator and denominator by y^3.

$$\sqrt[5]{\frac{x}{y^2}} = \sqrt[5]{\frac{x}{y^2} \cdot \frac{y^3}{y^3}} = \sqrt[5]{\frac{xy^3}{y^5}} = \frac{\sqrt[5]{xy^3}}{\sqrt[5]{y^5}} = \frac{\sqrt[5]{xy^3}}{y}.$$

d. $\sqrt[3]{\dfrac{2}{3x^4y^2}}.$

Since the index is 3, we want factors in the denominator to be cubes. This will be the case if the exponent for each factor is 3 or a multiple of 3. Thus we multiply 3 by 3^2, x^4 by x^2, and y^2 by y; that is, we multiply both numerator and denominator by 3^2x^2y.

$$\sqrt[3]{\frac{2}{3x^4y^2}} = \sqrt[3]{\frac{2}{3x^4y^2} \cdot \frac{3^2x^2y}{3^2x^2y}} = \sqrt[3]{\frac{2 \cdot 3^2x^2y}{3^3x^6y^3}} = \frac{\sqrt[3]{18x^2y}}{\sqrt[3]{(3x^2y)^3}} = \frac{\sqrt[3]{18x^2y}}{3x^2y}.$$

Another rule for radicals involves a root of a root of a number.

Rule 3. $\sqrt[m]{\sqrt[n]{a}} = \sqrt[mn]{a}.$

Note that the final index is the product of the individual indices. We can now find $\sqrt[3]{\sqrt{64}}$ in two ways. Using Rule 3 gives

$$\sqrt[3]{\sqrt{64}} = \sqrt[6]{64} = 2.$$

Without using Rule 3, we have

$$\sqrt[3]{\sqrt{64}} = \sqrt[3]{8} = 2.$$

EXAMPLE 4

Use of the rule $\sqrt[m]{\sqrt[n]{a}} = \sqrt[mn]{a}$.

a. $\sqrt[3]{\sqrt[4]{2}} = \sqrt[12]{2}$.

b. $\sqrt{\sqrt[3]{x}} = \sqrt[6]{x}$.

c. $\sqrt[4]{81} = \sqrt{\sqrt{81}} = \sqrt{9} = 3$.

Sometimes it is possible to **reduce the index** of a radical, as the following shows.

$$\sqrt[6]{x^3} = x^{3/6} = x^{1/2} = \sqrt{x}.$$

Here the index 6 was reduced to 2. You may also use the rule $\sqrt[mn]{a} = \sqrt[m]{\sqrt[n]{a}}$ to do such a problem.

$$\sqrt[6]{x^3} = \sqrt{\sqrt[3]{x^3}} = \sqrt{x}.$$

EXAMPLE 5

Reducing the index.

a. $\sqrt[4]{25} = \sqrt[4]{5^2} = 5^{2/4} = 5^{1/2} = \sqrt{5}$, or

$$\sqrt[4]{25} = \sqrt{\sqrt{25}} = \sqrt{5}.$$

b. $\sqrt[6]{16x^2}$.

We can't conveniently take the sixth root of the radicand, but we can take the square root. Since $6 = 3 \cdot 2$, we write

$$\sqrt[6]{16x^2} = \sqrt[3]{\sqrt{16x^2}} = \sqrt[3]{4x}.$$

Another approach is

$$\sqrt[6]{16x^2} = \sqrt[6]{(4x)^2} = (4x)^{2/6} = (4x)^{1/3} = \sqrt[3]{4x}.$$

c. $\sqrt[12]{8x^6y^9}$.

Here the radicand is the cube $(2x^2y^3)^3$ and 3 is a factor of the index 12. Thus we have

$$\sqrt[12]{8x^6y^9} = \sqrt[4]{\sqrt[3]{8x^6y^9}} = \sqrt[4]{2x^2y^3}.$$

We say that a radical is **simplified** when the following are true.

1. As many factors as possible are removed from the radicand (Example 2).
2. The denominator is rationalized (Example 3).
3. The index cannot be reduced (Example 5).

EXAMPLE 6

Simplify the following radicals.

a. $\sqrt{\frac{x^5}{z}} = \sqrt{\frac{x^5}{z} \cdot \frac{z}{z}} = \frac{\sqrt{x^5 z}}{\sqrt{z^2}} = \frac{\sqrt{x^5 z}}{z} = \frac{\sqrt{x^4(xz)}}{z} = \frac{x^2\sqrt{xz}}{z}$.

b. $\sqrt[4]{x^6 y^{10}} = \sqrt[4]{x^4 y^8(x^2 y^2)} = xy^2\sqrt[4]{x^2 y^2} = xy^2\sqrt[4]{(xy)^2}$
$= xy^2(xy)^{2/4} = xy^2(xy)^{1/2} = xy^2\sqrt{xy}.$

c. $\sqrt[6]{\frac{x^3}{y^9}} = \sqrt[6]{\frac{x^3}{y^9} \cdot \frac{y^3}{y^3}} = \sqrt[6]{\frac{x^3 y^3}{y^{12}}} = \frac{\sqrt[6]{x^3 y^3}}{\sqrt[6]{(y^2)^6}} = \frac{\sqrt[6]{(xy)^3}}{y^2} = \frac{\sqrt{xy}}{y^2}$.

d. $\sqrt[3]{x^{-6} y^6} = \sqrt[3]{\frac{y^6}{x^6}} = \sqrt[3]{\left(\frac{y^2}{x^2}\right)^3} = \frac{y^2}{x^2}$.

Exercise 8.2

In Problems **1–14**, *find the root.*

1. $\sqrt[4]{x^8}$. **2.** $\sqrt[3]{x^9}$. **3.** $\sqrt[3]{8x^{12}}$.

4. $\sqrt[6]{9^{12}}$. **5.** $\sqrt{9x^{16} y^{18}}$. **6.** $\sqrt[3]{x^3 y^3 z^6}$.

7. $\sqrt[3]{x^3 y^6 z^9}$. **8.** $\sqrt{(x^3 y^7)(x^7 y^{13})}$. **9.** $\sqrt[5]{\frac{x^{15}}{y^{20}}}$.

10. $\sqrt[4]{\frac{16x^8}{y^{16}}}$. **11.** $\sqrt{\sqrt{x^8}}$. **12.** $\sqrt[3]{\sqrt{x^{18}}}$.

13. $\sqrt[4]{\sqrt[3]{x^{12}}}$. **14.** $\sqrt{\sqrt[6]{x^{24}}}$.

In Problems **15–72**, *simplify the radical.*

15. $\sqrt{12}$. **16.** $\sqrt{18}$. **17.** $\sqrt{32}$.

18. $\sqrt{20}$. **19.** $\sqrt[3]{16}$. **20.** $\sqrt[3]{54}$.

21. $\sqrt{x^7}$.
22. $\sqrt[3]{x^7}$.
23. $\sqrt[3]{24x^6}$.
24. $\sqrt{8x^3}$.
25. $\sqrt[4]{x^9y^2}$.
26. $\sqrt[5]{x^{17}y^{20}}$.
27. $\sqrt[3]{x^6yz^4}$.
28. $\sqrt{x^5y^4z}$.
29. $\sqrt[3]{8a^3y^5}$.
30. $\sqrt[3]{24(a+b)^7}$.
31. $\sqrt[5]{x^{23}y^{10}z^6}$.
32. $\sqrt[4]{x^3y^3z^7}$.
33. $\sqrt{81xy^2z^3w^4}$.
34. $\sqrt[4]{32x^{17}}$.
35. $\sqrt{\frac{1}{2}}$.
36. $\sqrt{\frac{1}{5}}$.
37. $\sqrt[3]{\frac{2}{5}}$.
38. $\sqrt[3]{\frac{1}{3}}$.
39. $\sqrt[3]{\frac{x^2}{y^3}}$.
40. $\sqrt[4]{\frac{3}{x^4}}$.
41. $\sqrt{\frac{x}{y^4}}$.
42. $\sqrt[3]{\frac{y^2}{x^{12}}}$.
43. $\sqrt{\frac{2x}{y}}$.
44. $\sqrt{\frac{y}{x^9}}$.
45. $\sqrt[3]{\frac{2}{xy^2}}$.
46. $\sqrt[3]{\frac{2y}{xz}}$.
47. $\sqrt[4]{\frac{3}{2x^7yz^2}}$.
48. $\sqrt[4]{\frac{1}{xy^3z^5}}$.
49. $\sqrt[6]{x^2}$.
50. $\sqrt[8]{(xy)^4}$.
51. $\sqrt[4]{9}$.
52. $\sqrt[6]{8}$.
53. $\sqrt[4]{16x^4y^2}$.
54. $\sqrt[6]{27x^3y^3z^3}$.
55. $\sqrt[8]{\frac{x^4}{y^4}}$.
56. $\sqrt[6]{\frac{16x^2}{y^2}}$.
57. $\sqrt[12]{x^2y^2z^{10}}$.
58. $\sqrt[15]{x^3y^{15}z^6}$.
59. $\sqrt[4]{x^{10}y^2}$.
60. $\sqrt[12]{x^{15}y^{27}}$.
61. $\sqrt[6]{x^{20}x^{26}}$.
62. $\sqrt[8]{x^2z^{10}}$.
63. $\sqrt[3]{\sqrt{x^{12}y^5w^{25}}}$.
64. $\sqrt{\sqrt[3]{64x^{12}y^{11}w^7}}$.
65. $\sqrt[4]{\frac{16x^5}{y^8}}$.
66. $\sqrt[3]{\frac{x^6}{y^{-3}}}$.
67. $\sqrt[4]{\frac{1}{4}}$.
68. $\sqrt[6]{\frac{x^9}{y^{15}}}$.
69. $\sqrt[8]{\frac{x^4}{y^{12}}}$.
70. $\sqrt{x^4(2+x)}$.
71. $\sqrt[3]{\sqrt{x^3}}$.
72. $\sqrt[4]{\sqrt[3]{\sqrt{x^{48}}}}$.
73. Suppose consumers will buy (or demand) q units of product A when the price is p (in dollars) per unit, where p is given by

$$p = \frac{50}{\sqrt{q}}.$$

An equation such as this, which relates price per unit and quantity demanded for a product, is called a *demand equation*. For product A, if q quadruples, by what percentage does p decrease?

74. A psychologist is studying human response to electrical shock. She devises an experiment in which a human subject receives shocks of different intensities. After each shock, the subject estimates its magnitude by assigning a number M, called the *magnitude estimation*, to the shock. After performing the experiment several times, the psychologist estimated that M is given by

$$M = 5.5 \times 10^{-5} I^{1.5},$$

where I is the intensity of the shock (in microamperes). If the intensity of the shock doubles, by what factor does the magnitude estimation increase? Give your answer in terms of a simplified radical.

8.3 ADDITION, SUBTRACTION, AND MULTIPLICATION WITH RADICALS

You may combine radicals in addition or subtraction when the radicals have the *same index* and the *same radicand*. Such radicals are called **similar radicals**. The distributive law is used just as it was in combining *similar terms* in Chapter 3. For example,

$$8\sqrt{xy} - 4\sqrt{xy} + 7\sqrt{xy} = (8 - 4 + 7)\sqrt{xy} = 11\sqrt{xy}.$$

Sometimes, radicals that are *not* similar can be combined by simplifying them first. For example, in $\sqrt[3]{81} - \sqrt[3]{24}$ the radicands are different. But we can write

$$\begin{aligned} \sqrt[3]{81} - \sqrt[3]{24} &= \sqrt[3]{27 \cdot 3} - \sqrt[3]{8 \cdot 3} \\ &= 3\sqrt[3]{3} - 2\sqrt[3]{3} = \sqrt[3]{3}. \end{aligned}$$

EXAMPLE 1

Adding and subtracting radicals.

a. $$\begin{aligned} (16\sqrt[3]{x} + 15) - (12 - 2\sqrt[3]{x}) &= 16\sqrt[3]{x} + 15 - 12 + 2\sqrt[3]{x} \\ &= 18\sqrt[3]{x} + 3. \end{aligned}$$

b. $$\begin{aligned} 3\sqrt{75} - 2\sqrt{12} + \sqrt{7} &= 3\sqrt{25 \cdot 3} - 2\sqrt{4 \cdot 3} + \sqrt{7} \\ &= 3(5\sqrt{3}) - 2(2\sqrt{3}) + \sqrt{7} \\ &= 15\sqrt{3} - 4\sqrt{3} + \sqrt{7} = 11\sqrt{3} + \sqrt{7}. \end{aligned}$$

c. $\sqrt[3]{\frac{3}{4}} + \sqrt[3]{\frac{2}{9}} - \sqrt[3]{\frac{1}{36}} = \sqrt[3]{\frac{3}{4}\cdot\frac{2}{2}} + \sqrt[3]{\frac{2}{9}\cdot\frac{3}{3}} - \sqrt[3]{\frac{1}{6^2}\cdot\frac{6}{6}}$

$$= \frac{\sqrt[3]{6}}{2} + \frac{\sqrt[3]{6}}{3} - \frac{\sqrt[3]{6}}{6}$$

$$= \left(\frac{1}{2} + \frac{1}{3} - \frac{1}{6}\right)\sqrt[3]{6} = \frac{2}{3}\sqrt[3]{6}.$$

Radicals with the *same* index can be multiplied by means of the rule $\sqrt[n]{a} \cdot \sqrt[n]{b} = \sqrt[n]{ab}$. For example,

$$\sqrt{2x^3} \cdot \sqrt{8x} = \sqrt{(2x^3)(8x)} = \sqrt{16x^4} = 4x^2.$$

EXAMPLE 2

Multiplying radicals with the same index.

a. $\sqrt{yz} \cdot \sqrt{3z} = \sqrt{3yz^2} = z\sqrt{3y}.$

b. $\sqrt[4]{\frac{3}{2}} \cdot \sqrt[4]{6} = \sqrt[4]{\frac{3}{2} \cdot 6} = \sqrt[4]{9} = \sqrt[4]{3^2} = 3^{2/4} = 3^{1/2} = \sqrt{3}.$

c. $\sqrt{3}(\sqrt{3} - \sqrt{6}) = \sqrt{3} \cdot \sqrt{3} - \sqrt{3} \cdot \sqrt{6}$ [distributive law]

$$= 3 - \sqrt{18} = 3 - \sqrt{9 \cdot 2} = 3 - 3\sqrt{2}.$$

d. $(x\sqrt[3]{x^2y^3})^4 = x^4(\sqrt[3]{x^2y^3})^4$ [using $(ab)^n = a^n b^n$]

$= x^4\sqrt[3]{(x^2y^3)^4}$ [using $(\sqrt[n]{a})^m = \sqrt[n]{a^m}$]

$= x^4\sqrt[3]{x^8y^{12}}$ [using $(ab)^n = a^n b^n$]

$= x^4(x^2y^4\sqrt[3]{x^2})$ [simplifying]

$= x^6y^4\sqrt[3]{x^2}.$

When operating on radicals, be sure to simplify your answers, as we did in Example 2.

EXAMPLE 3

Multiplying radicals with the same index.

a. $(\sqrt{8} - \sqrt{3})(\sqrt{18} - \sqrt{48}).$

We'll simplify the radicals before multiplying.

$$\begin{aligned}&(2\sqrt{2}-\sqrt{3})(3\sqrt{2}-4\sqrt{3})\\&=2\sqrt{2}\cdot 3\sqrt{2}-2\sqrt{2}\cdot 4\sqrt{3}-\sqrt{3}\cdot 3\sqrt{2}+\sqrt{3}\cdot 4\sqrt{3} \qquad \text{[multiplying binomials]}\\&=6(\sqrt{2})^2-8\sqrt{6}-3\sqrt{6}+4(\sqrt{3})^2\\&=12-11\sqrt{6}+12\\&=24-11\sqrt{6}.\end{aligned}$$

b. $(\sqrt{5}-2\sqrt{7})(\sqrt{5}+2\sqrt{7})$.

This has the form $(a-b)(a+b)$. Thus

$$\begin{aligned}(\sqrt{5}-2\sqrt{7})(\sqrt{5}+2\sqrt{7})&=(\sqrt{5})^2-(2\sqrt{7})^2\\&=5-4(7)=5-28=-23.\end{aligned}$$

c. $(\sqrt{x}-\sqrt{y})^2$.

This is the square of a binomial.

$$\begin{aligned}(\sqrt{x}-\sqrt{y})^2&=(\sqrt{x})^2-2\sqrt{x}\sqrt{y}+(\sqrt{y})^2\\&=x-2\sqrt{xy}+y.\end{aligned}$$

Exercise 8.3

Perform the operations and simplify.

1. $4\sqrt[3]{3}-2\sqrt[3]{3}+\sqrt[3]{3}$.

2. $6\sqrt{7}-(2\sqrt{7}+3\sqrt{7})$.

3. $x^2\sqrt{2x}-3x^2\sqrt{2x}+4x^2\sqrt{2x}$.

4. $xy\sqrt{x}+4xy\sqrt{x}-2xy\sqrt{x}$.

5. $3\sqrt{75}-2\sqrt{12}$.

6. $5\sqrt{18}-(\sqrt{2}+1)$.

7. $5\sqrt{8}-(2\sqrt{18}-4\sqrt{32})$.

8. $\sqrt{75}-(\sqrt{27}-2\sqrt{3})$.

9. $2y\sqrt{16x}-3y\sqrt{9x}$.

10. $3\sqrt[3]{16}-\sqrt[3]{54}$.

11. $\sqrt[3]{128}-6\sqrt[3]{16}$.

12. $2(\sqrt[3]{54}-2\sqrt[3]{128})-2\sqrt[3]{16}$.

13. $30\sqrt{\frac{1}{15}}-72\sqrt{\frac{5}{12}}+50\sqrt{\frac{3}{5}}$.

14. $20\sqrt{\frac{2}{5}}-3\sqrt{40}-4\sqrt{\frac{5}{2}}$.

15. $4\sqrt[3]{\frac{1}{2}}+\sqrt[3]{32}-3\sqrt[3]{4}$.

16. $\sqrt{\frac{1}{6}}+\sqrt{\frac{2}{3}}+\sqrt{\frac{3}{2}}$.

17. $\sqrt[4]{4x^4}-3\sqrt[6]{8x^6}$.

18. $\sqrt[4]{4x^2}-3\sqrt[6]{8x^3}$.

19. $\sqrt{2}\sqrt{8}$.

20. $\sqrt{12}\sqrt{3}$.

21. $\sqrt{3}\sqrt{4}$.

22. $\sqrt{9}\sqrt{2}$.

23. $(2\sqrt{6})(3\sqrt{3})$.

24. $(5\sqrt{27})(2\sqrt{3})$.

25. $\sqrt[3]{3}\sqrt[3]{9}\sqrt[3]{12}$.

26. $\sqrt[3]{4}\sqrt[3]{16}\sqrt[3]{-1}$.

27. $\sqrt{2x}\sqrt{x}\sqrt{3x}$.

28. $\sqrt{5y}\sqrt{2y}\sqrt{y}$.

29. $(-\sqrt{3})^2$.

30. $(-\sqrt{5})^3$.

31. $(2\sqrt[3]{x})^4$.

32. $(\frac{1}{3}\sqrt{x})^3$.

33. $\sqrt{30}\sqrt{\frac{2}{3}}$.

34. $\sqrt{\frac{5}{8}}\sqrt{\frac{16}{3}}$.

35. $\sqrt{3}(2\sqrt{6}-4\sqrt{3})$.

36. $\sqrt{2}(\sqrt{2}+2\sqrt{18})$.

37. $(\sqrt{6}+\sqrt{2})(\sqrt{2}-2\sqrt{6})$.

38. $(\sqrt{3}-1)(\sqrt{3}+2)$.

39. $(2+\sqrt{7})(2-\sqrt{7})$.

40. $(\sqrt{7}-\sqrt{2})(\sqrt{7}+\sqrt{2})$.

41. $(\sqrt{5}+2)^2$.

42. $(1-\sqrt{5})^2$.

43. $(\sqrt{x}-1)(2\sqrt{x}+5)$.

44. $\sqrt{ab}\sqrt{a^2bc^2}\sqrt{abc^2}$.

45. $\sqrt{3xy^2}\sqrt{2xy}\sqrt{3xy^3}$.

46. $5\sqrt[4]{ab}(1-2\sqrt[4]{ab})$.

47. $(\sqrt{6}-5)^2$.

48. $(\sqrt{3}+4)^2$.

49. $(3\sqrt{8}+\sqrt{3})(8\sqrt{3}-\sqrt{8})$.

50. $(5\sqrt{2}+\sqrt{5})(2\sqrt{5}-\sqrt{2})$.

8.4 DIVISION WITH RADICALS

In Section 8.2 we rationalized the denominator for a radical such as $\sqrt{\frac{3}{7}}$. We also use the phrase *rationalizing the denominator* in another situation: when a fraction with a radical in its denominator, such as $\frac{2}{\sqrt[3]{5}}$, is rewritten so that there is no radical in the denominator.

To rationalize the denominator of $\frac{2}{\sqrt[3]{5}}$, we note that if the radicand were a perfect cube, then the denominator would simplify very easily by the rule $\sqrt[n]{a^n}=a$. We can change the radicand to 5^3 by first multiplying both numerator and denominator by $\sqrt[3]{5^2}$.

$$\begin{aligned}\frac{2}{\sqrt[3]{5}} &= \frac{2\sqrt[3]{5^2}}{\sqrt[3]{5}\sqrt[3]{5^2}}\\ &= \frac{2\sqrt[3]{5^2}}{\sqrt[3]{5^3}} \qquad [\text{using } \sqrt[n]{a}\sqrt[n]{b}=\sqrt[n]{ab}]\\ &= \frac{2\sqrt[3]{25}}{5}.\end{aligned}$$

To rationalize a square root in a denominator, such as in $\frac{4}{\sqrt{3}}$, we have

$$\frac{4}{\sqrt{3}}=\frac{4}{\sqrt{3}}\cdot\frac{\sqrt{3}}{\sqrt{3}}=\frac{4\sqrt{3}}{3}.$$

The process of simplifying a radical will now include rationalizing a denominator when a denominator contains a radical.

EXAMPLE 1

Rationalizing the denominator

a. $\frac{6}{\sqrt[5]{2}}=\frac{6}{\sqrt[5]{2}}\cdot\frac{\sqrt[5]{2^4}}{\sqrt[5]{2^4}}=\frac{6\sqrt[5]{16}}{\sqrt[5]{2^5}}=\frac{6\sqrt[5]{16}}{2}=3\sqrt[5]{16}.$

b. $\frac{3}{\sqrt{2x}}=\frac{3}{\sqrt{2x}}\cdot\frac{\sqrt{2x}}{\sqrt{2x}}=\frac{3\sqrt{2x}}{2x}.$

c. $\frac{7}{2\sqrt[5]{3x^3}}=\frac{7}{2\sqrt[5]{3x^3}}\cdot\frac{\sqrt[5]{3^4x^2}}{\sqrt[5]{3^4x^2}}=\frac{7\sqrt[5]{81x^2}}{2\sqrt[5]{3^5x^5}}=\frac{7\sqrt[5]{81x^2}}{2(3x)}=\frac{7\sqrt[5]{81x^2}}{6x}.$

Division of radicals with the *same* index may be handled by using the rule $\frac{\sqrt[n]{a}}{\sqrt[n]{b}}=\sqrt[n]{\frac{a}{b}}$. For example

$$\frac{\sqrt{7}}{\sqrt{3}}=\sqrt{\frac{7}{3}}=\sqrt{\frac{7}{3}\cdot\frac{3}{3}}=\frac{\sqrt{21}}{3}.$$

EXAMPLE 2

Division of radicals with same index.

a. $\frac{\sqrt[3]{3x}}{\sqrt[3]{2x}}=\sqrt[3]{\frac{3x}{2x}}=\sqrt[3]{\frac{3}{2}}=\sqrt[3]{\frac{3}{2}\cdot\frac{2^2}{2^2}}=\frac{\sqrt[3]{12}}{2}.$

b. $\frac{\sqrt{5xy}}{\sqrt{18xy^3}}=\frac{\sqrt{5xy}}{3y\sqrt{2xy}}=\frac{1}{3y}\sqrt{\frac{5xy}{2xy}}=\frac{1}{3y}\sqrt{\frac{5}{2}}=\frac{1}{3y}\sqrt{\frac{5}{2}\cdot\frac{2}{2}}=\frac{\sqrt{10}}{6y}.$

Sometimes a denominator of a fraction has two terms, one or both involving a square root, such as $2-\sqrt{3}$ or $\sqrt{5}+\sqrt{2}$. The denominator may be rationalized by multiplying by an expression that makes the denominator a difference of two squares. We use the fact that $(a+b)(a-b)=a^2-b^2$. For example,

$$\frac{4}{\sqrt{5}+\sqrt{2}} = \frac{4}{\sqrt{5}+\sqrt{2}} \cdot \frac{\sqrt{5}-\sqrt{2}}{\sqrt{5}-\sqrt{2}}$$

$$= \frac{4(\sqrt{5}-\sqrt{2})}{(\sqrt{5})^2-(\sqrt{2})^2} = \frac{4(\sqrt{5}-\sqrt{2})}{5-2}$$

$$= \frac{4(\sqrt{5}-\sqrt{2})}{3}.$$

EXAMPLE 3

Rationalize each denominator.

a. $\frac{2}{x-\sqrt{3}} = \frac{2}{x-\sqrt{3}} \cdot \frac{x+\sqrt{3}}{x+\sqrt{3}} = \frac{2(x+\sqrt{3})}{x^2-(\sqrt{3})^2} = \frac{2(x+\sqrt{3})}{x^2-3}.$

b. $\frac{\sqrt{2}}{\sqrt{2}-\sqrt{3}} = \frac{\sqrt{2}}{\sqrt{2}-\sqrt{3}} \cdot \frac{\sqrt{2}+\sqrt{3}}{\sqrt{2}+\sqrt{3}} = \frac{\sqrt{2}(\sqrt{2}+\sqrt{3})}{2-3}$

$$= \frac{2+\sqrt{6}}{-1} = -(2+\sqrt{6}) = -2-\sqrt{6}.$$

c. $\frac{\sqrt{5}-\sqrt{2}}{\sqrt{5}+\sqrt{2}} = \frac{\sqrt{5}-\sqrt{2}}{\sqrt{5}+\sqrt{2}} \cdot \frac{\sqrt{5}-\sqrt{2}}{\sqrt{5}-\sqrt{2}}$

$$= \frac{(\sqrt{5}-\sqrt{2})^2}{5-2} = \frac{5-2\sqrt{5}\sqrt{2}+2}{3}$$

$$= \frac{7-2\sqrt{10}}{3}.$$

Exercise 8.4

In Problems **1–8**, *rationalize each denominator.*

1. $\frac{3}{\sqrt{7}}.$

2. $\frac{5}{\sqrt{11}}.$

3. $\frac{4}{\sqrt{2x}}.$

4. $\frac{y}{\sqrt{2y}}.$

5. $\frac{1}{\sqrt[3]{2}}.$

6. $\frac{3}{\sqrt[4]{2}}.$

7. $\frac{1}{\sqrt[3]{3x}}.$

8. $\frac{4}{3\sqrt[3]{x^2}}.$

In Problems **9–18**, *perform each division and simplify.*

9. $\frac{\sqrt{32}}{\sqrt{2}}.$

10. $\frac{\sqrt{18}}{\sqrt{2}}.$

11. $\frac{\sqrt{2a^3}}{\sqrt{a}}.$

12. $\dfrac{\sqrt{3x^5}}{\sqrt{x}}$. **13.** $\dfrac{2x\sqrt[4]{x^7}}{\sqrt[4]{x^{12}}}$. **14.** $\dfrac{3a^2b\sqrt[5]{a^3}}{\sqrt[5]{a^5}}$.

15. $\dfrac{\sqrt[3]{6}}{\sqrt[3]{4x}}$. **16.** $\dfrac{\sqrt[3]{3y}}{\sqrt[3]{x}}$. **17.** $\dfrac{\sqrt[3]{2x}}{\sqrt[3]{5xy^2}}$.

18. $\dfrac{\sqrt{7xy}}{\sqrt{14xy^3}}$.

In Problems **19–28**, *rationalize each denominator.*

19. $\dfrac{1}{2+\sqrt{3}}$. **20.** $\dfrac{1}{1-\sqrt{2}}$. **21.** $\dfrac{\sqrt{2}}{\sqrt{3}-\sqrt{6}}$.

22. $\dfrac{5}{\sqrt{6}+\sqrt{7}}$. **23.** $\dfrac{2\sqrt{2}}{\sqrt{2}-\sqrt{3}}$. **24.** $\dfrac{2\sqrt{3}}{\sqrt{5}-\sqrt{2}}$.

25. $\dfrac{1}{x+\sqrt{5}}$. **26.** $\dfrac{3-\sqrt{5}}{\sqrt{4}+\sqrt{2}}$. **27.** $\dfrac{\sqrt{6}-\sqrt{3}}{\sqrt{6}+\sqrt{3}}$.

28. $\dfrac{\sqrt{3}-5\sqrt{2}}{\sqrt{3}+\sqrt{2}}$.

29. For a piano tuned to the equally tempered scale, the frequency of a half tone above any note can be obtained by multiplying the frequency of the note by $\sqrt[12]{2}$. The standard concert frequency for A above middle C is 440 vibrations per second. (a) What is the frequency for the note B that is two half tones above standard A? Give your anwer in a simplified radical form. (b) What is the frequency for the note D# that is six half tones below standard A? Give your answer in simplified radical form and then use Appendix A to approximate your answer to the nearest full vibration per second.

30. One result of Einstein's theory of relativity is that the mass of an object increases as the velocity of the object increases. If the mass at velocity v is m_v and the rest mass of the object is m_0, then the formula for m_v is

$$m_v = \frac{m_0}{\sqrt{1-\dfrac{v^2}{c^2}}},$$

where c is the velocity of light. If the object's velocity is half the speed of light, find its mass m_v in terms of its rest mass.

8.5 RADICAL EQUATIONS

The equation $\sqrt{x-7}=4$ has the variable x under the radical sign. It is called a **radical equation**. One way to solve it is to raise both sides to the same power to remove the radical. This step does *not* guarantee that the resulting equation is equivalent to the original. Thus we sometimes find a value of x that doesn't actually satisfy the original equation (that is, we get an extraneous solution). So it is important that all "solutions" be checked.

Let's solve our equation by squaring both sides.

$$\sqrt{x-7}=4,$$
$$(\sqrt{x-7})^2=4^2,$$
$$x-7=16,$$
$$x=23.$$

Substituting 23 for x in $\sqrt{x-7}=4$ gives $\sqrt{23-7}=4$, or $\sqrt{16}=4$, which is true. Thus the solution is $x=23$.

EXAMPLE 1

Solving radical equations.

a. $\sqrt{x+2}-x+4=0$.

It helps to rewrite the equation so that the radical is by itself on one side.

$$\sqrt{x+2}=x-4,$$
$$x+2=(x-4)^2 \qquad \text{[squaring both sides]},$$
$$x+2=x^2-8x+16,$$
$$0=x^2-9x+14,$$
$$0=(x-7)(x-2) \qquad \text{[factoring]},$$
$$x=7 \quad \text{or} \quad x=2.$$

Now we check these values in the *original* equation.

Replacing x by 7 gives $\sqrt{7+2}-7+4=0$, or $3-7+4=0$, which is true.
Replacing x by 2 gives $\sqrt{2+2}-2+4=0$, or $2-2+4=0$, which is *false*.

Thus the only solution is 7.

b. $\sqrt[3]{x-4}=3$.

$$\sqrt[3]{x-4}=3,$$
$$x-4=27 \qquad [\textit{cubing} \text{ both sides}],$$
$$x=31.$$

You may check that 31 is indeed the solution.

In some cases, you may have to raise both sides of an equation to the same power more than once. Example 2 shows this.

EXAMPLE 2

Solve $\sqrt{y-3} - \sqrt{y} = -3$.

When an equation has two terms involving radicals, you should first write the equation so that one radical is on each side.

$$\sqrt{y-3} = \sqrt{y} - 3,$$

$$y - 3 = (\sqrt{y} - 3)^2 \qquad \text{[squaring both sides]},$$

$$y - 3 = y - 6\sqrt{y} + 9,$$

$$6\sqrt{y} = 12,$$

$$\sqrt{y} = 2,$$

$$y = 4 \qquad \text{[squaring both sides]}.$$

Replacing y by 4 in the left side of the original equation gives $\sqrt{1} - \sqrt{4}$, which is -1. Since this does not equal the right side, -3, there is **no solution.**

Exercise 8.5

In Problems **1–18**, *solve each radical equation.*

1. $\sqrt{x-2} = 5$.

2. $\sqrt{x+7} = 9$.

3. $\sqrt{2y-5} = 6$.

4. $\sqrt{2x-6} - 16 = 0$.

5. $2\sqrt{2x+1} = 3\sqrt{3x-8}$.

6. $4\sqrt{8y+1} = 5\sqrt{5y+1}$.

7. $\sqrt{x^2+33} = x + 3$.

8. $\sqrt{4x-6} - \sqrt{x} = 0$.

9. $\sqrt{x} - \sqrt{x+1} = 1$.

10. $\sqrt{x-3} + 4 = 1$.

11. $\sqrt{x+2} = x - 4$.

12. $3\sqrt{x+4} = x - 6$.

13. $z + 2 = 2\sqrt{4z-7}$.

14. $x + \sqrt{x} - 2 = 0$.

15. $\sqrt{x+7} - \sqrt{2x} = 1$.

16. $\sqrt{3x} - \sqrt{5x+1} = -1$.

17. $\sqrt[3]{x^2+2} = 3$.

18. $\sqrt[4]{2x+1} = 3$.

19. The number of species of plants counted by researchers on a plot may depend on the size of the plot. For example, in Fig. 8-1 we see that on 1-m^2 plots there are three species (A, B, and C on the left plot and A, B, and D on the right plot); on a 2-m^2 plot there are four species (A, B, C, and D).

In a study* of rooted plants in one geographic region, it was estimated that the

---

* Adapted from R. W. Poole, *An Introduction to Quantitative Ecology*. (New York: McGraw-Hill, 1974).

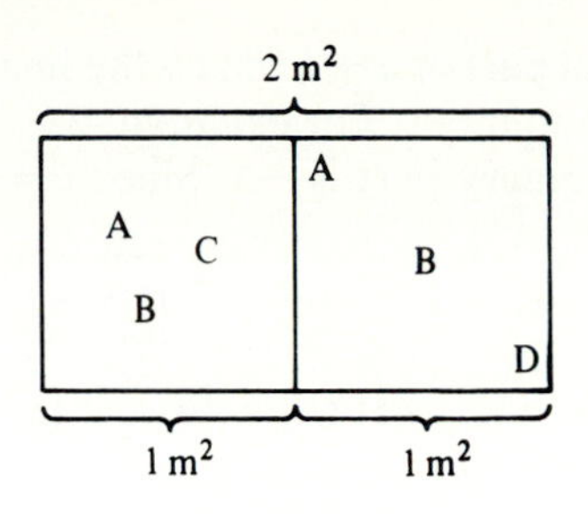

FIGURE 8-1

average number of species, S, occurring on plots of size M (in square meters) is given by

$$S = 12\sqrt[4]{M}.$$

On what size plots does an average of 36 species occur?

20. One result of Einstein's theory of relativity is that the mass of an object increases as the velocity of the object increases. If the mass at velocity v is m_v and the rest mass of the object is m_0, then the formula for m_v is

$$m_v = \frac{m_0}{\sqrt{1 - \frac{v^2}{c^2}}},$$

where c is the velocity of light. At what velocity would the mass m_v be twice the rest mass? (Assume $v > 0$.)

21. Police have used the formula $s = \sqrt{30fd}$ to estimate the speed s (in miles per hour) of a car if it skidded d feet when stopping. The literal number f is the coefficient of friction determined by the kind of road (such as concrete, asphalt, gravel, or tar) and whether the road is wet or dry. Some values of f are given in Table 8-2. At 40 mi/h, about how many feet will a car skid on a dry concrete road? Give your answer to the nearest foot.

TABLE 8-2

	Concrete	Tar
Wet	.4	.5
Dry	.8	1.0

8.6 OPERATIONS WITH COMPLEX NUMBERS

In this section you will learn how to add, subtract, multiply, and divide complex numbers. Recall from Section 6-4 that a complex number has the form $a + bi$, where a and b are real numbers and $i = \sqrt{-1}$:

$$\text{complex number: } a + bi.$$

We call a the **real part** of $a + bi$ and b the **imaginary part**. Note that the imaginary part, b, is a real number. For example, $2 - 3i$ is a complex number. Its real part is 2 and its imaginary part is -3. Since $i = \sqrt{-1}$, i has the following property.

$$i^2 = -1.$$

Many of the rules that apply to real numbers, such as the associative, commutative, and distributive laws, also apply to complex numbers.

To add or subtract complex numbers, we use the following rules.

ADDITION AND SUBTRACTION OF COMPLEX NUMBERS

$$(a + bi) + (c + di) = (a + c) + (b + d)i.$$

$$(a + bi) - (c + di) = (a - c) + (b - d)i.$$

Notice in the first rule that the sum of two complex numbers is also a complex number. Its real part is the sum of the real parts of the given numbers; its imaginary part is the sum of their imaginary parts. A similar statement can be made for subtraction.

EXAMPLE 1

Adding and subtracting complex numbers.

a. $(2 + 3i) + (4 + i) = (2 + 4) + (3 + 1)i = 6 + 4i.$

b. $(6 + 3i) - (2 + 5i) = (6 - 2) + (3 - 5)i = 4 - 2i.$

When adding or subtracting complex numbers, we can apply our usual rules for removing parentheses and combining similar terms. We treat i like any other literal number. Thus we can do Example 1a as

$$(2 + 3i) + (4 + i) = 2 + 3i + 4 + i = 6 + 4i$$

and Example 1b as

$$(6 + 3i) - (2 + 5i) = 6 + 3i - 2 - 5i = 4 - 2i.$$

EXAMPLE 2

a. $(-2 - 3i) + (2 - 4i) = -2 - 3i + 2 - 4i = -7i.$

b. $(5 - 2i) - (-2i) = 5 - 2i + 2i = 5.$

We multiply the complex numbers $a + bi$ and $c + di$ the same way we would multiply binomials. By the distributive law,

$$(a + bi)(c + di) = a(c + di) + bi(c + di)$$
$$= ac + adi + bci + bdi^2.$$

We now replace i^2 by -1 and combine similar terms.

$$(a + bi)(c + di) = ac + adi + bci + bd(-1)$$
$$= (ac - bd) + (ad + bc)i.$$

In general, we have the following rule.

> Multiply complex numbers as you would algebraic expressions. Then replace i^2 by -1 and simplify.

EXAMPLE 3

Multiplying complex numbers.

a. $(3 + 2i)(2 - 5i) = 3(2 - 5i) + 2i(2 - 5i)$

$$= 6 - 15i + 4i - 10i^2$$
$$= 6 - 11i - 10(-1)$$
$$= 16 - 11i.$$

b. $i(-2 + 3i) = -2i + 3i^2 = -2i - 3 = -3 - 2i.$

c. $(8i)(4i) = 32i^2 = -32.$

d. $(4 + 5i)^2.$

We treat this as a square of a binomial.

$$(4 + 5i)^2 = 4^2 + 2(4)(5i) + (5i)^2$$
$$= 16 + 40i + 25i^2$$
$$= 16 + 40i - 25$$
$$= -9 + 40i.$$

EXAMPLE 4

Simplify.

a. $i^3.$

Since $i^2 = -1$, we have

$$i^3 = i^2 i = (-1)i = -i.$$

b. i^4.

$$i^4 = i^2 i^2 = (-1)(-1) = 1.$$

With each complex number we can associate another complex number called its *conjugate*.

> The **conjugate** of $a + bi$ is $a - bi$.

Notice that $a + bi$ and its conjugate $a - bi$ have the same real parts, but their imaginary parts are *opposites*.

EXAMPLE 5

Complex Number	*Conjugate*
$2 + 3i$	$2 - 3i$
$5 - 6i$	$5 + 6i$
$-2i$	$2i$
$8 = 8 + 0i$	$8 - 0i = 8$

The product of a complex number and its conjugate has an interesting property. Consider $2 + 3i$ and its conjugate $2 - 3i$. These expressions are in the forms $x + y$ and $x - y$, where 2 plays the role of x and $3i$ plays the role of y. Thus their product is a difference of squares.

$$\begin{aligned}(2 + 3i)(2 - 3i) &= 2^2 - (3i)^2 \\ &= 4 - 9i^2 = 4 + 9 = 13.\end{aligned}$$

Notice that the product is a *real number*!

> **PRODUCT OF CONJUGATES**
>
> $(a + bi)(a - bi) = a^2 + b^2$, a real number.

*The product of the conjugates $5 + 2i$ and $5 - 2i$ is a **sum** of squares of real numbers:*

$$(5 + 2i)(5 - 2i) = 5^2 + 2^2 = 29.$$

Don't write the product as a difference: $(5 + 2i)(5 - 2i) \neq 5^2 - 2^2 = 21$.

We divide complex numbers by making use of conjugates. For example, to find

$$\frac{2-5i}{4+2i},$$

we multiply both the numerator and denominator by the conjugate of the denominator, $4-2i$. The resulting denominator, which is the product of conjugates, must be a real number.

$$\frac{2-5i}{4+2i}=\frac{(2-5i)(4-2i)}{(4+2i)(4-2i)}=\frac{8-4i-20i+10i^2}{4^2+2^2}$$

$$=\frac{8-24i-10}{16+4}=\frac{-2-24i}{20}$$

$$=-\frac{2}{20}-\frac{24}{20}i=-\frac{1}{10}-\frac{6}{5}i.$$

We shall write our answers in the form $a+bi$, as above. Here is a general rule.

> To divide two complex numbers, multiply both the numerator and denominator by the conjugate of the denominator and then simplify.

EXAMPLE 6

Dividing complex numbers.

a. $\frac{2-i}{3+2i}=\frac{(2-i)(3-2i)}{(3+2i)(3-2i)}=\frac{6-4i-3i+2i^2}{3^2+2^2}$

$$=\frac{6-7i-2}{9+4}=\frac{4-7i}{13}=\frac{4}{13}-\frac{7}{13}i.$$

b. $\frac{10-4i}{2}$.

Since the denominator 2 is real, we do not have to multiply numerator and denominator by the conjugate of 2.

$$\frac{10-4i}{2}=\frac{10}{2}-\frac{4}{2}i=5-2i.$$

c. $\frac{2}{1-i}=\frac{2(1+i)}{(1-i)(1+i)}=\frac{2+2i}{1^2+1^2}=\frac{2+2i}{2}=\frac{2}{2}+\frac{2}{2}i=1+i.$

d. $\frac{6}{5i}=\frac{6(-5i)}{5i(-5i)}=\frac{-30i}{-25i^2}=\frac{-30i}{25}=-\frac{6}{5}i.$

If the denominator is a pure imaginary number, as it is in $\frac{6}{5i}$, we can simply multiply numerator and denominator by i.

$$\frac{6}{5i} = \frac{6(i)}{5i(i)} = \frac{6i}{5i^2} = \frac{6i}{-5} = -\frac{6}{5}i.$$

Whenever you must perform operations with expressions involving square roots of negative numbers, first rewrite the square roots in terms of i. Then perform the given operations.

EXAMPLE 7

a. $$\begin{aligned}\sqrt{6} - \sqrt{-6} - \sqrt{-24} &= \sqrt{6} - i\sqrt{6} - i\sqrt{24} \\ &= \sqrt{6} - i\sqrt{6} - 2i\sqrt{6} \\ &= \sqrt{6} - 3i\sqrt{6}.\end{aligned}$$

b. $$\begin{aligned}2\sqrt{-4}(5 - \sqrt{-9}) &= 2(2i)(5 - 3i) = 4i(5 - 3i) \\ &= 20i - 12i^2 = 20i + 12 = 12 + 20i.\end{aligned}$$

c. $$\begin{aligned}\frac{2 - \sqrt{-9}}{1 - \sqrt{-4}} &= \frac{2 - 3i}{1 - 2i} = \frac{(2 - 3i)(1 + 2i)}{(1 - 2i)(1 + 2i)} \\ &= \frac{2 + 4i - 3i - 6i^2}{1^2 + 2^2} = \frac{2 + i + 6}{5} \\ &= \frac{8 + i}{5} = \frac{8}{5} + \frac{1}{5}i.\end{aligned}$$

$$\sqrt{-8} \cdot \sqrt{-2} = i\sqrt{8} \cdot i\sqrt{2} = \sqrt{8} \cdot \sqrt{2} \cdot i^2 = \sqrt{16}(-1) = -4.$$

Don't write $\sqrt{-8} \cdot \sqrt{-2} = \sqrt{(-8)(-2)} = \sqrt{16} = 4$, *which is* ***false!*** *Be certain to write* $\sqrt{-8}$ *and* $\sqrt{-2}$ *in terms of i* ***before*** *you multiply.**

Exercise 8.6

In Problems **1–44**, *perform the operations.*

1. $(3 - 5i) + (-6 + 4i)$.

2. $(-2 + 6i) + (7 + 4i)$.

3. $(7 - 3i) - (9 - 6i)$.

4. $(4 + 3i) - (-6 + 5i)$.

5. $2i + 3i - 3$.

6. $2i - 6i + 4i$.

---

* The rule $\sqrt{a}\sqrt{b} = \sqrt{ab}$ does not apply when both a and b are negative.

7. $(2 - i) + (3 + i) - (4 - i)$.

8. $4(2 + i) - 3(6 + 2i) + 7i$.

9. $(2 + i)(3 + i)$.

10. $(1 + i)(6 - 2i)$.

11. $(4 + 3i)(5 - 2i)$.

12. $(-2 - i)(-3 + i)$.

13. i^5.

14. $-i^6$.

15. $(2i^2)(3i)^2$.

16. $i(2i)(3i)(4i)$.

17. $2i(4 - 5i)$.

18. $(3i)^2(1 + i)$.

19. $(3 + 2i)^2$.

20. $(8 - 4i)^2$.

21. $(7 + 3i)(7 - 3i)$.

22. $(-4 - 2i)(-4 + 2i)$.

23. $\dfrac{4 + 3i}{2}$.

24. $\dfrac{8 - 10i}{5}$.

25. $\dfrac{3 - i}{4 + i}$.

26. $\dfrac{1 - i}{1 + i}$.

27. $\dfrac{2 - i}{3 - i}$.

28. $\dfrac{1 - i}{2 + i}$.

29. $\dfrac{2}{1 - 4i}$.

30. $\dfrac{-3i}{2i}$.

31. $\dfrac{3}{2i}$.

32. $\dfrac{3}{2(1 - i)}$.

33. $\dfrac{4 - 2i}{-i}$.

34. $\dfrac{1}{i} + \dfrac{2}{i}$.

35. $\sqrt{-4} + 2\sqrt{-8} + 3\sqrt{12}$.

36. $5 - 2\sqrt{-16} + (3 - \sqrt{-2})$.

37. $\sqrt{-2}\sqrt{-20}$.

38. $(-\sqrt{-4})(\sqrt{-5})$.

39. $(6 + i)(-7 - \sqrt{-4})$.

40. $(2 - \sqrt{-3})(3 + \sqrt{-12})$.

41. $(3 - \sqrt{-25})^2$.

42. $(2i - \sqrt{-64})^2$.

43. $\dfrac{6}{2 - \sqrt{-25}}$.

44. $\dfrac{\sqrt{4}}{\sqrt{-4}}$.

45. In electrical theory, if two impedances, Z_1 and Z_2, are connected in series, then the resulting impedance is Z, where

$$Z = Z_1 + Z_2.$$

These impedances can be represented by complex numbers. (a) Find Z if $Z_1 = 1 + i$ and $Z_2 = 2 + i$. (b) If Z_1 and Z_2 are connected in parallel, then

$$Z = \frac{Z_1 Z_2}{Z_1 + Z_2}.$$

Find Z for the values of Z_1 and Z_2 given in part a.

8.7 REVIEW

IMPORTANT TERMS

fractional exponent *(8.1)*
reducing the index *(8.2)*
similar radicals *(8.3)*
real part *(8.6)*
conjugate *(8.6)*
rationalizing the denominator *(8.2)*
simplified radical *(8.2)*
radical equation *(8.5)*
imaginary part *(8.6)*

REVIEW PROBLEMS

In Problems **1–8**, *evaluate the expressions.*

1. $100^{1/2}$.
2. $64^{1/3}$.
3. $4^{3/2}$.
4. $(25)^{-3/2}$.
5. $(32)^{-2/5}$.
6. $\left(\frac{9}{100}\right)^{3/2}$.
7. $\left(\frac{1}{16}\right)^{5/4}$.
8. $\left(-\frac{27}{64}\right)^{2/3}$.

In Problems **9–20**, *simplify. Use only positive exponents in your answers.*

9. $(9z^6)^{1/2}$.
10. $(16y^8)^{3/4}$.
11. $\left(\frac{27t^3}{8}\right)^{2/3}$.
12. $\left(\frac{1000}{a^9}\right)^{-2/3}$.
13. $(ab^2c^3)^{3/4}$.
14. $(2x^{3/4}y^{1/2})(xy^{3/2})$.
15. $(-3x^{1/2}y^{2/3})^3$.
16. $[(x-4)^{1/5}]^{10}$.
17. $2x^{1/2}y^{-3}x^{1/3}$.
18. $(4xy^3)^{1/2}(-2x^{3/2}y)^4$.
19. $(x^{1/2}+y^{1/2})(x^{1/2}-y^{1/2})$.
20. $\left(\frac{x^{2/3}y^4}{x^{1/6}y^{2/3}}\right)^{3/2}$.

In Problems **21–56**, *perform the operations and simplify.*

21. $\sqrt{32}$.
22. $\sqrt[3]{24}$.
23. $\sqrt[3]{2x^3}$.
24. $\sqrt{4x}$.
25. $\sqrt{16x^4}$.
26. $\sqrt[4]{\frac{x}{16}}$.
27. $\sqrt{7}\sqrt{4}\sqrt{14}$.
28. $\frac{2}{\sqrt{x^3}}$.
29. $\sqrt{\sqrt[3]{t^4}}$.
30. $\frac{\sqrt{3}\sqrt{6}}{\sqrt{2}}$.
31. $\sqrt[4]{\frac{3}{2x^2y^3}}$.
32. $\frac{\sqrt[3]{t^5}}{\sqrt[3]{t^2}}$.
33. $2\sqrt{8}-(5\sqrt{2}-\sqrt{18})$.
34. $(\sqrt{3}-\sqrt{2})(\sqrt{3}+2\sqrt{2})$.
35. $\sqrt{2}(1-\sqrt{6})$.
36. $\sqrt{75k^4}$.

37. $(\sqrt[5]{2})^{10}$.

38. $\sqrt{x}\sqrt{x^2y^3}\sqrt{xy^2}$.

39. $\frac{2}{\sqrt{7}}$.

40. $\frac{8}{\sqrt[3]{4}}$.

41. $\frac{3}{\sqrt[4]{x}}$.

42. $\sqrt[5]{\sqrt[3]{x^{10}}}$.

43. $\sqrt[4]{81x^6}$.

44. $(\sqrt[5]{x^2y})^{10}$.

45. $\sqrt[3]{\sqrt{\sqrt[3]{x^{36}}}}$.

46. $\sqrt[4]{2}\sqrt[4]{24}$.

47. $\sqrt{x}\sqrt{3x}$.

48. $\sqrt[6]{x^7y^{13}z^{12}}$.

49. $\sqrt[6]{\frac{x^6}{y^9}}$.

50. $\sqrt[9]{x^3y^6}$.

51. $\frac{3}{\sqrt[3]{xy^2}}$.

52. $\frac{1}{\sqrt{x}\sqrt{y}}$.

53. $\frac{\sqrt[3]{3x^2}}{\sqrt[3]{2x}}$.

54. $\sqrt[5]{\frac{xy^2}{x^2y}}$.

55. $\frac{1}{\sqrt{6}-2}$.

56. $\frac{4\sqrt{6}}{\sqrt{6}+\sqrt{4}}$.

In Problems **57–64**, *solve the equations.*

57. $\sqrt{2x+5}=5$.

58. $\sqrt{3x-4}=\sqrt{2x+5}$.

59. $\sqrt[3]{11x+9}=4$.

60. $\sqrt{x^2+5x+25}=x+4$.

61. $\sqrt{y}+6=5$.

62. $\sqrt{z^2+9}=5$.

63. $\sqrt{x-1}+\sqrt{x+6}=7$.

64. $\sqrt{2x+1}=x-7$.

In Problems **65–78**, *perform the operations.*

65. $(8-3i)+2(6+4i)$.

66. $9i^2-3i+2i^3$.

67. $(4+3i)-(6+i)+10i$.

68. $3i(1+2i)$.

69. $(1+2i)(3+4i)$.

70. $(1-i)^2$.

71. $(2+i)(2-i)$.

72. $(2-5i)+i(6i)$.

73. $\frac{4+2i}{3+i}$.

74. $\frac{3i}{6-2i}$.

75. $-\sqrt{-3}+\sqrt{-12}+2$.

76. $\sqrt{-8}(1-\sqrt{-2})$.

77. $(1+\sqrt{-4})(1+\sqrt{-9})$.

78. $\frac{2}{2+\sqrt{-25}}$.

79. The mass, m, of an object moving with speed v is given by

$$m=\frac{m_0}{\sqrt{1-\frac{v^2}{c^2}}},$$

where m_0 is the rest mass of the object and c is the speed of light. If $m=\frac{3}{2}m_0$, solve for v. (Assume $v>0$.)

9 Graphs and Straight Lines

9.1 RECTANGULAR COORDINATE SYSTEM

On television and in newspapers, you've probably seen how information can be given by means of graphs. In this chapter you'll see how graphs are used in algebra.

Our graphs will be drawn on a **rectangular** (or *Cartesian*) **coordinate plane**. This is a plane (think of a sheet of paper) on which two number lines are placed, one horizontal and one vertical, as in Fig. 9-1. The origins of the number lines meet at a point called the **origin** of the coordinate plane.

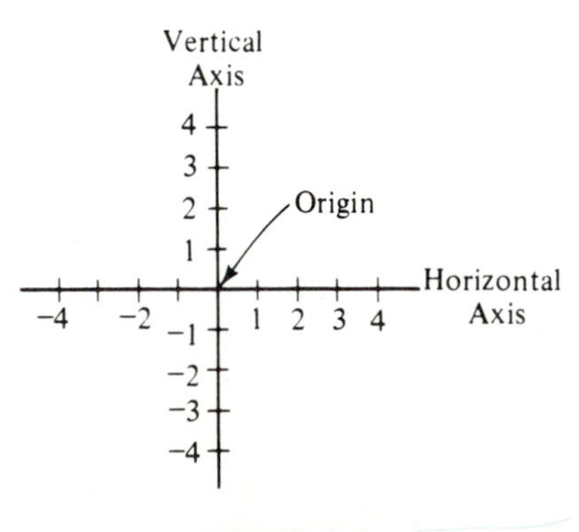

FIGURE 9-1

The number lines in Fig. 9-1 are called **coordinate axes**. The *horizontal axis* has its positive numbers to the right of the origin, while the *vertical axis* has them above the origin. The unit distance on the vertical axis does not have to be the same as that on the horizontal axis.

Every point in the plane can be labeled to indicate its position. For example, let's label point P in Fig. 9-2(a). First, from P we draw a perpendicular to the

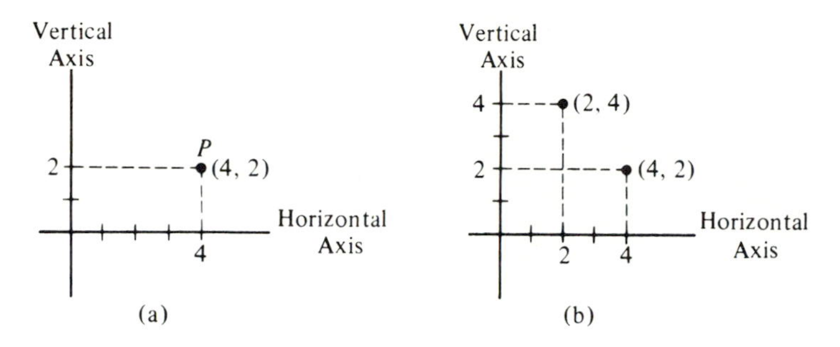

FIGURE 9-2

horizontal axis. It meets this axis at 4. Next, we draw a perpendicular from P to the vertical axis. It meets this axis at 2. Thus P determines two numbers, 4 and 2, and in Fig. 9-2(a) we label P with the **ordered pair** $(4, 2)$. The first number of the ordered pair is called the **first coordinate**, or **abscissa**, of P. The second number is the **second coordinate**, or **ordinate**. Together, 4 and 2 are called the **rectangular coordinates** of P.

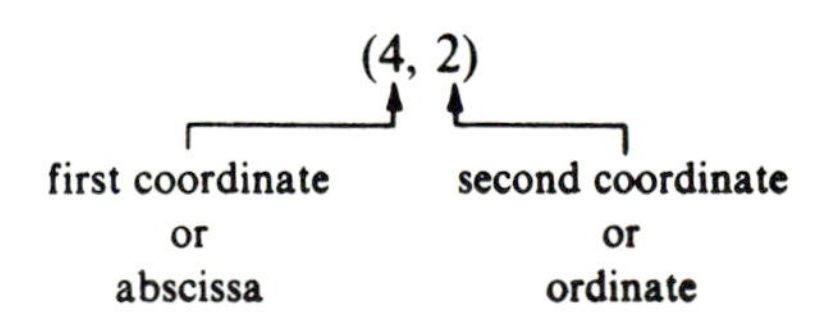

Thus we can think of a point in the plane as an ordered pair, and vice versa. The word *ordered* is important because

$$(4, 2) \neq (2, 4) \qquad \text{[see Fig. 9-2(b) above].}$$

Figure 9-3 shows the coordinates of several points. Notice a few things.

- The origin has coordinates $(0, 0)$.
- Any point on the horizontal axis has a *second* coordinate of 0.
- Any point on the vertical axis has a *first* coordinate of 0.

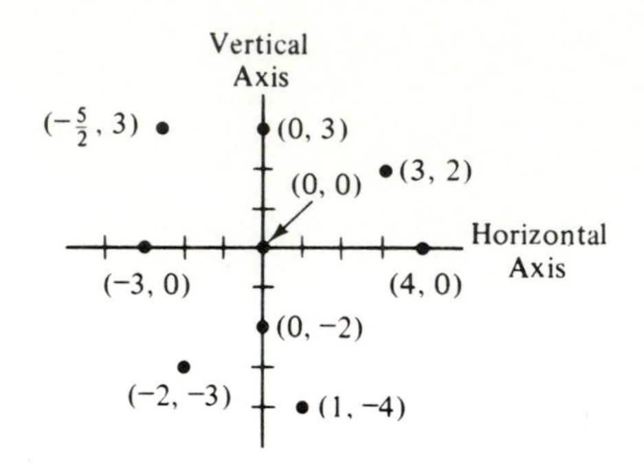

FIGURE 9-3

The coordinate axes divide the plane into four parts, called **quadrants.** They are numbered as in Fig. 9-4. Any point (a_1, b_1) in Quadrant I (or the *first* quadrant) has positive coordinates. Any point (a_2, b_2) in Quadrant II has a negative first coordinate and a positive second coordinate. Look at Fig. 9-4 and see what must be true for points in Quadrants III and IV.

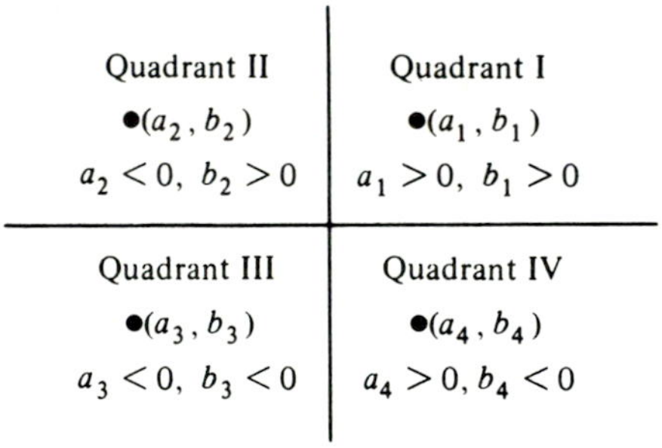

Quadrant II	Quadrant I
•(a_2, b_2)	•(a_1, b_1)
$a_2 < 0,\ b_2 > 0$	$a_1 > 0,\ b_1 > 0$
Quadrant III	Quadrant IV
•(a_3, b_3)	•(a_4, b_4)
$a_3 < 0,\ b_3 < 0$	$a_4 > 0, b_4 < 0$

FIGURE 9-4

EXAMPLE 1

a. $(4, -2)$ lies in Quadrant IV, since $4 > 0$ and $-2 < 0$.

b. $(1, 8)$ lies in Quadrant I, since $1 > 0$ and $8 > 0$.

c. $(-3, -1)$ lies in Quadrant III.

d. $(-\frac{1}{2}, \frac{5}{8})$ lies in Quadrant II.

e. A point on an axis does not lie in any quadrant.

Exercise 9.1

In Problems **1–6**, *give the coordinates of each indicated point. See Fig.* 9-5.

1. P. **2.** Q. **3.** R. **4.** S. **5.** T. **6.** U.

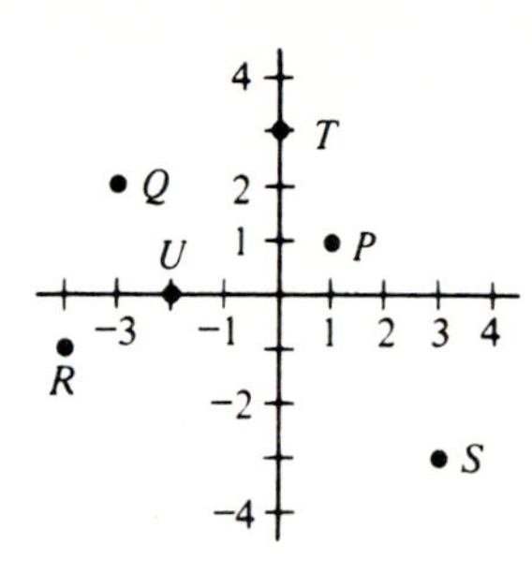

FIGURE 9-5

In Problems **7–18**, *construct a rectangular coordinate plane. Then locate and label (that is, plot) each of the points.*

7. $(0,0)$. **8.** $(2,3)$. **9.** $(1,4)$. **10.** $(0,-2)$.

11. $(-2,0)$. **12.** $(-1,-2)$. **13.** $(-1,2)$. **14.** $(2,-3)$.

15. $(-4,-3)$. **16.** $(-1,1)$. **17.** $(0,-\frac{1}{2})$. **18.** $(\frac{3}{2},-\frac{1}{2})$.

In Problems **19–30**, *give the quadrant in which the point lies.*

19. $(3,2)$. **20.** $(-3,2)$. **21.** $(2,-3)$. **22.** $(-4,-8)$.

23. $(-6,1)$. **24.** $(4,1)$. **25.** $(-1,-2)$. **26.** $(5,-5)$.

27. $(-\frac{3}{5},\frac{2}{3})$. **28.** $(\frac{4}{7},-\frac{2}{5})$. **29.** $(\sqrt{2},\sqrt{3})$. **30.** $(-.1,-.2)$.

9.2 GRAPHS OF EQUATIONS

Up to now we have worked with equations with one variable. Now we'll look at some in two variables.

For example, $y = x^2$ is an equation in the variables x and y. Here a solution is a *pair* of numbers: a value of x *and* a value of y that make the equation true. For instance, $x = 2$ and $y = 4$ is a solution:

$$y = x^2,$$
$$4 = 2^2,$$
$$4 = 4. \checkmark$$

Table 9-1 shows more solutions. We found them by first choosing values of x and then substituting them into $y = x^2$ to get corresponding values of y. For example,

$$\text{if} \quad x = -\frac{1}{2}, \quad \text{then} \quad y = x^2 = \left(-\frac{1}{2}\right)^2 = \frac{1}{4},$$

so a solution is

$$x = -\frac{1}{2}, \qquad y = \frac{1}{4}.$$

By no means are all solutions listed. That's impossible to do because there are infinitely many.

TABLE 9-1

x	$y\ (=x^2)$	
0	0	$\leftrightarrow$ (0,0)
$\frac{1}{2}$	$\frac{1}{4}$	$\leftrightarrow$ $(\frac{1}{2}, \frac{1}{4})$
$-\frac{1}{2}$	$\frac{1}{4}$	$\leftrightarrow$ $(-\frac{1}{2}, \frac{1}{4})$
1	1	$\leftrightarrow$ (1, 1)
−1	1	$\leftrightarrow$ (−1, 1)
2	4	$\leftrightarrow$ (2, 4)
−2	4	$\leftrightarrow$ (−2, 4)
3	9	$\leftrightarrow$ (3, 9)
−3	9	$\leftrightarrow$ (−3, 9)

Notice in Table 9-1 that we associated ordered pairs with the solutions. In each ordered pair the first number is the x-value of a solution, while the second is the y-value.

Now, from the last section we know that ordered pairs can be thought of as points in a plane. In Fig. 9-6(a) we've *plotted* (or located) the ordered pairs in Table 9-1. *Each point represents a solution of* $y = x^2$.

The horizontal axis is labeled the ***x*-axis** (or, more simply, x) and the vertical axis is the ***y*-axis**. This is natural because the first coordinate of each point is an x-value and the second coordinate is a y-value. Sometimes we say that the first coordinate is the *x-coordinate*, and the second coordinate is the *y-coordinate*. When we look geometrically at solutions of an equation in x and y, we usually choose the horizontal axis as the x-axis.

If we were able to plot all the solutions of $y = x^2$, we would get a picture like Fig. 9-6(b). It is called the *graph* of $y = x^2$. **The *graph of an equation* is the geometric representation of its solutions**. To get Fig. 9-6(b), we connected the points in Fig. 9-6(a) by a smooth curve. It should be clear that the graph extends upward indefinitely (as indicated by arrows), and every point on the graph gives a solution of the equation.

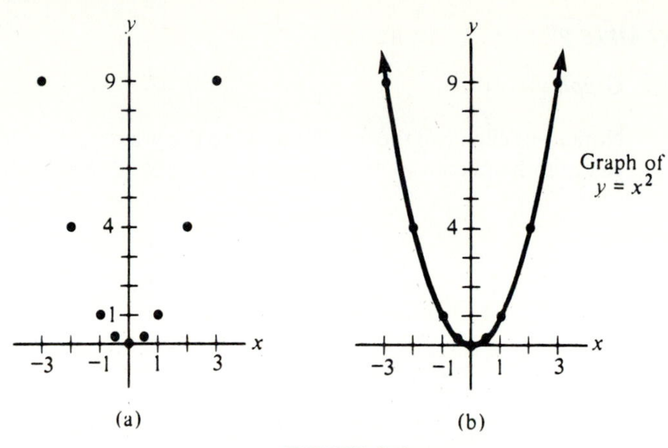

FIGURE 9-6

Usually, the more points that we plot originally, the better is our graph. Always plot enough points so that the general shape of the graph is clear. When in doubt, plot more points.

EXAMPLE 1

Graph the equation $y = 2x - 1$.

We choose some values for x and find the corresponding y-values.

$$\text{If} \quad x = 0, \quad \text{then} \quad y = 2x - 1 = 2(0) - 1 = -1.$$

Thus $(0, -1)$ lies on the graph.

$$\text{If} \quad x = 1, \quad \text{then} \quad y = 2x - 1 = 2(1) - 1 = 1.$$

Thus $(1, 1)$ lies on the graph. Continuing in this way, we get the table in Fig. 9-7. Plotting those points and connecting them, we get a straight line.

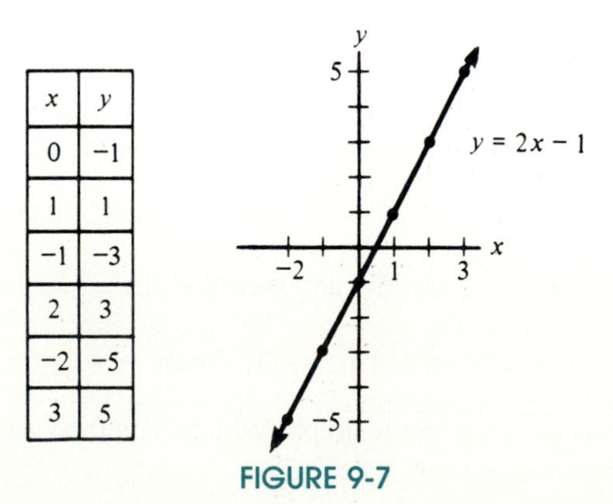

x	y
0	−1
1	1
−1	−3
2	3
−2	−5
3	5

FIGURE 9-7

EXAMPLE 2

Graph $y = 2x^3$.

Notice our choice of the unit distance on the y-axis in Fig. 9-8. It makes the graphing easier to handle.

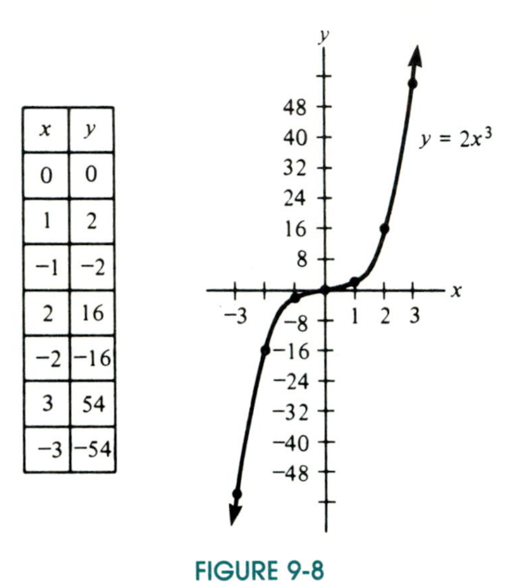

x	y
0	0
1	2
−1	−2
2	16
−2	−16
3	54
−3	−54

FIGURE 9-8

EXAMPLE 3

Graph $x = \frac{1}{2}y^2$.

We'll show two ways to do this problem: a messy way and a better way.

Messy Way.

If we choose a value of x such as 1, then to find y we have to solve the quadratic equation $1 = \frac{1}{2}y^2$. Although nothing is wrong with this, it can be a little messy. Solving gives $y = \pm\sqrt{2}$. Thus the points $(1, \sqrt{2})$ and $(1, -\sqrt{2})$ lie on the graph.

Better Way.

We choose values of y and then find the corresponding values of x. For example,

$$\text{if} \quad y = 1, \quad \text{then} \quad x = \frac{1}{2}y^2 = \frac{1}{2}(1)^2 = \frac{1}{2}.$$

Thus $(\frac{1}{2}, 1)$ lies on the graph. With this method we do not have to solve a quadratic equation. See Fig. 9-9.

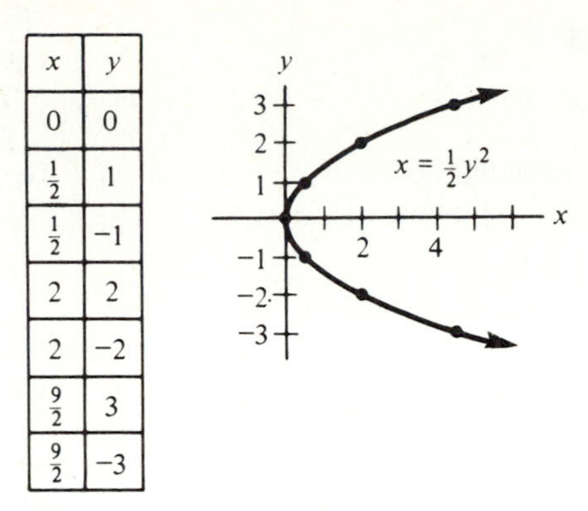

x	y
0	0
$\frac{1}{2}$	1
$\frac{1}{2}$	−1
2	2
2	−2
$\frac{9}{2}$	3
$\frac{9}{2}$	−3

FIGURE 9-9

Moral.

Sometimes it is better to choose values for one variable rather than for the other.

EXAMPLE 4

Graph $y = 2$.

We can think of $y = 2$ as an equation in the variables x and y. You may say to yourself, "But I don't see any x!" Well, you can if you write the equation in the form $y = 2 + 0x$. From this form it is clear that no matter what x is, any solution must have a y-value of 2. For example,

$$\text{if} \quad x = 4, \quad \text{then} \quad y = 2 + 0x = 2 + 0(4) = 2.$$

Thus $(4, 2)$ lies on the graph. In fact, every point on the graph will be of the form $(x, 2)$, where x is a real number. See Fig. 9-10. *The graph of $y = 2$ is a horizontal line.*

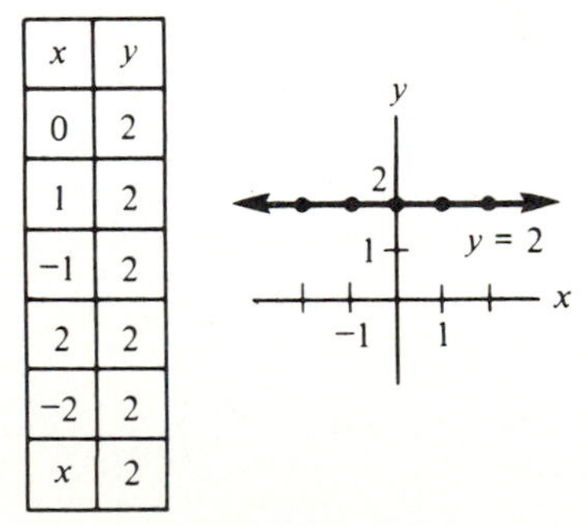

x	y
0	2
1	2
−1	2
2	2
−2	2
x	2

FIGURE 9-10

EXAMPLE 5

Graph $x^2 + y^2 = 4$.

Let's solve this equation for y in terms of x. Then we'll have a better equation from which to get points.

$$x^2 + y^2 = 4,$$

$$y^2 = 4 - x^2 \qquad \text{[form } u^2 = k\text{]},$$

$$y = \pm\sqrt{4 - x^2} \qquad \text{[square-root method]}.$$

If $x = 0$, then $y = \pm\sqrt{4-(0)^2} = \pm 2$. Thus, both $(0, 2)$ and $(0, -2)$ lie on the graph.

If $x = 1$, then $y = \pm\sqrt{4-(1)^2} = \pm\sqrt{3} \approx \pm 1.73$. (The symbol $\approx$ means approximately equals, and the value 1.73 comes from the table in Appendix A.)

Similarly, if $x = -1$, then $y = \pm\sqrt{4-(-1)^2} = \pm\sqrt{3} \approx \pm 1.73$.

Other values of x and y are given in Fig. 9-11. Notice that we do not choose any values of x greater than 2 or less than -2. This is because we don't want $4 - x^2$ to

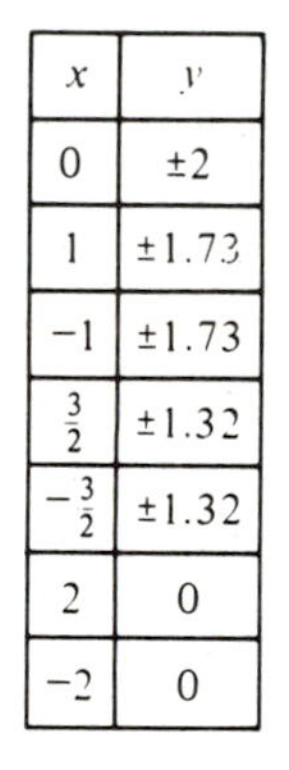

x	y
0	±2
1	±1.73
−1	±1.73
$\frac{3}{2}$	±1.32
$-\frac{3}{2}$	±1.32
2	0
−2	0

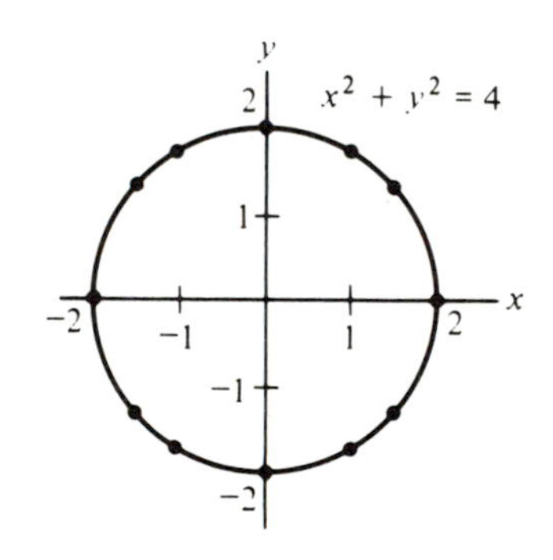

FIGURE 9-11

be negative, for then y (which is $\pm\sqrt{4 - x^2}$) would be imaginary. We don't plot points with imaginary coordinates in a rectangular coordinate plane.

From Fig. 9-11, we see that *the graph is a circle with center at* $(0, 0)$ *and radius* 2.

Exercise 9.2

In Problems **1–28**, *graph each equation.*

1. $y = x$.

2. $y = -x$.

3. $y = 3x - 2$.

4. $y = 2x + 3$.

5. $x = y + 1$.

6. $x + y + 1 = 0$.

7. $y = -1$.

8. $x = 2$.

9. $y = 2 - \frac{x}{2}$.

10. $y = \frac{x}{4} + 1$.

11. $y = 2x^2$.

12. $y = \frac{1}{2}x^2$.

13. $x = y^2$.

14. $x = y^2 - 2$.

15. $x = 3$.

16. $y = 3$.

17. $y = x^3$.

18. $y = \frac{x^3}{3}$.

19. $y = 0$.

20. $x = 0$.

21. $y = -x^3$.

22. $y = 2 - x^3$.

23. $y = x^2 - 4x + 3$.

24. $y = x^2 - 2x - 3$.

25. $y = 2x - x^2$.

26. $y = 4x - x^2$.

27. $x^2 + y^2 = 9$.

28. $y^2 = 25 - x^2$.

29. Graph $s = t^2 - 4t$. Choose t for the horizontal axis.

30. Graph $p + q = 1$. Choose p for the horizontal axis.

9.3 SLOPE OF A LINE AND POINT-SLOPE FORM

Some of the graphs we drew in the last section were straight lines. One feature of a straight line is its "steepness." For example, in Fig. 9-12 line L_1 rises faster as it goes from left to right than does line L_2. We say that L_1 is steeper.

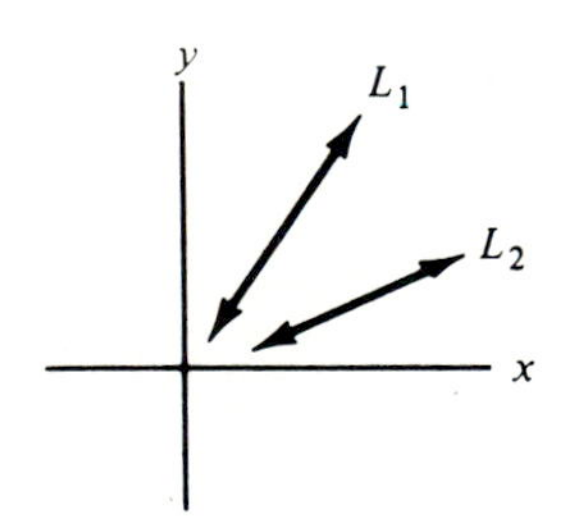

FIGURE 9-12

There's a way to measure steepness of a line. We choose two points on the line and see how the y-values change as the x-values change. In Fig. 9-13, as x changes from 2 to 4 (an increase of 2 units), notice that y changes from 1 to 5 (an increase of 4 units). Let's find how y changes for each 1-unit increase in x. To do this we divide the change in y (vertical change) by the change in x (horizontal change):

$$\frac{\text{change in } y}{\text{change in } x} = \frac{5-1}{4-2} = \frac{4}{2} = \frac{2}{1} = 2.$$

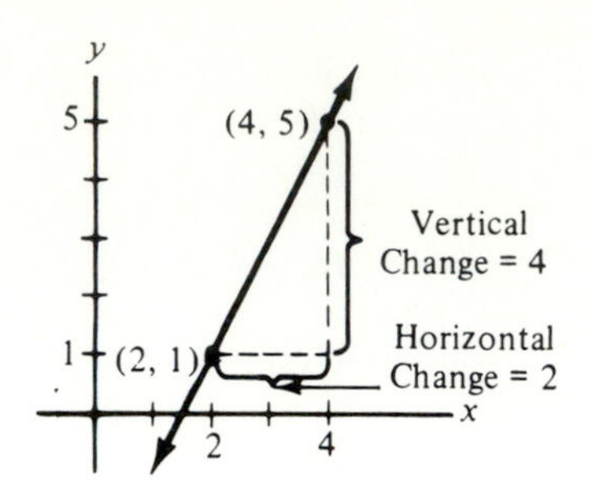

FIGURE 9-13

This means that for each 1-unit increase in x, there is a 2-unit increase in y. We say that the *slope* of the line is 2. Choosing two other different points on this line would also give a slope of 2. In general we have the following.

> **SLOPE OF A LINE**
>
> Suppose that (x_1, y_1) and (x_2, y_2) are two different points on a nonvertical line. Then the **slope** of the line is the number m given by
>
> $$m = \frac{y_2 - y_1}{x_2 - x_1} \qquad \left(= \frac{\text{change in } y}{\text{change in } x}\right).$$

EXAMPLE 1

Find the slope of the line passing through $(3, 4)$ *and* $(5, 7)$.

In the slope formula, let's choose $(3, 4)$ as (x_1, y_1) and $(5, 7)$ as (x_2, y_2). Then the slope is

$$m = \frac{y_2 - y_1}{x_2 - x_1} = \frac{7 - 4}{5 - 3} = \frac{3}{2}.$$

This means that if x increases by 1 unit, then y *increases* by $\frac{3}{2}$ units. In other words, if x increases by 2 units, then y *increases* by 3 units (see Fig. 9-14). Thus the line must

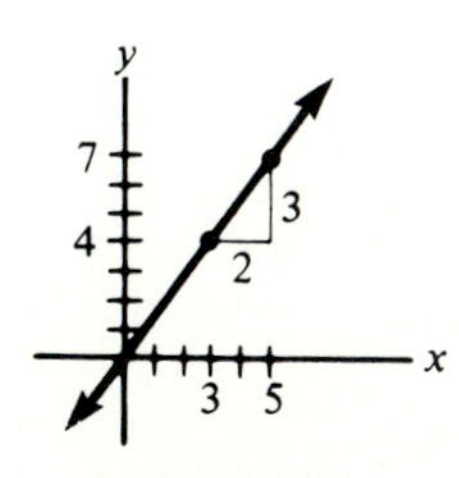

FIGURE 9-14

rise from left to right. When finding the slope, it doesn't matter which point we pick to be (x_1, y_1). If we choose $(5,7)$ as (x_1, y_1) and $(3,4)$ as (x_2, y_2), then

$$m = \frac{y_2 - y_1}{x_2 - x_1} = \frac{4-7}{3-5} = \frac{-3}{-2} = \frac{3}{2},$$

as before.

Always align the subscripts correctly when you use the slope formula. Remember,

$$m = \frac{y_2 - y_1}{x_2 - x_1}. \qquad \textbf{\textit{Don't write}}\ m = \frac{y_2 - y_1}{x_1 - x_2}.$$

EXAMPLE 2

Find the slope of the line through $(4,5)$ *and* $(6,1)$.

Let $(4,5) = (x_1, y_1)$ and $(6,1) = (x_2, y_2)$. Then

$$m = \frac{y_2 - y_1}{x_2 - x_1} = \frac{1-5}{6-4} = \frac{-4}{2} = -2.$$

Here the slope is negative, -2. This means that if x increases 1 unit, then y *decreases* 2 units. Thus the line must *fall* from left to right (see Fig. 9-15).

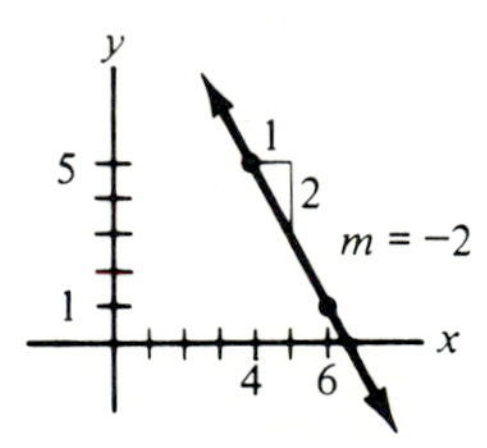

FIGURE 9-15

From Examples 1 and 2, we see that the slope tells us about the "rising" or "falling" nature of a line.

Positive slope: Line *rises* from left to right (Fig. 9-14).
Negative slope: Line *falls* from left to right (Fig. 9-15).

EXAMPLE 3

a. The slope of the *horizontal* line through $(3,4)$ and $(5,4)$ is [see Fig. 9-16(a)]

$$m = \frac{y_2 - y_1}{x_2 - x_1} = \frac{4-4}{5-3} = \frac{0}{2} = 0.$$

This means that the change in y is 0 for any change in x. In fact, the slope of *every* horizontal line is 0.

Zero slope: horizontal line.

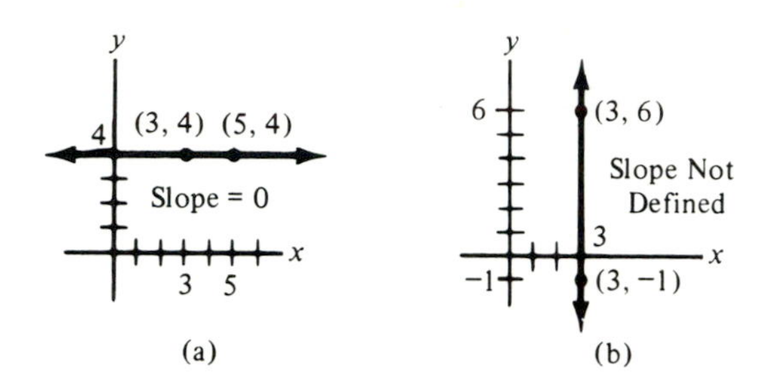

FIGURE 9-16

b. If we use the slope formula on the *vertical* line through $(3, -1)$ and $(3, 6)$ [see Fig. 9-16(b)], we get

$$m = \frac{y_2 - y_1}{x_2 - x_1} = \frac{6 - (-1)}{3 - 3} = \frac{7}{0}, \qquad \text{which is not defined.}$$

Here there is no change in x at all. In fact, the slope of *every* vertical line is undefined.

Undefined slope: vertical line.

Figure 9-17 shows lines with different slopes. Notice that **the closer the slope is to 0, the more nearly horizontal is the line. The greater the absolute value of the slope, the more nearly vertical is the line.**

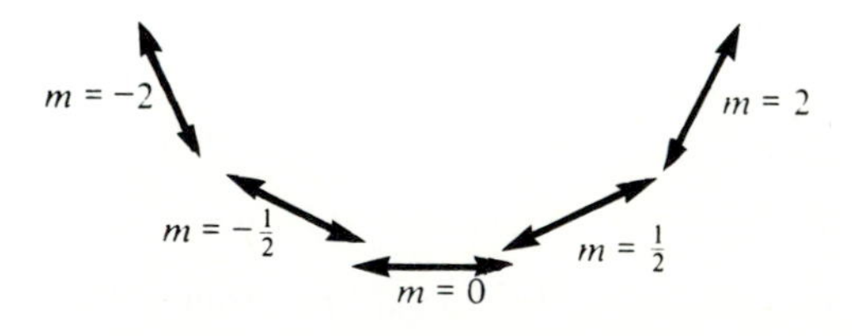

FIGURE 9-17

Now that you know what slope is, let's do something with it. Assume that you are given the slope of a line and the coordinates of one point on it. Then you can find an equation whose graph is that line. Here's how.

Suppose that the given line has a slope of 3 and the point $(1, 2)$ lies on it (see Fig. 9-18). Let (x, y) be *any* other point on the line. Then by applying the slope

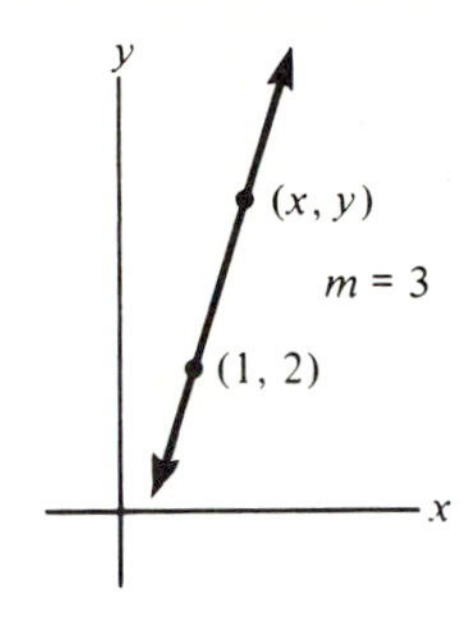

FIGURE 9-18

formula to the points $(1, 2)$ and (x, y), we have

$$\frac{y-2}{x-1} = 3.$$

Multiplying both sides by $x - 1$ gives

$$y - 2 = 3(x - 1). \tag{1}$$

This equation is called a *point-slope form* of an equation of the line. We can write Eq. 1 as

$$\begin{aligned} y - 2 &= 3x - 3, \\ y &= 3x - 1. \end{aligned}$$

The coordinates of every point on the line make the equation $y = 3x - 1$ true. We say that these coordinates *satisfy* the equation. Also, every point whose coordinates satisfy $y = 3x - 1$ is on the line. For example, if $x = 2$, then $y = 3x - 1 = 3(2) - 1 = 5$. Thus $(2, 5)$ is on the line.

Let's get more general. In the above discussion, we replace 3 by m and $(1, 2)$ by (x_1, y_1). Thus, in the manner of Eq. 1, we have the following.

POINT-SLOPE FORM

$$y - y_1 = m(x - x_1)$$

is the **point-slope form** of an equation of the line passing through (x_1, y_1) and having slope m.

EXAMPLE 4

Find an equation of the line with slope -3 *and that passes through* $(2, -1)$.

Using the point-slope form, we set $m = -3$ and $(x_1, y_1) = (2, -1)$.

$$y - y_1 = m(x - x_1),$$
$$y - (-1) = -3(x - 2),$$
$$y + 1 = -3x + 6.$$

We can write our answer as

$$y = -3x + 5. \tag{2}$$

If we want to sketch the line, we need one more point besides $(2, -1)$. If $x = 0$, then from Eq. 2 we get $y = -3(0) + 5 = 5$ (see Fig. 9-19).

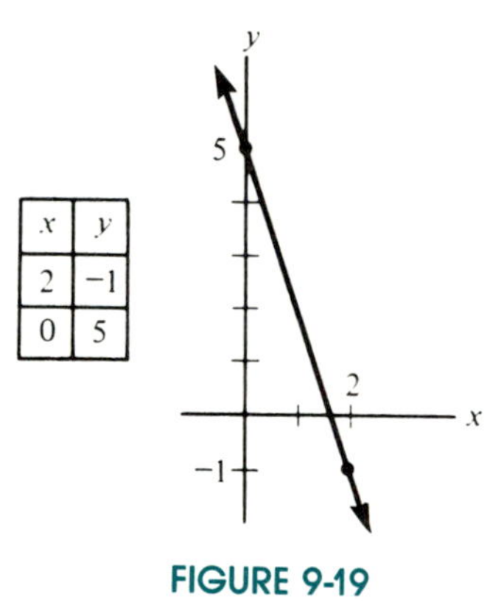

FIGURE 9-19

You can also find an equation of a line if you know only two points on it. First find the slope.* Then use a point-slope form with either point as (x_1, y_1).

EXAMPLE 5

Find an equation of the line through $(4, -2)$ *and* $(-3, 8)$.

We first find the slope. Let $(x_1, y_1) = (4, -2)$ and $(x_2, y_2) = (-3, 8)$. Then

$$m = \frac{y_2 - y_1}{x_2 - x_1} = \frac{8 - (-2)}{-3 - 4} = -\frac{10}{7}.$$

---

* We assume that the line is nonvertical and thus has a slope. Equations of vertical lines are discussed in the next section.

Using the point-slope form with $(4, -2)$ as (x_1, y_1) and $m = -\frac{10}{7}$, we obtain

$$y - y_1 = m(x - x_1),$$

$$y - (-2) = -\frac{10}{7}(x - 4),$$

$$y + 2 = -\frac{10}{7}x + \frac{40}{7}.$$

Simplifying gives

$$y = -\frac{10}{7}x + \frac{26}{7},$$

which is an equation of the line. If we had chosen $(-3, 8)$ as (x_1, y_1), we would get the point-slope form

$$y - 8 = -\frac{10}{7}(x + 3).$$

This also simplifies to $y = -\frac{10}{7}x + \frac{26}{7}$.

Exercise 9.3

In Problems **1–8**, *find the slope of the line passing through the given points.*

1. $(6, 3)$, $(8, 6)$.

2. $(-3, 4)$, $(0, 1)$.

3. $(-2, 2)$, $(3, -2)$.

4. $(2, -4)$, $(3, -4)$.

5. $(-2, 4)$, $(-2, 8)$.

6. $(0, 0)$, $(-3, 4)$.

7. $(3, 6)$, $(-1, 6)$.

8. $(0, 6)$, $(1, 1)$.

In Problems **9–16**, *find an equation of the line that has the given properties.*

9. Passes through $(1, 1)$ and has slope 1.

10. Passes through $(2, 4)$ and has slope 3.

11. Passes through $(-2, 5)$ and has slope $-\frac{1}{4}$.

12. Passes through $(3, 5)$ and has slope 0.

13. Passes through $(3, -6)$ and has slope $\frac{1}{3}$.

14. Has slope $\frac{3}{5}$ and passes through $(0, 2)$.

15. Has slope -5 and passes through the origin.

16. Has slope $\frac{1}{2}$ and passes through $(-\frac{1}{2}, -\frac{3}{2})$.

In Problems **17–24**, *find an equation of the line passing through the given points.*

17. $(2,4)$, $(8,7)$.

18. $(2,-5)$, $(3,4)$.

19. $(0,0)$, $(5,-5)$.

20. $(-2,5)$, $(3,5)$.

21. $(3,-1)$, $(2,-9)$.

22. $(2,3)$, $(0,0)$.

23. $(2,-7)$, $(3,-7)$.

24. $(-1,-1)$, $(-3,-3)$.

25. A straight line passes through $(1,2)$ and $(-3,8)$. Find the point on it that has a first coordinate of 5.

26. A straight line has slope -3 and passes through $(4,-1)$. Find the point on it that has a second coordinate of -2.

9.4 MORE EQUATIONS OF STRAIGHT LINES

In the last section we discussed a point-slope form of an equation of a line. We'll look at other forms in this section.

In Fig. 9-20 the line intersects the y-axis at the point $(0,b)$, which is called the **y-intercept** of the line. Sometimes we simply say that the *number b* is the y-intercept.

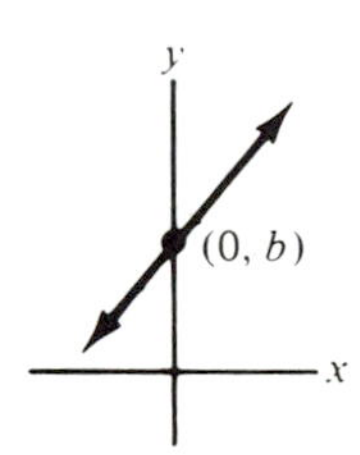

FIGURE 9-20

Suppose that we know the slope m and y-intercept b of a line. Then $(0,b)$ is on the line. Using the point-slope form, we have

$$y - y_1 = m(x - x_1),$$
$$y - b = m(x - 0).$$

If we solve for y, we'll get an important form of an equation of the line.

$$y - b = mx,$$
$$y = mx + b,$$

which is called the *slope-intercept form*.

SLOPE-INTERCEPT FORM

$$y = mx + b$$

is the **slope-intercept form** of an equation of the line with slope m and y-intercept b.

EXAMPLE 1

Find an equation of the line with slope 3 *and y-intercept* 4.

$$y = mx + b,$$
$$y = 3x + 4.$$

EXAMPLE 2

Discuss and draw the graph of $y = -\frac{2}{3}x + 7$.

The form of $y = -\frac{2}{3}x + 7$ is $y = mx + b$, with $m = -\frac{2}{3}$ and $b = 7$. Thus the graph is a straight line with slope $-\frac{2}{3}$ and y-intercept 7. To draw the graph, all we really need is two points on it. We can use the y-intercept for one of them, as in Fig. 9-21. Observe that since the slope is $-\frac{2}{3}$, then as x increases by 3 units, y *decreases* by 2 units.

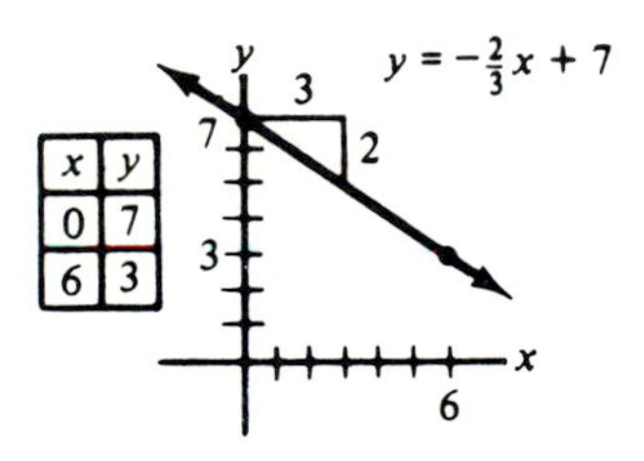

FIGURE 9-21

Let's turn to horizontal and vertical lines. In Fig. 9-22(a), the vertical line through $(2, 3)$ consists of all points (x, y) where $x = 2$. Thus an equation of the line is $x = 2$. More generally, we have the following.

VERTICAL LINE

An equation of the *vertical* line through (a, b) is

$$x = a.$$

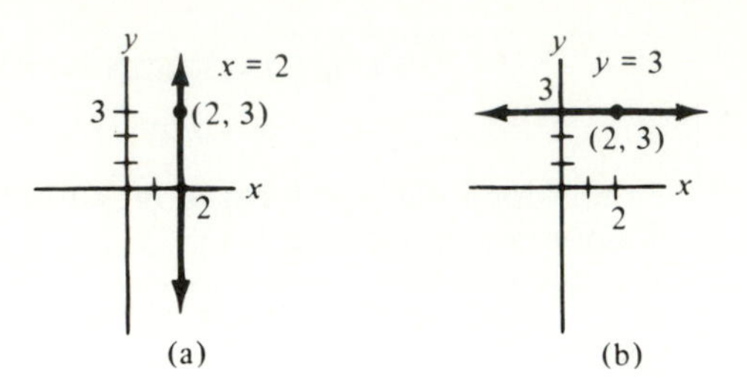

FIGURE 9-22

Similarly, in Fig. 9-22(b), an equation for the horizontal line through $(2, 3)$ is $y = 3$. In general, we have the following.

HORIZONTAL LINE

An equation of the *horizontal* line through (a, b) is

$$\boldsymbol{y = b}.$$

It turns out that every straight line is the graph of an equation that can be put in the form

$$Ax + By + C = 0,$$

where A, B, and C are constants and A and B are not both zero. This is called a **general linear equation in *x* and *y***. We say that x and y are **linearly related**. For example, we can get a general linear equation for the line $y = 7x - 2$ by getting one side of this equation to be 0.

$$y = 7x - 2,$$

$$-7x + y + 2 = 0,$$

$$(-7)x + (1)y + (2) = 0 \qquad [A = -7,\ B = 1,\ C = 2].$$

On the other hand, the graph of every general linear equation is a straight line. For example, by solving the equation

$$3x + 4y - 5 = 0$$

for y, we can get a slope-intercept form:

$$3x + 4y - 5 = 0,$$

$$4y = -3x + 5,$$

$$y = -\frac{3}{4}x + \frac{5}{4}.$$

This is the slope-intercept form of a line with slope $-\frac{3}{4}$ and y-intercept $\frac{5}{4}$.

EXAMPLE 3

Find the slope of the line $16x - 8y + 3 = 0$.

The equation is in general linear form. We solve it for y to get the slope-intercept form.

$$16x - 8y + 3 = 0,$$
$$-8y = -16x - 3,$$
$$y = 2x + \frac{3}{8}.$$

The slope is 2.

EXAMPLE 4

Sketch the graph of $2x - 3y + 6 = 0$.

Method 1.

Since this is a general linear equation, its graph is a straight line. All we have to do is get two points on the graph. To make things easy, let $x = 0$. Then $y = 2$. This gives the y-intercept $(0, 2)$. Now, let $y = 0$. Then $x = -3$. Thus $(-3, 0)$ lies on the line. The graph is given in Fig. 9-23(a). The point $(-3, 0)$ is the point where the line intersects the x-axis and is called the **x-intercept** of the line.

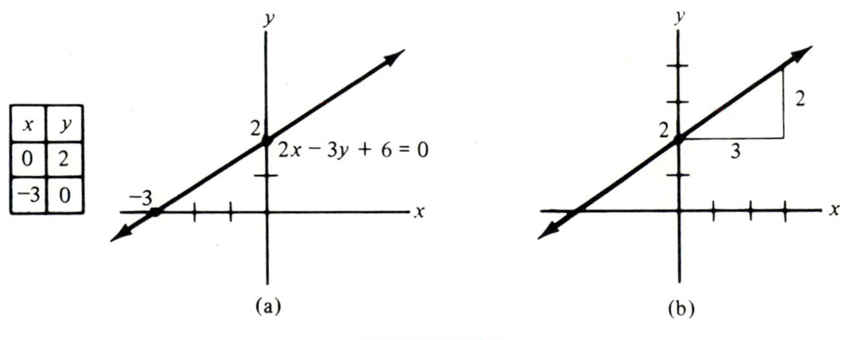

FIGURE 9-23

Method 2.

Here we use the slope and y-intercept to sketch the line. First we solve for y to get the slope-intercept form.

$$2x - 3y + 6 = 0,$$
$$-3y = -2x - 6,$$
$$y = \frac{2}{3}x + 2.$$

Hence the line intersects the y-axis at 2 and has slope $\frac{2}{3}$. Thus as x increases by 3 units, y increases by 2 units. Using this information, we sketch the graph as in Fig. 9-23(b).

EXAMPLE 5

Fahrenheit temperature F and Celsius temperature C are linearly related. Use the facts that 32°F = 0°C and 212°F = 100°C to find an equation that relates F and C. (Express C in terms of F.) Also, find C when F = 50.

Since F and C are linearly related, the graph of the equation is a straight line. In Fig. 9-24 we used F for the horizontal axis and C for the vertical. (We could have just as well reversed our choices.) When $F = 32$, then $C = 0$, so $(32, 0)$ is on the line. Likewise $(212, 100)$ is on it.

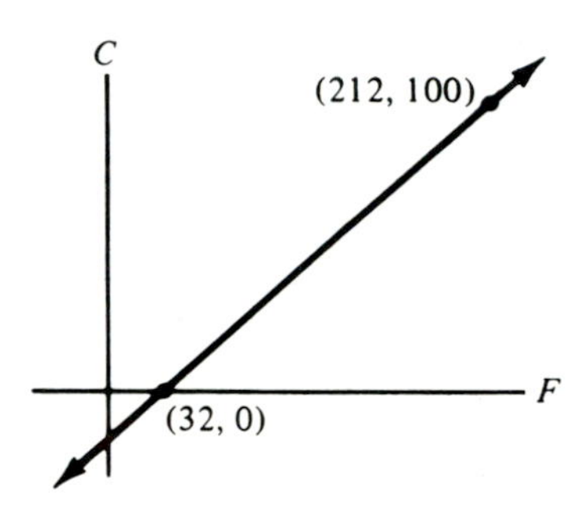

FIGURE 9-24

The slope m of the line is given by

$$m = \frac{C_2 - C_1}{F_2 - F_1} = \frac{100 - 0}{212 - 32} = \frac{100}{180} = \frac{5}{9}.$$

Using the point-slope form with the point $(32, 0)$, we get

$$C - C_1 = m(F - F_1),$$

$$C - 0 = \frac{5}{9}(F - 32),$$

$$C = \frac{5}{9}(F - 32).$$

Here C is in terms of F. We can use this equation to find C when $F = 50$.

$$C = \frac{5}{9}(F - 32) = \frac{5}{9}(50 - 32) = \frac{5}{9}(18) = 10.$$

Table 9-2 gives a summary of forms of equations of straight lines.

TABLE 9-2

FORMS OF EQUATIONS OF LINES	
Point-slope form	$y - y_1 = m(x - x_1)$.
Slope-intercept form	$y = mx + b$.
Vertical line	$x = a$.
Horizontal line	$y = b$.
General linear form	$Ax + By + C = 0$.

Exercise 9.4

In Problems **1–8**, *find an equation of the line that has the given properties.*

1. Has slope 2 and y-intercept 4.

2. Has slope 7 and y-intercept -5.

3. Has slope $-\frac{1}{2}$ and y-intercept -3.

4. Has slope 0 and y-intercept $-\frac{1}{2}$.

5. Is horizontal and passes through $(-3, -2)$.

6. Is vertical and passes through $(-1, 4)$.

7. Passes through $(2, -3)$ and is vertical.

8. Passes through the origin and is horizontal.

In Problems **9–18**, *write each line in slope-intercept form and find the slope and y-intercept. Sketch each line.*

9. $y = 2x - 1$.

10. $y = -3x + 2$.

11. $2y = -6x + 4$.

12. $y = x$.

13. $y - 4x = 0$.

14. $x = 2y + 1$.

15. $x + 3y + 3 = 0$.

16. $3x - 1 = 5y$.

17. $y = 1$.

18. $y - 2 = 3(x - 1)$.

In Problems **19–22**, *find a general linear form for the given line.*

19. $y = x - 6$.

20. $4x = 5 - 9y$.

21. $x = -2y + 2$.

22. $\frac{x}{2} + 3 = \frac{y}{3}$.

23. A straight line has slope 2 and y-intercept 1. Does the point $(-1, -1)$ lie on the line?

24. Graph the lines $y = x + k$ for $k = -1$, 0, and 1.

25. Suppose that s and t are linearly related such that $s = 40$ when $t = 12$, and $s = 25$ when $t = 18$. Find an equation that gives s in terms of t. Also, find s when $t = 24$.

26. The force F suspended on a spring and the stretch S of the spring that the force produces are linearly related. When $F = 1$ (pound), then $S = .4$ (inch). When $F = 4$, then $S = 1.6$. Find an equation that gives S in terms of F. Also, find the stretch when a force of 2 lb is suspended.

27. Suppose that the total cost to produce 10 units of a certain product is \$40, and for 20 units the total cost is \$70. Assume that total cost c is linearly related to the number q of units produced. Find an equation that gives c in terms of q. Also, find the total cost of producing 35 units.

28. In a circuit the voltage V and current i are linearly related. When $i = 4$ (amperes), then $V = 2$ (volts); when $i = 12$, then $V = 6$. Find an equation that gives V in terms of i. Also, find the voltage when the current is 10.

29. In studies of production, an *isocost line* is a line whose points represent all combinations of two factors of production that can be bought for the same amount of money. Suppose a farmer has \$20,000 and wants to use it all to buy x tons of fertilizer (costing \$200 per ton) and y acres of land (costing \$2000 per acre). Here, x and y cannot be negative. Find an equation of the isocost line that describes the different combinations that can be bought for \$20,000.

30. A cancer patient is to receive drug and radiation therapies. Each cubic centimeter of the drug to be used contains 200 curative units, and each minute of radiation exposure gives 300 curative units. The patient requires 2400 curative units. If d cubic centimeters of the drug and r minutes of radiation are administered, determine an equation relating d and r. Graph the equation for $d \geq 0$ and $r \geq 0$; label the horizontal axis as d.

31. For sheep maintained at high environmental temperatures, respiratory rate r (per minute) increases as wool length l (in centimeters) decreases.* Suppose sheep with a wool length of 2 cm have an (average) respiratory rate of 160, and those with a wool length of 4 cm have a respiratory rate of 125. Assume that r and l are linearly related. (a) Find an equation that gives r in terms of l. (b) Find the respiratory rate of sheep with a wool length of 1 cm.

32. Biologists have found that the number of chirps made per minute by crickets of a certain species is related to the temperature. The relationship is very close to being linear. At 68°F, those crickets chirp about 124 times a minute. At 80°F, they chirp about 172 times a minute. (a) Find an equation that gives Fahrenheit temperature t in terms of the number of chirps c per minute. (b) If you count chirps for only 15 s, how can you quickly estimate the temperature?

33. In a certain learning experiment involving repetition and memory†, the proportion p of items recalled was estimated to be linearly related to effective study time t (in seconds), where t is between 5 and 9. For an effective study time of five seconds, the proportion of items recalled was .32. For each one-second increase in study time, the

---

* Adapted from G. E. Folk, Jr., *Textbook of Environmental Physiology*, 2nd ed. (Philadelphia: Lea and Febiger, 1974).

† "Repetition and Learning," ed. G. H. Bower, *The Psychology of Learning and Motivation* (New York: Academic Press, 1976), 10:77.

proportion recalled increased by .059. (a) Find an equation that gives p in terms of t. (b) What proportion of items was recalled with nine seconds of effective study time?

34. The value of a piece of machinery decreases each year by 10% of its original value. If the original value is $8000, find an equation that gives the value V of the machinery in terms of the number of years t after the purchase. Here t must be between 0 and 10. Graph the equation. (Choose t for the horizontal axis and V for the vertical.) What is the slope of the resulting line? This method of considering the value of equipment is called *straight-line depreciation.*

35. The results of testing an experimental diet for pigs suggested that the (average) live weight w (in kilograms) of a pig was linearly related to the number of days d after the diet began. (Here d is between 0 and 100, inclusive.) Suppose the weight of a pig beginning the diet was 20 kg and the pig gained 6.6 kg every 10 days. (a) Find an equation that gives w in terms of d. (b) Find the weight of the pig 50 days after the diet began.

36. For reasons of comparison, a teacher wants to rescale the scores on a set of test papers so that the maximum score is still 100 but the mean (average) is 80 instead of 56. (a) Find a linear equation that will do this. (*Hint*: You want 56 to become an 80 and 100 to remain 100. Consider the points (56, 80) and (100, 100) and, more generally, (x, y) where x is the old score and y is the new score. Find the slope and use a point-slope form. Express y in terms of x.) (b) If 60 on the new scale is the lowest passing score, what was the lowest passing score on the original scale?

9.5 PARALLEL AND PERPENDICULAR LINES

You can tell when two lines are parallel by comparing their slopes.

> Two lines with slopes m_1 and m_2 are **parallel** if, and only if, $m_1 = m_2$.

EXAMPLE 1

Show that the lines $y = 2x - 3$ and $y = 2x + 2$ are parallel.

Both lines have a slope of 2, so they are parallel. See Fig. 9-25.

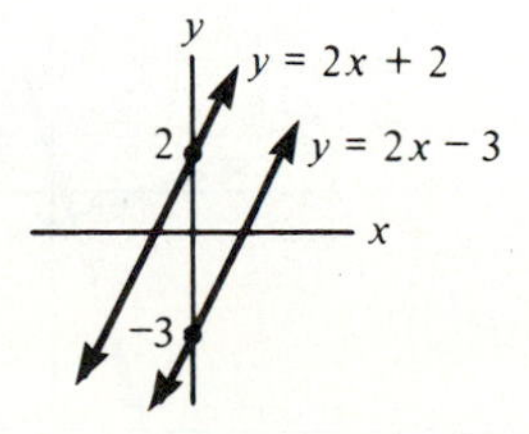

FIGURE 9-25

Here's a rule for perpendicular lines.

> Two lines with slopes m_1 and m_2 are **perpendicular** if, and only if, $m_1 = -\frac{1}{m_2}$.

Thus two lines are perpendicular when the slope of one line is the negative reciprocal of the slope of the other.

EXAMPLE 2

Show that the line $y = \frac{1}{2}x + 3$ is perpendicular to the line $y = -2x + 7$.

The line $y = \frac{1}{2}x + 3$ has slope $m_1 = \frac{1}{2}$, while $y = -2x + 7$ has slope $m_2 = -2$. Now, $\frac{1}{2}$ is the negative reciprocal of -2 (that is, $\frac{1}{2} = -\frac{1}{-2}$). Thus the lines are perpendicular. See Fig. 9-26.

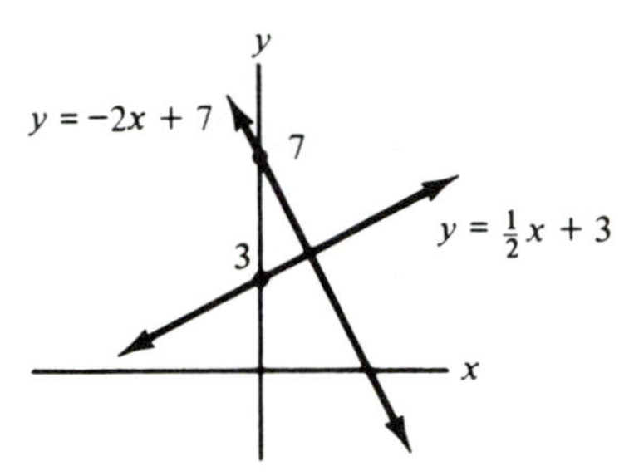

FIGURE 9-26

EXAMPLE 3

Find an equation of the line that passes through $(3, -2)$ and that is

a. *parallel*
b. *perpendicular*

to the line $y = 3x + 1$. See Fig. 9-27.

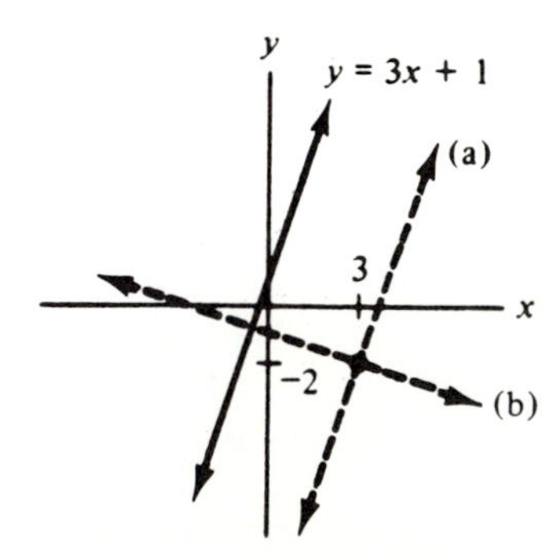

FIGURE 9-27

a. The slope of $y = 3x + 1$ is 3. Thus the line through $(3, -2)$ that is *parallel* to $y = 3x + 1$ also has slope 3. Using the point-slope form, we get

$$y - (-2) = 3(x - 3),$$
$$y + 2 = 3x - 9,$$
$$y = 3x - 11.$$

b. The slope of a line *perpendicular* to $y = 3x + 1$ must be $-\frac{1}{3}$ (= the negative reciprocal of 3). Using the point-slope form, we get

$$y - (-2) = -\tfrac{1}{3}(x - 3),$$
$$y + 2 = -\tfrac{1}{3}x + 1,$$
$$y = -\tfrac{1}{3}x - 1.$$

For the two rules concerning parallel and perpendicular lines, we assumed that each line had a slope. But what can be said about vertical lines (their slopes are undefined)?

Two vertical lines are always parallel.

Any vertical line and any horizontal line are perpendicular.

EXAMPLE 4

a. The lines $x = 2$ and $x = 3$ are both vertical. Thus they are parallel. See Fig. 9-28(a).

b. The line $x = 2$ is vertical and the line $y = 3$ is horizontal. Thus the lines are perpendicular. See Fig. 9-28(b).

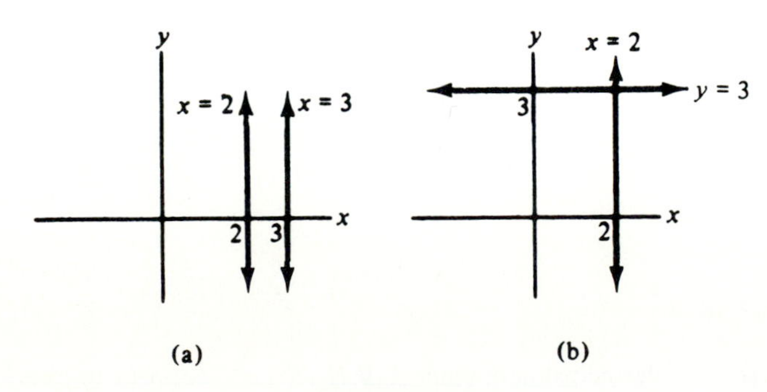

FIGURE 9-28

Exercise 9.5

In Problems **1–10**, *determine if the lines are parallel, perpendicular, or neither.*

1. $y = 7x + 2, \quad y = 7x - 3.$

2. $y = 4x + 3, \quad y = 5 + 4x.$

3. $y = 5x + 2, \quad -5x + y - 3 = 0.$

4. $y = x, \quad y = -x.$

5. $x + 2y + 1 = 0, \quad y = 2x.$

6. $x + 2y = 0, \quad x + y - 4 = 0.$

7. $y = 3, \quad y = -\frac{1}{3}.$

8. $x = 3, \quad x = -4.$

9. $3x + y = 4, \quad 3x - y + 1 = 0.$

10. $x - 1 = 0, \quad y = 0.$

In Problems **11–20**, *find an equation of the line satisfying the given conditions. Give the answer in slope-intercept form if possible.*

11. Passing through $(-1, 3)$ and parallel to $y = 4x - 5$.

12. Passing through $(2, -8)$ and parallel to $x = -4$.

13. Passing through $(2, 1)$ and parallel to $y = 2$.

14. Passing through $(3, -4)$ and parallel to $y = 3 + 2x$.

15. Perpendicular to $y = 3x - 5$ and passing through $(3, 4)$.

16. Perpendicular to $y = -4$ and passing through $(1, 1)$.

17. Passing through $(7, 4)$ and perpendicular to $y = -4$.

18. Passing through $(-5, 4)$ and perpendicular to $2y = -x + 1$.

19. Passing through $(-7, -5)$ and parallel to $2x + 3y + 6 = 0$.

20. Passing through $(-2, 1)$ and parallel to the y-axis.

21. A coordinate map of a college campus gives the coordinates (x, y) of three major buildings as follows: computation center, $(3.5, -1)$; engineering lab, $(.5, 0)$; and library, $(-1, -4.5)$. Find the equations (in slope-intercept form) of the straight-line paths connecting (a) the engineering lab with the computation center, and (b) the engineering lab with the library. Show that these two paths are perpendicular.

9.6 REVIEW

IMPORTANT TERMS

rectangular coordinate plane *(9.1)*

coordinate axes *(9.1)*

origin *(9.1)*

ordered pair *(9.1)*

first coordinate *(9.1)*

second coordinate *(9.1)*

abscissa *(9.1)*
ordinate *(9.1)*
quadrant *(9.1)*
graph of equation *(9.2)*
slope *(9.3)*
point-slope form *(9.3)*
y-intercept *(9.4)*
x-intercept *(9.4)*
slope-intercept form *(9.4)*
linearly related *(9.4)*
general linear equation *(9.4)*
parallel lines *(9.5)*
perpendicular lines *(9.5)*

REVIEW PROBLEMS

In Problems **1–8**, *graph the given equation.*

1. $y = 5x - 4$.

2. $y = -2x + 4$.

3. $x = 3$.

4. $y = 5$.

5. $x = y^2$.

6. $2x + 3y = 12$.

7. $y = 4 - x^2$.

8. $x^2 + y^2 = 36$.

In Problems **9–12**, *find the slope of the line passing through the given points.*

9. $(2, 3)$, $(-1, 4)$.

10. $(1, -1)$, $(2, 3)$.

11. $(2, 1)$, $(5, 1)$.

12. $(-2, 3)$, $(3, -2)$.

In Problems **13–26**, *find an equation of the line satisfying the given conditions. Give the answer in slope-intercept form if possible.*

13. Passes through $(2, -3)$ and has slope -2.

14. Passes through $(-6, 2)$ and has slope $\frac{1}{3}$.

15. Passes through $(-2, 3)$ and $(4, 5)$.

16. Passes through $(1, -1)$ and $(3, 0)$.

17. Passes through $(-2, 2)$ and has slope 0.

18. Passes through $(1, 2)$ and is vertical.

19. Passes through the origin and is vertical.

20. Passes through $(4, -1)$ and is parallel to the x-axis.

21. Has slope 3 and y-intercept -4.

22. Has slope -1 and passes through $(0, 2)$.

23. Passes through $(1, 2)$ and is perpendicular to $-3y + 5x = 7$.

24. Passes through $(-2, 4)$ and is horizontal.

25. Parallel to $y = 3 - 5x$ and passes through $(1, 2)$.

26. Has slope 1 and passes through $(1, 0)$.

In Problems **27–32**, *determine if the lines are parallel, perpendicular, or neither.*

27. $x + 4y + 2 = 0, \quad 8x - 2y - 2 = 0.$

28. $y - 2 = 2(x - 1), \quad 2x + 4y - 3 = 0.$

29. $x - 3 = 2(y + 4), \quad y = 4x + 2.$

30. $3x + 5y + 4 = 0, \quad 6x + 10y = 0.$

31. $y = \frac{1}{2}x + 5, \quad 2x = 4y - 3.$

32. $y = 7x, \quad y = 7.$

In Problems **33** *and* **34**, *write the given line in slope-intercept form and a general linear form. Find the slope.*

33. $3x - 2y = 4.$

34. $x = -3y + 4.$

35. Suppose a and b are linearly related so that $a = 1$ when $b = 2$, and $a = 2$ when $b = 1$. Find a general linear form of an equation that relates a and b. Also, find a when $b = 3$.

36. When the temperature T (in degrees Celsius) of a cat is reduced, the cat's heart rate r (in beats per minute) decreases. Under laboratory conditions, a cat at a temperature of 37°C had a heart rate of 220, and at a temperature of 32°C its heart rate was 150. If r is linearly related to T, where T is between 26 and 38, (a) determine an equation for r in terms of T, and (b) determine the heart rate at a temperature of 28°C.

10 Systems of Equations

10.1 METHODS OF ELIMINATION

When a situation must be described mathematically, it is not unusual for a *set* of equations to arise. For example, suppose that the manager of a factory is setting up a production schedule for two models of a new product. The first model requires 4 widgets and 9 klunkers. The second requires 5 widgets and 14 klunkers. From its suppliers, the factory gets 335 widgets and 850 klunkers each day. How many of each model should the manager plan to make each day so that all the widgets and klunkers are used?

It's a good idea to construct a table that summarizes the important information. Table 10-1 shows the number of widgets and klunkers required for each model, as well as the total number available.

TABLE 10-1

	First Model	Second Model	Total Available
Widgets	4	5	335
Klunkers	9	14	850

Suppose we let x be the number of first models made each day and y be the number of second models. Then these require a total of $4x + 5y$ widgets and $9x + 14y$ klunkers. Since 335 widgets and 850 klunkers are available, we have

$$\begin{cases} 4x + 5y = 335, & (1) \\ 9x + 14y = 850. & (2) \end{cases}$$

We call this set of equations a **system** of two linear equations in the variables (or unknowns) x and y. The problem is to find values of x and y for which *both* equations are true *simultaneously*. These are called *solutions* of the system.

Since Eqs. 1 and 2 are linear, their graphs are straight lines, call them L_1 and L_2. Now, the coordinates of any point on a line satisfy the equation of that line; that is, they make the equation true. Thus the coordinates of any point of intersection of L_1 and L_2 will satisfy both equations. This means that a point of intersection gives a solution of the system.

If L_1 and L_2 are drawn on the same plane, they will appear in one of three ways.

1. L_1 and L_2 may meet at exactly one point, say (x_0, y_0). See Fig. 10-1(a). Thus the system has the solution $x = x_0$ and $y = y_0$.

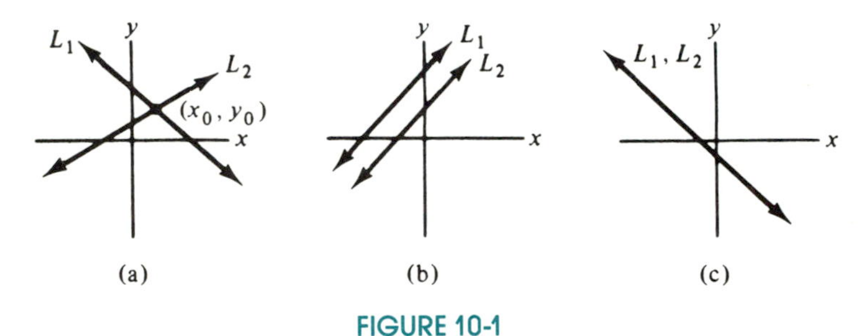

FIGURE 10-1

2. L_1 and L_2 may be parallel and have no points in common. See Fig. 10-1(b). Thus there is no solution.
3. L_1 and L_2 may be the same line. See Fig. 10-1(c). Thus the coordinates of any point on the line are a solution of the system. Consequently there are infinitely many solutions.

We can solve a system with algebra, as well as by graphing. With algebra we use operations to obtain an equation with only one variable in it. That is, one of the variables is *eliminated*. After solving this equation, we can easily get the value of the other variable.

Let's solve our widget and klunker problem by eliminating y from the system

$$\begin{cases} 4x + 5y = 335, & (1) \\ 9x + 14y = 850. & (2) \end{cases}$$

One way involves getting the coefficients of the y-terms in each equation to be the same except for sign. We can multiply Eq. 1 by 14 (that is, multiply both sides of Eq. 1 by 14), and multiply Eq. 2 by -5. This gives

$$\begin{cases} 56x + 70y = 4690, & (3) \\ -45x - 70y = -4250. & (4) \end{cases}$$

The left and right sides of Eq. 3 are equal, so each can be *added* to the corresponding side of Eq. 4. This gives

$$11x = 440,$$

which has only one variable, as planned. Solving this gives

$$x = 40.$$

To find y, we'll replace x by 40 in either one of the *original* equations, such as Eq. 1, and then solve for y.

$$\begin{aligned} 4(40) + 5y &= 335, \\ 160 + 5y &= 335, \\ 5y &= 175, \\ y &= \frac{175}{5} = 35. \end{aligned}$$

We can check our answer by substituting $x = 40$ and $y = 35$ into *both* of the original equations. In Eq. 1 we get $4(40) + 5(35) = 335$, or $335 = 335$. In Eq. 2 we get $9(40) + 14(35) = 850$, or $850 = 850$. Thus the solution is

$$x = 40 \quad \text{and} \quad y = 35.$$

Each day the manager should plan to make 40 of the first model and 35 of the second.

The method we used is called **elimination by addition**. Although we chose to eliminate y, we could eliminate x as follows. Multiplying Eq. 1 by 9 and Eq. 2 by -4 gives

$$\begin{cases} 36x + 45y = 3015, & (5) \\ -36x - 56y = -3400, & (6) \end{cases}$$

Adding Eq. 5 to Eq. 6 (that is, adding each side of Eq. 5 to the corresponding side of Eq. 6) gives

$$\begin{aligned} -11y &= -385, \\ y &= 35. \end{aligned}$$

Finally, replacing y in Eq. 1 by 35 gives $x = 40$, as expected.

EXAMPLE 1

Use elimination by addition to solve the system

$$\begin{cases} 3x - 4y = 13, \\ 3y + 2x = 3. \end{cases}$$

Aligning the x- and y-terms for convenience gives

$$\begin{cases} 3x - 4y = 13, & (7) \\ 2x + 3y = 3. & (8) \end{cases}$$

Let's eliminate y. We multiply Eq. 7 by 3 and Eq. 8 by 4.

$$\begin{cases} 9x - 12y = 39, & (9) \\ 8x + 12y = 12. & (10) \end{cases}$$

Adding Eq. 9 to Eq. 10 gives

$$17x = 51,$$

$$\boxed{x = \frac{51}{17} = 3.}$$

Replacing x by 3 in Eq. 7 gives

$$3(3) - 4y = 13,$$

$$-4y = 4,$$

$$\boxed{y = -1.}$$

The solution is $x = 3$ and $y = -1$. Figure 10-2 shows a graph of the system.

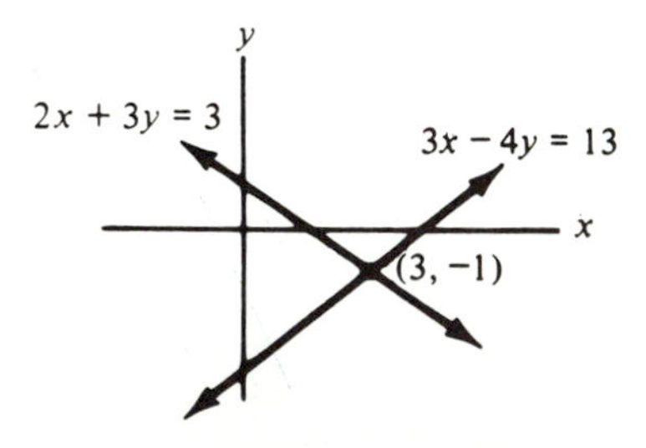

FIGURE 10-2

There's another way to solve the system in Example 1:

$$\begin{cases} 3x - 4y = 13, & (11) \\ 3y + 2x = 3. & (12) \end{cases}$$

We first choose one of the equations and solve it for one variable in terms of the other. Let's solve Eq. 11 for x in terms of y:

$$3x - 4y = 13,$$
$$3x = 13 + 4y,$$
$$x = \frac{13}{3} + \frac{4}{3}y. \qquad (13)$$

Next we *substitute* the right side of Eq. 13 for x in Eq. 12:

$$3y + 2\left(\frac{13}{3} + \frac{4}{3}y\right) = 3. \qquad (14)$$

Thus x has been eliminated. (Note that we did not substitute the right side of Eq. 13 for x in Eq. 11, because Eq. 13 was derived from Eq. 11.) Solving Eq. 14, we have

$$3y + \frac{26}{3} + \frac{8}{3}y = 3,$$
$$9y + 26 + 8y = 9 \quad \text{[clearing of fractions]},$$
$$17y = -17,$$
$$y = -1.$$

Replacing y in Eq. 13 by -1 gives $x = 3$, as before. This method is called **elimination by substitution**.

EXAMPLE 2

Use the method of elimination by substitution to solve

$$\begin{cases} 4x + 2y + 4 = 0, & (15) \\ 2x + \ \ y - 8 = 0. & (16) \end{cases}$$

The variable for which it is easier to solve is y in Eq. 16:

$$y = -2x + 8.$$

Substituting $-2x + 8$ for y in Eq. 15 gives

$$4x + 2(-2x + 8) + 4 = 0,$$
$$4x - 4x + 16 + 4 = 0,$$
$$20 = 0. \qquad (17)$$

Equation 17 is *never* true. When an impossible equation like this occurs, there is **no**

solution to the system. The reason in our case is clear from Fig. 10-3. The graphs of Eqs. 15 and 16 are different parallel lines.

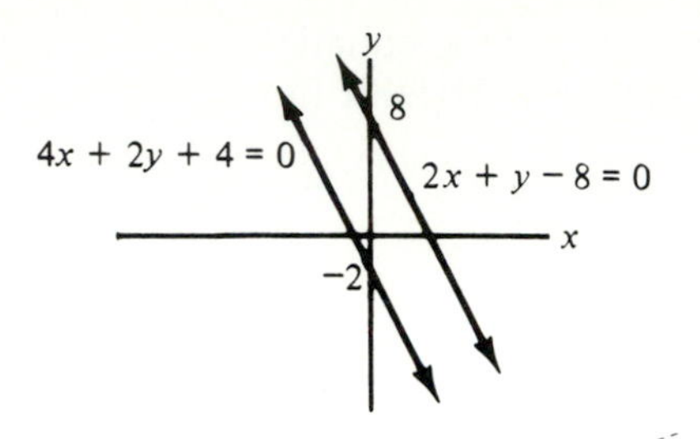

FIGURE 10-3

EXAMPLE 3

Solve the system

$$\begin{cases} 2x + y = 1, & (18) \\ 4x + 2y = 2. & (19) \end{cases}$$

We'll eliminate y by addition. Multiplying Eq. 18 by -2 gives

$$\begin{cases} -4x - 2y = -2, & (20) \\ 4x + 2y = 2. & (21) \end{cases}$$

Adding Eq. 20 to Eq. 21, we have

$$0 = 0,$$

which is *always* true. We might have expected this result, since each term in Eq. 19 is two times the corresponding term in Eq. 18. Equation 19 is said to be a *multiple* of Eq. 18. Because of this, the graphs of Eqs. 18 and 19 are the same line (see Fig.

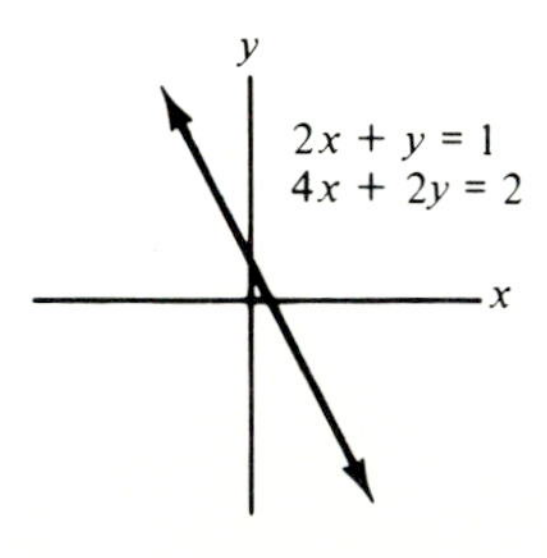

FIGURE 10-4

10-4). Thus any solution to $2x + y = 1$ is a solution to the system. The solution of the system is the coordinates of any point on the line $2x + y = 1$.

Since we can write that line as $y = 1 - 2x$, we can easily find some specific solutions. If $x = 0$, then $y = 1$. Thus $x = 0$, $y = 1$ is one solution. Similarly, some others are $x = 1$, $y = -1$ and $x = -3$, $y = 7$.

EXAMPLE 4

In a laboratory a student must combine a 25% hydrogen peroxide solution (25% by volume is hydrogen peroxide) with a 40% hydrogen peroxide solution to get 2 liters of a 30% solution. How many liters of each solution should the student mix?

Let x be the number of liters of the 25% solution and y be the number of liters of the 40% solution that should be mixed. Then

$$x + y = 2. \tag{22}$$

See Fig. 10-5. In 2 liters of a 30% solution, there is $.30(2) = .6$ liter of hydrogen

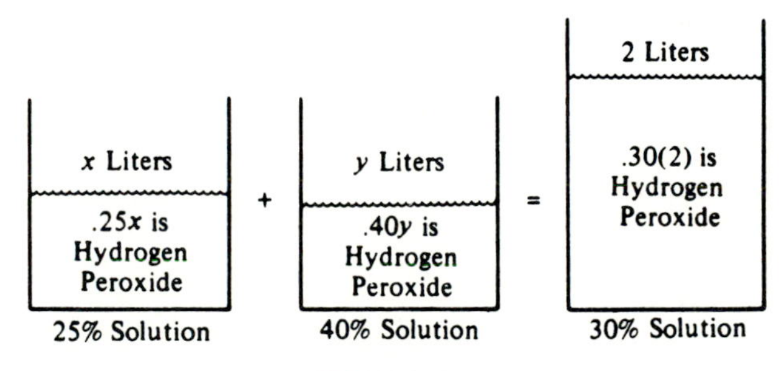

FIGURE 10-5

peroxide. This hydrogen peroxide comes from two places: $.25x$ liter of it comes from the 25% solution and $.40y$ liter comes from the 40% solution. Thus

$$.25x + .40y = .6. \tag{23}$$

Equations 22 and 23 form a system. Solving Eq. 22 for x gives $x = 2 - y$. Substituting $2 - y$ for x in Eq. 23 gives

$$.25(2 - y) + .40y = .6,$$

$$.5 - .25y + .40y = .6 \quad \text{[distributive law]},$$

$$.15y = .1,$$

$$y = \frac{.1}{.15} = \frac{10}{15} = \frac{2}{3}.$$

From Eq. 22, $x = 2 - y = 2 - \frac{2}{3} = \frac{4}{3}$. Thus $\frac{4}{3}$ liters of the 25% solution and $\frac{2}{3}$ liter of the 40% solution must be mixed.

Exercise 10.1

In Problems **1–4**, *use elimination by addition to solve the system.*

1. $\begin{cases} x + 2y = 1, \\ 3x + y = -2. \end{cases}$

2. $\begin{cases} 2x + 3y = -10, \\ 3x - 2y = -2. \end{cases}$

3. $\begin{cases} 4x - 3y = 6, \\ 3x + 2y = 13. \end{cases}$

4. $\begin{cases} 5x + 3y = 10, \\ 3x + 4y = 6. \end{cases}$

In Problems **5–8**, *use elimination by substitution to solve the system.*

5. $\begin{cases} 4x + y = 6, \\ 3x + 2y = 2. \end{cases}$

6. $\begin{cases} x + 5y = -14, \\ -2x - 7y = 16. \end{cases}$

7. $\begin{cases} 5x + 7y = 13, \\ -x + 5y = 7. \end{cases}$

8. $\begin{cases} 4x - 3y = 13, \\ 3x + y = 0. \end{cases}$

In Problems **9–20**, *solve the system by addition or substitution.*

9. $\begin{cases} 3x + y = 7, \\ 2x + 2y = -2. \end{cases}$

10. $\begin{cases} 2x - y = -11, \\ y + 5x = -7. \end{cases}$

11. $\begin{cases} 3x - 4y - 13 = 0, \\ 2x + 3y - 3 = 0. \end{cases}$

12. $\begin{cases} 5x - 3y = 2, \\ -10x + 6y = 4. \end{cases}$

13. $\begin{cases} 2q = 36 - 5p, \\ 8p = 3q - 54. \end{cases}$

14. $\begin{cases} x = 3 - y, \\ 3x + 2y = 19. \end{cases}$

15. $\begin{cases} 4x + 12y = 6, \\ 2x + 6y = 3. \end{cases}$

16. $\begin{cases} 2u - v - 1 = 0, \\ -u + 2v - 7 = 0. \end{cases}$

17. $\begin{cases} 2x + y = 4, \\ 10y - 41 = -20x. \end{cases}$

18. $\begin{cases} 3y = 4x - 2, \\ y + \frac{2}{3} = \frac{4}{3}x. \end{cases}$

19. $\begin{cases} \frac{2}{3}x + \frac{1}{2}y = 2, \\ \frac{3}{8}x + \frac{5}{6}y = -\frac{11}{2}. \end{cases}$

20. $\begin{cases} 3x + 2y - 3 = x, \\ y - 3x = -2. \end{cases}$

21. A chemical manufacturer wishes to fill an order for 700 gal of a 24% acid solution. Solutions of 20% and 30% are in stock. How many gallons of each must be mixed to fill the order?

22. A chemical manufacturer wishes to obtain 500 gal of a 25% acid solution by mixing a 30% solution with an 18% solution. How many gallons of each should be mixed?

23. Universal Control Co. makes industrial control units. Their new models are the Argon I and the Argon II. To make each Argon I unit, they use 6 doodles and 3 skeeters. To make each Argon II unit, they use 10 doodles and 8 skeeters. The company receives a total of 760 doodles and 500 skeeters each day from its supplier. How many units of each model of the Argon can the company make each day? Assume that all the parts are used.

24. On a trip on a raft, it took $\frac{3}{4}$ h to travel 12 mi downstream. The return trip took $1\frac{1}{2}$ h. Find the speed of the raft in still water and the speed of the current.

25. An airplane travels 900 mi in 3 h with the aid of a tail wind. It takes 3 h 36 min for the return trip flying against the same wind. Find the speed of the airplane in still air and the speed of the wind.

26. A private parcel service charges a certain rate per pound for the first 5 lb of a package's weight. For each additional pound it charges another rate. An 8-lb package will be delivered for $3.55. A 15-lb package will be delivered for $6.00. Find the two rates.

27. A company that manufactures dining room sets produces two styles: early American and contemporary. The company knows that 20% more early American styles than contemporary styles can be sold. The profit on each early American sold is $250. On each contemporary sold, the profit is $350. This year the company wants a total profit of $130,000. How many units of each style must be sold?

28. A 10,000-gal railroad tank car is to be filled with solvent from two storage tanks, A and B. Solvent from A is pumped at the rate of 20 gal/min. Solvent from B is pumped at 30 gal/min. Usually both pumps operate at the same time. However, because of a blown fuse the pump on A is delayed 10 min. How many gallons from each storage tank will be used to fill the car?

29. In news reports, profits of a company this year (T) are often compared with those of last year (L), but actual values of T and L are not always given. This year a company had profits of $20 million more than last year. The profits were up 25%. Determine T and L from these data.

30. A company has taxable income of $312,000. The federal tax is 25% of that portion left after the state tax has been paid. The state tax is 10% of that portion left after the federal tax has been paid. Find the federal and state taxes.

31. United Products Co. manufactures calculators and has plants in the cities of Exton and Whyton. At the Exton plant, fixed costs are $7000 per month, and the manufacturing cost of each calculator is $7.50. At the Whyton plant, fixed costs are $8800 per month, and each calculator costs $6.00 to produce. Next month United Products must produce 1500 calculators. How many must be made at each plant if the total cost at each plant is to be the same?

32. A company pays its salespeople on a basis of a certain percentage of the first $100,000 in sales, plus a certain percentage of any amount over $100,000 in sales. If one salesperson earned $8500 on sales of $175,000 and another salesperson $14,800 on sales of $280,000, find the two percentages.

*A **supply equation** for a product relates the number of units q that manufacturers will sell when the price is p (in dollars) per unit. A **demand equation** for a product relates the number of units q that consumers will buy when the price is p (in dollars) per unit. Suppose you form a system using the supply and demand equations for a product. The values of q and p that* *satisfy the system are given special names. The q-value is the* ***equilibrium quantity****. The p-value is the* ***equilibrium price****. In Problems* **33** *and* **34**, *find the equilibrium quantities and prices.*

33. Supply equation: $3q - 200p = -1800$.

Demand equation: $3q + 100p = 1800$.

34. Supply equation: $p = \frac{3}{100}q + 2$.

Demand equation: $p = -\frac{7}{100}q + 12$.

A ***total revenue equation*** *for a manufacturer of a product relates the total amount of money, y (in dollars), that the manufacturer receives when q units of the product are sold. The manufacturer's* ***total cost equation*** *relates the total amount of money, y (in dollars), that it costs when q units are produced. Suppose you form a system using the total revenue and total cost equations. The q-value of a solution to this system is called the* ***break-even quantity****. It corresponds to the point at which the manufacturer has neither a profit nor a loss. In Problems* **35** *and* **36**, *find the break-even quantities.*

35. Total revenue equation: $y = 14q$.

Total cost equation: $y = \frac{40}{3}q + 1200$.

36. Total revenue equation: $y = 3q$.

Total cost equation: $y = 2q + 4500$.

10.2 SYSTEMS OF LINEAR EQUATIONS IN THREE UNKNOWNS

A linear equation in three unknowns, such as x, y, and z, is an equation that can be put in the form $Ax + By + Cz = D$, where A, B, C, and D are constants.* To solve a system of three linear equations in three unknowns, we use the methods of elimination in Sec. 10.1. For example, we might perform the following steps.

1. Select two *pairs* of the given equations and eliminate the *same* variable from each pair.
2. From Step 1 we have two equations in the same two variables. We solve these by elimination.
3. In Step 2 we found values of two of the variables. We substitute these in one of the original equations to find the value of the third variable.

We shall follow these steps in the next example.

* Also, A, B, and C are not all zero.

EXAMPLE 1

Solve the system

$$\begin{cases} 4x - y - 3z = 1, & (1) \\ 2x + y + 2z = 5, & (2) \\ 8x + y - z = 5. & (3) \end{cases}$$

Step 1.

As our first pair of equations, let's select Eqs. 1 and 2:

$$\begin{cases} 4x - y - 3z = 1, \\ 2x + y + 2z = 5. \end{cases}$$

Adding these to eliminate y gives

$$6x - z = 6. \qquad (4)$$

As our second pair of equations, let's select Eqs. 2 and 3:

$$\begin{cases} 2x + y + 2z = 5, \\ 8x + y - z = 5. \end{cases}$$

We must also eliminate y from these. Multiplying the second equation by -1 gives

$$\begin{cases} 2x + y + 2z = 5, \\ -8x - y + z = -5. \end{cases}$$

Adding these, we have

$$-6x + 3z = 0. \qquad (5)$$

Step 2.

From Step 1 we have two equations in two variables, Eqs. 4 and 5:

$$\begin{cases} 6x - z = 6, & (4) \\ -6x + 3z = 0. & (5) \end{cases}$$

We now solve this system. Adding these equations and solving for z, we obtain

$$2z = 6,$$

$$\boxed{z = 3.}$$

Substituting 3 for z in Eq. 4 gives

$$6x - 3 = 6,$$

$$6x = 9,$$

$$\boxed{x = \frac{9}{6} = \frac{3}{2}.}$$

Step 3.

In Step 2 we found $z = 3$ and $x = \frac{3}{2}$. Substituting these in Eq. 1, we find y.

$$4x - y - 3z = 1,$$

$$4\left(\frac{3}{2}\right) - y - 3(3) = 1,$$

$$-y = 4,$$

$$\boxed{y = -4.}$$

The solution is $x = \frac{3}{2}$, $y = -4$, and $z = 3$.

Exercise 10.2

In Problems **1–8**, *solve the system.*

1. $\begin{cases} x + y + z = 6, \\ x - y + z = 2, \\ 2x - y + 3z = 6. \end{cases}$

2. $\begin{cases} 2x - y + 3z = 12, \\ x + 2y - 3z = -10, \\ x + y - z = -3. \end{cases}$

3. $\begin{cases} x - z = 14, \\ y + z = 21, \\ x - y + z = -10. \end{cases}$

4. $\begin{cases} x + y = -6, \\ z = 4, \\ -x + y + 2z = 16. \end{cases}$

5. $\begin{cases} 2x + y + 6z = 3, \\ x - y + 4z = 1, \\ 3x + 2y - 2z = 2. \end{cases}$

6. $\begin{cases} 5x - 7y + 4z = 2, \\ 3x + 2y - 2z = 3, \\ 2x - y + 3z = 4. \end{cases}$

7. $\begin{cases} 2x - 3y + z = -2, \\ 3x + 3y - z = 2, \\ x - 6y + 3z = -2. \end{cases}$

8. $\begin{cases} x + y + z = -1, \\ 3x + y + z = 1, \\ 4x - 2y + 2z = 0. \end{cases}$

9. A company makes three types of patio furniture: chairs, rockers, and chaise lounges. Each requires wood, plastic, and aluminum, as in Table 10-2. The company has in stock 400 units of wood, 600 units of plastic, and 1500 units of aluminum. For its end-of-season production run, the company wants to use up all the stock. To do this, how many chairs, rockers, and chaise lounges should it make?

TABLE 10-2

	WOOD	PLASTIC	ALUMINUM
Chair	1 unit	1 unit	2 units
Rocker	1 unit	1 unit	3 units
Chaise lounge	1 unit	2 units	5 units

10. Table 10-3 shows how alloys A, B, and C are composed (by weight). How much of A, B, and C must be mixed to produce 100 kg of an alloy that is 53% copper and 19% zinc?

TABLE 10-3

	A	B	C
Copper	50%	60%	40%
Zinc	30%	20%	
Nickel	20%	20%	60%

11. A coffee wholesaler blends together three types of coffee that sell for \$2.20, \$2.30, and \$2.60 per pound to obtain 100 lb of coffee worth \$2.40 per pound. If the wholesaler uses the same amount of the two higher-priced coffees, how much of each type must be used in the blend?

12. A total of \$35,000 was invested at three interest rates: 7%, 8%, and 9%. The interest for the first year was \$2830, which was not reinvested. The second year the amount originally invested at 9% earned 10%, and the other rates remained the same. The total interest the second year was \$2960. How much was invested at each rate?

13. The graph of $y = ax^2 + bx + c$ passes through the points $(2, 0)$, $(0, 0)$, and $(-1, 3)$. Find a, b, and c. (*Hint*: The coordinates of each point must satisfy the equation, so replace x by 2 and y by 0 to get an equation in a, b, and c. Do the same for the other two points. In this way obtain three equations in the unknowns a, b, and c.)

14. Repeat Problem 13 if the graph passes through the points $(2, 5)$, $(-3, 5)$, and $(1, 1)$.

In studies of the electrical circuit shown in Fig. 10-6, *Kirchhoff's laws lead to the following system*:

$$\begin{cases} R_1 i_1 + R_2 i_1 + R_3 i_3 + \varepsilon_1 = 0, \\ R_3 i_3 + R_4 i_2 + \varepsilon_2 = 0, \\ i_1 + i_2 = i_3. \end{cases}$$

Here the R's are resistances (in ohms, Ω*), the i's are currents (in amperes), and the* ε*'s are potential differences (in volts, V). Problems* **15** *and* **16** *relate to this system.*

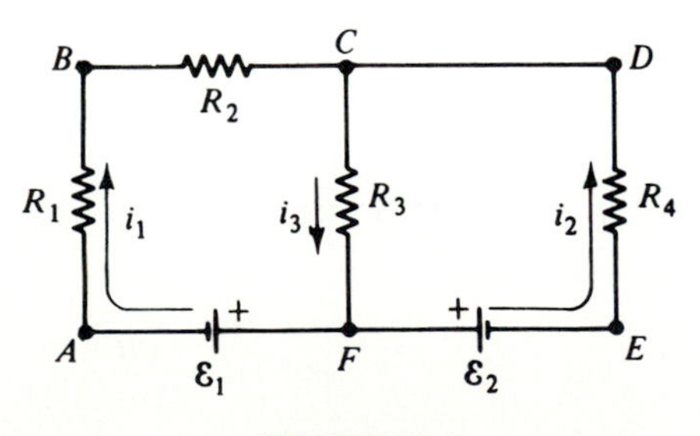

FIGURE 10-6

15. Find i_1, i_2, and i_3 (in amperes) if $R_1 = 2$, $R_2 = 3$, $R_3 = 5$, and $R_4 = 3\ \Omega$ and $\varepsilon_1 = 15$ and $\varepsilon_2 = 2$ V. (It is possible to get negative values of i in your answer.)

16. Find i_1, i_2, and i_3 (in amperes) if $R_1 = 1$, $R_2 = 1$, $R_3 = 4$, and $R_4 = 6\ \Omega$ and $\varepsilon_1 = 2$ and $\varepsilon_2 = 60$ V. (It is possible to get negative values of i in your answer.)

10.3 DETERMINANTS

Besides elimination there is another way to solve systems of linear equations. This method involves *determinants*. In this section you will learn what determinants are and how to evaluate them. In the next section you will use them to solve linear systems.

A **determinant** is simply a number associated with a square array of numbers. We denote it by placing vertical bars before and after the array. For example,

$$\begin{vmatrix} 2 & 1 \\ 5 & 7 \end{vmatrix} \text{ is a determinant.}$$

The numbers inside the bars are called **entries**. The horizontal rows are numbered consecutively, with row 1 at the top. The vertical columns are also numbered consecutively, with column 1 on the left.

$$\begin{array}{r c} & \begin{array}{cc} \text{column 1} & \text{column 2} \\ \downarrow & \downarrow \end{array} \\ \begin{array}{r} \text{row 1}\rightarrow \\ \text{row 2}\rightarrow \end{array} & \begin{vmatrix} 2 & 1 \\ 5 & 7 \end{vmatrix} \end{array}$$

Since there are two rows and two columns, we say that this determinant has **order** 2, or is a 2×2 (read "2 by 2") determinant.

$$\begin{vmatrix} 2 & 1 \\ 5 & 7 \end{vmatrix} \text{ has order 2.}$$

Similarly,

$$\begin{vmatrix} 1 & 0 & -1 \\ 3 & 2 & 4 \\ 2 & 1 & 0 \end{vmatrix} \text{ has order 3,}$$

since it has three rows and three columns.

To evaluate a 2×2 determinant, we use the following rule (which is actually a definition).

$$\begin{vmatrix} a & b \\ c & d \end{vmatrix} = ad - bc.$$

Notice that ad is the product of the entries on the diagonal that goes from the upper left corner to the lower right corner, while bc is the product of the entries on the other diagonal. The value of the determinant is the difference of these products.

$$\begin{vmatrix} a & b \\ c & d \end{vmatrix} = ad - bc.$$

EXAMPLE 1

Evaluating determinants.

a. $\begin{vmatrix} 2 & 1 \\ 5 & 7 \end{vmatrix} = (2)(7) - (1)(5) = 14 - 5 = 9.$

b. $\begin{vmatrix} -6 & -2 \\ 4 & 3 \end{vmatrix} = (-6)(3) - (-2)(4) = -18 + 8 = -10.$

To evaluate determinants of order 3, we use *minors* and *cofactors*. Consider the 3×3 determinant

$$\begin{vmatrix} a_1 & b_1 & c_1 \\ a_2 & b_2 & c_2 \\ a_3 & b_3 & c_3 \end{vmatrix}.$$

With a given entry, we can associate a determinant of order 2. This 2×2 determinant is found by deleting the entries in the row and column in which the given entry lies. For example, with the entry a_2 we delete the entries in row 2 and column 1.

$$\begin{vmatrix} a_1 & b_1 & c_1 \\ a_2 & b_2 & c_2 \\ a_3 & b_3 & c_3 \end{vmatrix}.$$

This leaves the 2×2 determinant

$$\begin{vmatrix} b_1 & c_1 \\ b_3 & c_3 \end{vmatrix},$$

which is called the **minor** of a_2. Similarly, the minor of b_2 is

$$\begin{vmatrix} a_1 & c_1 \\ a_3 & c_3 \end{vmatrix},$$

and for c_2 it is

$$\begin{vmatrix} a_1 & b_1 \\ a_3 & b_3 \end{vmatrix}.$$

With each entry in

$$\begin{vmatrix} a_1 & b_1 & c_1 \\ a_2 & b_2 & c_2 \\ a_3 & b_3 & c_3 \end{vmatrix},$$

we also associate a number that depends on the row and column of the entry. That number is

$$(-1)^{\text{row no.}+\text{column no.}}.$$

For example, since a_2 is in row 2 and column 1, with a_2 we associate the number

$$(-1)^{2+1} = (-1)^3 = -1.$$

With b_2 we associate the number

$$(-1)^{2+2} = (-1)^4 = 1,$$

and with c_2 the number

$$(-1)^{2+3} = (-1)^5 = -1.$$

Notice that these numbers are either 1 or -1.

For an entry in a given row and column of

$$\begin{vmatrix} a_1 & b_1 & c_1 \\ a_2 & b_2 & c_2 \\ a_3 & b_3 & c_3 \end{vmatrix},$$

the **cofactor** of the entry is the product

$$(-1)^{\text{row no.}+\text{column no.}} \text{ (minor of entry)}.$$

Thus the cofactor of a_2 is

$$(-1)^{2+1}\begin{vmatrix} b_1 & c_1 \\ b_3 & c_3 \end{vmatrix},$$

and the cofactor of b_2 is

$$(-1)^{2+2}\begin{vmatrix} a_1 & c_1 \\ a_3 & c_3 \end{vmatrix}.$$

EXAMPLE 2

For $\begin{vmatrix} 2 & 0 & 3 \\ -2 & 2 & 4 \\ -1 & 6 & 5 \end{vmatrix}$, *find the cofactors of* 4 *and* -1.

Since 4 is in row 2 and column 3, its cofactor is

$$(-1)^{2+3}\begin{vmatrix} 2 & 0 \\ -1 & 6 \end{vmatrix} = (-1)^5[(2)(6) - (0)(-1)] = -1 \cdot [12 - 0] = -12.$$

The entry -1 is in row 3 and column 1, so its cofactor is

$$(-1)^{3+1}\begin{vmatrix} 0 & 3 \\ 2 & 4 \end{vmatrix} = (-1)^4[(0)(4)-(3)(2)] = 1\cdot[0-6] = -6.$$

To find the value of a 3×3 determinant, we use the following rule.

> To evaluate a **determinant of order 3**, choose *any* row (or column) and multiply each entry in that row (or column) by its cofactor. The *sum* of these products is the value of the determinant.

EXAMPLE 3

Find the value of $\begin{vmatrix} 2 & -1 & 3 \\ 4 & 0 & -5 \\ -2 & 1 & 1 \end{vmatrix}$.

We shall apply the rule to the first row. (This is called *expanding along the first row.*)

$$\begin{aligned}
&\begin{vmatrix} 2 & -1 & 3 \\ 4 & 0 & -5 \\ -2 & 1 & 1 \end{vmatrix} \\
&= 2\begin{pmatrix}\text{cofactor}\\ \text{of } 2\end{pmatrix} + (-1)\begin{pmatrix}\text{cofactor}\\ \text{of } -1\end{pmatrix} + 3\begin{pmatrix}\text{cofactor}\\ \text{of } 3\end{pmatrix} \\
&= 2(-1)^{1+1}\begin{vmatrix} 0 & -5 \\ 1 & 1 \end{vmatrix} + (-1)(-1)^{1+2}\begin{vmatrix} 4 & -5 \\ -2 & 1 \end{vmatrix} + 3(-1)^{1+3}\begin{vmatrix} 4 & 0 \\ -2 & 1 \end{vmatrix} \\
&= 2(1)[0-(-5)] + (-1)(-1)[4-10] + 3(1)[4-0] \\
&= 2(1)(5) + (-1)(-1)(-6) + 3(1)(4) \\
&= 10 - 6 + 12 = 16.
\end{aligned}$$

If we had expanded along the second column, we would have obtained

$$\begin{aligned}
&\begin{vmatrix} 2 & -1 & 3 \\ 4 & 0 & -5 \\ -2 & 1 & 1 \end{vmatrix} \\
&= (-1)\begin{pmatrix}\text{cofactor}\\ \text{of } -1\end{pmatrix} + 0\begin{pmatrix}\text{cofactor}\\ \text{of } 0\end{pmatrix} + 1\begin{pmatrix}\text{cofactor}\\ \text{of } 1\end{pmatrix} \\
&= (-1)(-1)^{1+2}\begin{vmatrix} 4 & -5 \\ -2 & 1 \end{vmatrix} + 0 + 1(-1)^{3+2}\begin{vmatrix} 2 & 3 \\ 4 & -5 \end{vmatrix} \\
&= (-1)(-1)[4-10] + 0 + 1(-1)[-10-12] \\
&= (-1)(-1)(-6) + 0 + 1(-1)(-22) \\
&= -6 + 0 + 22 = 16,
\end{aligned}$$

as before. The computation is easier here since we did not have to find the cofactor of 0. In general it's a good idea to expand along a row or column with the greatest number of zeros.

EXAMPLE 4

Find $\begin{vmatrix} 1 & 2 & -1 \\ 3 & -6 & 2 \\ 1 & 4 & 1 \end{vmatrix}$.

Expanding along the first row, we have

$$\begin{aligned} &\begin{vmatrix} 1 & 2 & -1 \\ 3 & -6 & 2 \\ 1 & 4 & 1 \end{vmatrix} \\ &= 1(-1)^{1+1}\begin{vmatrix} -6 & 2 \\ 4 & 1 \end{vmatrix} + 2(-1)^{1+2}\begin{vmatrix} 3 & 2 \\ 1 & 1 \end{vmatrix} + (-1)(-1)^{1+3}\begin{vmatrix} 3 & -6 \\ 1 & 4 \end{vmatrix} \\ &= 1(1)[-6-8] + 2(-1)[3-2] + (-1)(1)[12-(-6)] \\ &= 1(1)(-14) + 2(-1)(1) + (-1)(1)(18) \\ &= -14 - 2 - 18 = -34. \end{aligned}$$

Exercise 10.3

In Problems **1–8**, *find the value of the determinant.*

1. $\begin{vmatrix} 2 & 1 \\ 3 & 2 \end{vmatrix}$.

2. $\begin{vmatrix} 1 & 2 \\ 3 & 4 \end{vmatrix}$.

3. $\begin{vmatrix} 2 & -6 \\ 5 & 3 \end{vmatrix}$.

4. $\begin{vmatrix} 3 & 2 \\ -5 & -4 \end{vmatrix}$.

5. $\begin{vmatrix} \frac{1}{2} & \frac{1}{3} \\ \frac{1}{4} & -4 \end{vmatrix}$.

6. $\begin{vmatrix} -2 & -3 \\ -4 & -6 \end{vmatrix}$.

7. $\begin{vmatrix} -3 & 1 \\ -a & b \end{vmatrix}$.

8. $\begin{vmatrix} -2 & -a \\ -a & 2 \end{vmatrix}$.

In Problems **9–16**, *find the indicated value for the determinant*

$$\begin{vmatrix} 1 & 2 & 3 \\ 4 & 5 & 6 \\ 7 & 8 & 9 \end{vmatrix}.$$

9. The minor of 1.

10. The minor of 2.

11. The minor of 6.

12. The minor of 5.

13. The cofactor of 5.

14. The cofactor of 6.

15. The cofactor of 8.

16. The cofactor of 7.

In Problems **17–28**, *find the value of the determinant.*

17. $\begin{vmatrix} 2 & 1 & 3 \\ 2 & 0 & 1 \\ -4 & 0 & 6 \end{vmatrix}$.

18. $\begin{vmatrix} 1 & 0 & -1 \\ 0 & 1 & 0 \\ 1 & -1 & 1 \end{vmatrix}$.

19. $\begin{vmatrix} 2 & 1 & 5 \\ -3 & 4 & -1 \\ 0 & 6 & -1 \end{vmatrix}$.

20. $\begin{vmatrix} 5 & 0 & 8 \\ 6 & 0 & 8 \\ 7 & 0 & 9 \end{vmatrix}$.

21. $\begin{vmatrix} 1 & 0 & 0 \\ 0 & 2 & 0 \\ 0 & 0 & 3 \end{vmatrix}$.

22. $\begin{vmatrix} 1 & 2 & 3 \\ 4 & 5 & 6 \\ 7 & 8 & 9 \end{vmatrix}$.

23. $\begin{vmatrix} 1 & 2 & 3 \\ 4 & 5 & 4 \\ 3 & 2 & 1 \end{vmatrix}$.

24. $\begin{vmatrix} 3 & 2 & 1 \\ 1 & -2 & 3 \\ -1 & 3 & 2 \end{vmatrix}$.

25. $\begin{vmatrix} 2 & -1 & 3 \\ 1 & 2 & -3 \\ 1 & 1 & -1 \end{vmatrix}$.

26. $\begin{vmatrix} 1 & 2 & -3 \\ 4 & 5 & 4 \\ 3 & -2 & 1 \end{vmatrix}$.

27. $\begin{vmatrix} \frac{1}{2} & \frac{2}{3} & -\frac{1}{2} \\ -1 & \frac{1}{3} & \frac{2}{3} \\ 3 & -4 & 1 \end{vmatrix}$.

28. $\begin{vmatrix} -\frac{1}{3} & \frac{1}{4} & 4 \\ \frac{3}{2} & \frac{3}{8} & -2 \\ -\frac{1}{8} & \frac{9}{2} & 1 \end{vmatrix}$.

10.4 CRAMER'S RULE

In this section you will see how determinants are used to solve systems of two linear equations in two unknowns and three linear equations in three unknowns. This method will give us a way to solve for one unknown without having to solve for the others.

To begin, let's use elimination to solve the system

$$\begin{cases} a_1 x + b_1 y = c_1, & (1) \\ a_2 x + b_2 y = c_2. & (2) \end{cases}$$

To eliminate y, we first multiply Eq. 1 by b_2 and Eq. 2 by $-b_1$.

$$\begin{cases} a_1b_2x + b_1b_2y = b_2c_1, \\ -a_2b_1x - b_1b_2y = -b_1c_2. \end{cases}$$

Adding these equations, we have

$$a_1b_2x - a_2b_1x = b_2c_1 - b_1c_2.$$

Now we solve for x.

$$(a_1b_2 - a_2b_1)x = b_2c_1 - b_1c_2 \qquad \text{[factoring]},$$

$$x = \frac{b_2c_1 - b_1c_2}{a_1b_2 - a_2b_1}.$$

Here we assume that the denominator $a_1b_2 - a_2b_1$ is not 0. We can find y by eliminating x from the system. Multiplying Eq. 1 by $-a_2$ and Eq. 2 by a_1 and then adding the resulting equations, we have

$$\begin{cases} -a_1a_2x - a_2b_1y = -a_2c_1, \\ a_1a_2x + a_1b_2y = a_1c_2, \end{cases}$$

$$a_1b_2y - a_2b_1y = a_1c_2 - a_2c_1.$$

Solving for y gives

$$(a_1b_2 - a_2b_1)y = a_1c_2 - a_2c_1 \qquad \text{[factoring]},$$

$$y = \frac{a_1c_2 - a_2c_1}{a_1b_2 - a_2b_1}.$$

Thus

$$x = \frac{b_2c_1 - b_1c_2}{a_1b_2 - a_2b_1} \quad \text{and} \quad y = \frac{a_1c_2 - a_2c_1}{a_1b_2 - a_2b_1}.$$

Notice that both x and y are expressed as fractions with the *same* denominator. Also, the numerators and denominators look like the values of determinants of order 2. In fact, we can write x and y as

$$x = \frac{\begin{vmatrix} c_1 & b_1 \\ c_2 & b_2 \end{vmatrix}}{\begin{vmatrix} a_1 & b_1 \\ a_2 & b_2 \end{vmatrix}} \quad \text{and} \quad y = \frac{\begin{vmatrix} a_1 & c_1 \\ a_2 & c_2 \end{vmatrix}}{\begin{vmatrix} a_1 & b_1 \\ a_2 & b_2 \end{vmatrix}}.$$

Check this out! These expressions give us convenient formulas for solving the original system,

$$\begin{cases} a_1x + b_1y = c_1, \\ a_2x + b_2y = c_2. \end{cases}$$

The determinants in the formulas can be easily remembered. The first column of the determinant in the denominators consists of the coefficients of the x-terms in the system; the second column consists of the coefficients of the y-terms. The numerator for x is the same as its denominator, except that the "x-column" is replaced by the constant terms of the system. Similarly, the numerator for y can be found by replacing the "y-column" of the denominator by the column of constant terms. Usually we use the letter D to stand for the determinants in the denominators.

This method of using determinants to solve a system of linear equations is called *Cramer's rule.*

CRAMER'S RULE

The system

$$\begin{cases} a_1 x + b_1 y = c_1, \\ a_2 x + b_2 y = c_2 \end{cases}$$

has the solution

$$x = \frac{\begin{vmatrix} c_1 & b_1 \\ c_2 & b_2 \end{vmatrix}}{D}, \qquad y = \frac{\begin{vmatrix} a_1 & c_1 \\ a_2 & c_2 \end{vmatrix}}{D},$$

where $D = \begin{vmatrix} a_1 & b_1 \\ a_2 & b_2 \end{vmatrix}$. *We assume that* $D \neq 0$.

EXAMPLE 1

Use Cramer's rule to solve

$$\begin{cases} 2x + y + 5 = 0, \\ \phantom{2x + {}} 3y + x = 6. \end{cases}$$

First, we rewrite the system so that it has the form given in Cramer's rule.

$$\begin{cases} 2x + y = -5, \\ x + 3y = 6. \end{cases}$$

In both equations the x- and y-terms are aligned and the constant terms are on the right sides. Now we evaluate D.

$$D = \begin{vmatrix} 2 & 1 \\ 1 & 3 \end{vmatrix} = (2)(3) - (1)(1) = 6 - 1 = 5.$$

By Cramer's rule,

$$x = \frac{\begin{vmatrix} -5 & 1 \\ 6 & 3 \end{vmatrix}}{D} = \frac{(-5)(3) - (1)(6)}{5} = \frac{-15 - 6}{5} = -\frac{21}{5}$$

and

$$y = \frac{\begin{vmatrix} 2 & -5 \\ 1 & 6 \end{vmatrix}}{D} = \frac{(2)(6) - (-5)(1)}{5} = \frac{12 + 5}{5} = \frac{17}{5}.$$

You cannot use Cramer's rule to solve a system if $D = 0$. In that situation, use the method of elimination. It can be shown that the system will have either no solution or infinitely many solutions.

Cramer's rule can be extended to handle systems of three linear equations in three unknowns.

CRAMER'S RULE

The system

$$\begin{cases} a_1x + b_1y + c_1z = d_1, \\ a_2x + b_2y + c_2z = d_2, \\ a_3x + b_3y + c_3z = d_3 \end{cases}$$

has the solution

$$x = \frac{D_x}{D}, \qquad y = \frac{D_y}{D}, \qquad z = \frac{D_z}{D},$$

where

$$D = \begin{vmatrix} a_1 & b_1 & c_1 \\ a_2 & b_2 & c_2 \\ a_3 & b_3 & c_3 \end{vmatrix}$$

and

$$D_x = \begin{vmatrix} d_1 & b_1 & c_1 \\ d_2 & b_2 & c_2 \\ d_3 & b_3 & c_3 \end{vmatrix}, \qquad D_y = \begin{vmatrix} a_1 & d_1 & c_1 \\ a_2 & d_2 & c_2 \\ a_3 & d_3 & c_3 \end{vmatrix}, \qquad D_z = \begin{vmatrix} a_1 & b_1 & d_1 \\ a_2 & b_2 & d_2 \\ a_3 & b_3 & d_3 \end{vmatrix}.$$

We assume that $D \neq 0$.

Again, there is a pattern for the determinants. For D, notice that columns 1, 2, and 3 consist of the coefficients of the x-, y-, and z-terms, respectively, in the given system. The numerator D_x is found by replacing the x-column of D by the constant terms. The numerators D_y and D_z are found similarly.

EXAMPLE 2

Use Cramer's rule to solve

$$\begin{cases} 2x + y + z = 0, \\ 4x + 3y + 2z = 2, \\ 2x - y - 3z = 0. \end{cases}$$

This system is in the form stated in Cramer's rule. We have

$$D = \begin{vmatrix} 2 & 1 & 1 \\ 4 & 3 & 2 \\ 2 & -1 & -3 \end{vmatrix},$$

$$D_x = \begin{vmatrix} 0 & 1 & 1 \\ 2 & 3 & 2 \\ 0 & -1 & -3 \end{vmatrix}, \qquad D_y = \begin{vmatrix} 2 & 0 & 1 \\ 4 & 2 & 2 \\ 2 & 0 & -3 \end{vmatrix}, \qquad D_z = \begin{vmatrix} 2 & 1 & 0 \\ 4 & 3 & 2 \\ 2 & -1 & 0 \end{vmatrix}.$$

We shall evaluate D by expanding along the first row.

$$\begin{aligned} D &= 2(-1)^{1+1}\begin{vmatrix} 3 & 2 \\ -1 & -3 \end{vmatrix} + (1)(-1)^{1+2}\begin{vmatrix} 4 & 2 \\ 2 & -3 \end{vmatrix} + (1)(-1)^{1+3}\begin{vmatrix} 4 & 3 \\ 2 & -1 \end{vmatrix} \\ &= 2(1)(-7) + (1)(-1)(-16) + (1)(1)(-10) \\ &= -14 + 16 - 10 = -8. \end{aligned}$$

For D_x we expand along column 1.

$$\begin{aligned} D_x &= 0 + 2(-1)^{2+1}\begin{vmatrix} 1 & 1 \\ -1 & -3 \end{vmatrix} + 0 \\ &= 0 + 2(-1)(-2) + 0 = 4. \end{aligned}$$

For D_y we expand along column 2.

$$\begin{aligned} D_y &= 0 + 2(-1)^{2+2}\begin{vmatrix} 2 & 1 \\ 2 & -3 \end{vmatrix} + 0 \\ &= 0 + 2(1)(-8) + 0 = -16. \end{aligned}$$

For D_z we expand along column 3.

$$\begin{aligned} D_z &= 0 + 2(-1)^{2+3}\begin{vmatrix} 2 & 1 \\ 2 & -1 \end{vmatrix} + 0 \\ &= 0 + 2(-1)(-4) + 0 = 8. \end{aligned}$$

Thus

$$x = \frac{D_x}{D} = \frac{4}{-8} = -\frac{1}{2},$$

$$y = \frac{D_y}{D} = \frac{-16}{-8} = 2,$$

$$z = \frac{D_z}{D} = \frac{8}{-8} = -1.$$

Exercise 10.4

In Problems **1–18**, *use Cramer's rule to solve the system.*

1. $\begin{cases} 2x + 3y = 5, \\ x - 2y = -1. \end{cases}$

2. $\begin{cases} x - 3y = -11, \\ 4x + 3y = 9. \end{cases}$

3. $\begin{cases} 2x - y = 4, \\ 3x + y = 5. \end{cases}$

4. $\begin{cases} 3x + y = 6, \\ 7x - 2y = 5. \end{cases}$

5. $\begin{cases} -2x = 4 - 3y, \\ y = 6x - 1. \end{cases}$

6. $\begin{cases} x + 2y - 6 = 0, \\ y - 1 = 3x. \end{cases}$

7. $\begin{cases} 3(x+2)=5, \\ 6(x+y)=-8. \end{cases}$

8. $\begin{cases} w-2z=4, \\ 3w-4z=6. \end{cases}$

9. $\begin{cases} \frac{3}{2}x-\frac{1}{4}z=1, \\ \frac{1}{3}x+\frac{1}{2}z=2. \end{cases}$

10. $\begin{cases} .6x-.7y=.33, \\ 2.1x-.9y=.69. \end{cases}$

11. $\begin{cases} x+y+z=6, \\ x-y+z=2, \\ 2x-y+3z=6. \end{cases}$

12. $\begin{cases} 2x-y+3z=12, \\ x+y-z=-3, \\ x+2y-3z=-10. \end{cases}$

13. $\begin{cases} x+y-z=6, \\ 2x-3y-2z=2, \\ x-y-5z=18. \end{cases}$

14. $\begin{cases} x-y-3z=-4, \\ 2x-y-4z=-7, \\ x+y-z=-2. \end{cases}$

15. $\begin{cases} 2x-3y+4z=0, \\ x+y-3z=4, \\ 3x+2y-z=0. \end{cases}$

16. $\begin{cases} 3r-t=7, \\ 4r-s+3t=9, \\ 3s+2t=15. \end{cases}$

17. $\begin{cases} 2r-3s+t=-2, \\ r-6s+3t=-2, \\ 3r+3s-2t=2. \end{cases}$

18. $\begin{cases} x-z=14, \\ y+z=21, \\ x-y+z=-10. \end{cases}$

19. Show that Cramer's rule does *not* apply to

$$\begin{cases} x+y=2, \\ 3+x=-y, \end{cases}$$

but that from geometrical considerations there is no solution.

10.5 NONLINEAR SYSTEMS

A system of equations in which at least one equation is not linear is called a **nonlinear system**. We often solve nonlinear systems using elimination by substitution, which was one of our methods for solving linear systems.

EXAMPLE 1

Solve the nonlinear system

$$\begin{cases} x^2-2x+y-7=0, & (1) \\ 3x-y+1=0. & (2) \end{cases}$$

Solving Eq. 2 for y gives

$$y = 3x + 1. \tag{3}$$

Substituting $3x + 1$ for y in Eq. 1 and simplifying, we have

$$x^2 - 2x + (3x + 1) - 7 = 0,$$

$$x^2 + x - 6 = 0 \qquad \text{[quadratic equation]},$$

$$(x + 3)(x - 2) = 0.$$

$$x + 3 = 0, \quad x = -3. \qquad \bigg| \qquad x - 2 = 0, \quad x = 2.$$

From Eq. 3, if $x = -3$, then $y = -8$; if $x = 2$, then $y = 7$. You should check that each pair of values satisfies the given system. The solutions are

$$x = -3,\ y = -8 \quad \text{and} \quad x = 2,\ y = 7.$$

We can see these solutions geometrically in the graph of the system in Fig. 10-7. The solutions correspond to the intersection points $(-3, -8)$ and $(2, 7)$.

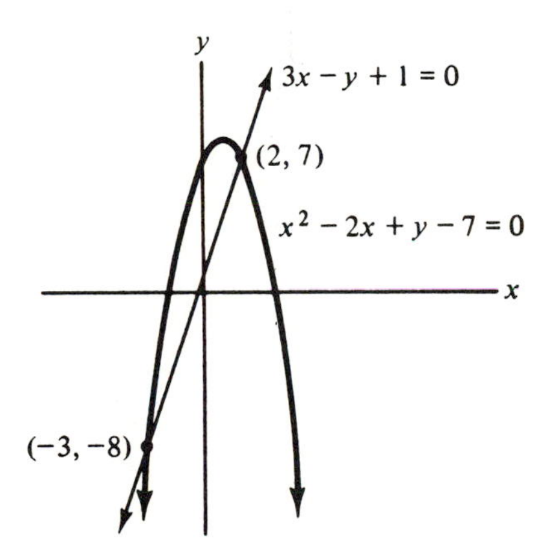

FIGURE 10-7

EXAMPLE 2

Solve the nonlinear system

$$\begin{cases} y = \sqrt{x + 2}, \\ x + y = 4. \end{cases}$$

Solving the second equation for y gives

$$y = 4 - x. \tag{4}$$

Substituting in the first equation gives

$$\begin{aligned}
4 - x &= \sqrt{x+2}, \\
16 - 8x + x^2 &= x + 2 \qquad \text{[squaring both sides]}, \\
x^2 - 9x + 14 &= 0, \\
(x-2)(x-7) &= 0.
\end{aligned}$$

$$x - 2 = 0, \quad x = 2. \qquad \Big| \qquad x - 7 = 0, \quad x = 7.$$

Thus $x = 2$ or $x = 7$. From Eq. 4, if $x = 2$, then $y = 2$; if $x = 7$, then $y = -3$. The pair $x = 2$ and $y = 2$ satisfies the original equations. But the pair $x = 7$ and $y = -3$ does *not* satisfy the first equation. Check this out! The solution is $x = 2$, $y = 2$.

Exercise 10.5

Solve each nonlinear system.

1. $\begin{cases} y = x + 1, \\ x^2 + y^2 = 1. \end{cases}$

2. $\begin{cases} y = 4 - x^2, \\ 3x + y = 0. \end{cases}$

3. $\begin{cases} y^2 = 4 - x, \\ y = x + 2. \end{cases}$

4. $\begin{cases} y = x^2, \\ x + y - 6 = 0. \end{cases}$

5. $\begin{cases} y^2 - x^2 = 28, \\ x - y = 14. \end{cases}$

6. $\begin{cases} y^2 - x = 0, \\ 3x - 2y - 1 = 0. \end{cases}$

7. $\begin{cases} p = 4q - q^2 + 8, \\ p = q^2 - 2q. \end{cases}$

8. $\begin{cases} q^2 + p^2 - 2pq = 1, \\ 3q - p = 5. \end{cases}$

9. $\begin{cases} x^2 - y = 8, \\ y - x^2 = 0. \end{cases}$

10. $\begin{cases} y = x^3, \\ x - y = 0. \end{cases}$

11. $\begin{cases} y = \sqrt{x + 14}, \\ x - y = -2. \end{cases}$

12. $\begin{cases} y^4 - x^2 = 2x + 4, \\ y = \sqrt{x + 1}. \end{cases}$

13. $\begin{cases} w^2 = z^2 + 14, \\ z = w^2 - 16. \end{cases}$

14. $\begin{cases} z = w + 6, \\ w = 3\sqrt{z + 4}. \end{cases}$

15. $\begin{cases} y = \dfrac{x^2}{x - 1} + 1, \\ y = \dfrac{1}{x - 1}. \end{cases}$

16. $\begin{cases} y = \dfrac{4}{x}, \\ 3y = 2x + 2. \end{cases}$

10.6 REVIEW

IMPORTANT TERMS

system of linear equations *(10.1)*
elimination by substitution *(10.1)*
entry of determinant *(10.3)*
minor *(10.3)*
Cramer's rule *(10.4)*
elimination by addition *(10.1)*
determinant *(10.3)*
order of determinant *(10.3)*
cofactor *(10.3)*
nonlinear system of equations *(10.5)*

REVIEW PROBLEMS

In Problems **1–12**, *solve the system by elimination.*

1. $\begin{cases} 4x + 5y = 3, \\ 3x + 4y = 2. \end{cases}$

2. $\begin{cases} 2x + 3y = 5, \\ x - 2y = -1. \end{cases}$

3. $\begin{cases} 3s + t - 4 = 0, \\ 12s + 4t - 2 = 0. \end{cases}$

4. $\begin{cases} u - 3v + 11 = 0, \\ 4u + 3v - 9 = 0. \end{cases}$

5. $\begin{cases} 3x + \frac{1}{2}y = 2, \\ \frac{1}{2}x - \frac{1}{4}y = 0. \end{cases}$

6. $\begin{cases} \frac{1}{3}x - \frac{1}{2}y = 4, \\ \frac{1}{4}x - \frac{3}{8}y = 3. \end{cases}$

7. $\begin{cases} 6x = 3 - 9y, \\ 12y = 4 - 8x. \end{cases}$

8. $\begin{cases} 4x = 7 + 12y, \\ 5x = 15y - 2. \end{cases}$

9. $\begin{cases} x - y = 2, \\ x + z = 1, \\ y - z = 3. \end{cases}$

10. $\begin{cases} 2x - 4z = 8, \\ x - 2y - 2z = 14, \\ 3x + y + z = 0. \end{cases}$

11. $\begin{cases} 4r - s + 2t = 2, \\ 8r - 3s + 4t = 1, \\ r + 2s + 2t = 8. \end{cases}$

12. $\begin{cases} 3u - 2v + w = -2, \\ 2u + v + w = 1, \\ u + 3v - w = 3. \end{cases}$

In Problems **13–20**, *find the value of the determinant.*

13. $\begin{vmatrix} 2 & -1 \\ 4 & 7 \end{vmatrix}$.

14. $\begin{vmatrix} 5 & 8 \\ 3 & 0 \end{vmatrix}$.

15. $\begin{vmatrix} -1 & -2 \\ -3 & 4 \end{vmatrix}$.

16. $\begin{vmatrix} -2 & -6 \\ 3 & 5 \end{vmatrix}$.

17. $\begin{vmatrix} 1 & 2 & -1 \\ 0 & 1 & 4 \\ 1 & 2 & 2 \end{vmatrix}$.

18. $\begin{vmatrix} 2 & 0 & 3 \\ 1 & 4 & 6 \\ -1 & 2 & -1 \end{vmatrix}$.

19. $\begin{vmatrix} 2 & 1 & -1 \\ 1 & 1 & 3 \\ -1 & 1 & -1 \end{vmatrix}$.

20. $\begin{vmatrix} 0 & 1 & 2 \\ -2 & 0 & 3 \\ 4 & 2 & 0 \end{vmatrix}$.

In Problems **21–24**, *solve the system by Cramer's rule.*

21. $\begin{cases} 3x - y = 1, \\ 2x + 3y = 8. \end{cases}$

22. $\begin{cases} 2x = 5y, \\ 4x + 3y = 0. \end{cases}$

23. $\begin{cases} x + y + z = 0, \\ x - z = 0, \\ x - y + 2z = 5. \end{cases}$

24. $\begin{cases} 3x + y + 4z = 1, \\ x + z = 0, \\ 2y + z = 2. \end{cases}$

In Problems **25–28**, *solve each nonlinear system.*

25. $\begin{cases} x^2 - y + 2x = 7, \\ x^2 + y = 5. \end{cases}$

26. $\begin{cases} x^2 + y^2 = 5, \\ x + y - 3 = 0. \end{cases}$

27. $\begin{cases} y - 3 = \sqrt{x + 2}, \\ x + y = 3. \end{cases}$

28. $\begin{cases} y = \dfrac{18}{x + 4}, \\ x - y + 7 = 0. \end{cases}$

29. Two alloys of copper are to be mixed so that the result is 15 kg of a 45% alloy (by weight). One alloy is 20% copper, and the other is 50% copper. How many kilograms of each should be used?

30. The advertising department of a company performed a product rating survey for the company's new product. A total of 250 people were interviewed. The department reported that 62.5% more people liked the product than disliked it. The report did not indicate that 16% of those interviewed had no comment. How many of those surveyed liked the product? How many disliked it? How many had no comment?

31. A company pays skilled workers in its assembly department \$8 per hour. Semiskilled workers in that department are paid \$4 per hour. Shipping clerks are paid \$5 per hour. Because of an increase in orders, the company wants to employ a total of 70 workers in the assembly and shipping departments. It will pay a total of \$370 per hour to these employees. Because of a union contract, twice as many semiskilled workers as skilled workers must be employed. How many semiskilled workers, skilled workers, and shipping clerks should the company employ?

11 Inequalities

11.1 NONLINEAR INEQUALITIES

In Section 4.5 you learned how to solve linear inequalities in one variable. Now we'll show you how to solve inequalities that are not linear. Let's begin with $x^2 + 3x - 4 > 0$. First we factor the left side:

$$(x + 4)(x - 1) > 0.$$

Here we have a product of two factors that is positive. This can happen only if either both factors are positive *or* both factors are negative. Thus there are two cases to consider.

Case 1. *Both factors positive.*

Here $x + 4 > 0$ *and* $x - 1 > 0$. Thus $x > -4$ *and* $x > 1$. This means that x must lie not only to the right of -4, but also to the right of 1. See Fig. 11-1. Both conditions are met when $x > 1$. Thus the solution in Case 1 is $x > 1$.

$x > 1$

-4 1 $x > 1$

$x > -4$

FIGURE 11-1

Case 2. *Both factors negative.*

Here $x + 4 < 0$ *and* $x - 1 < 0$. Thus $x < -4$ *and* $x < 1$, so x must lie to the left of both -4 and 1. See Fig. 11-2. Both conditions are met when $x < -4$. The solution in Case 2 is $x < -4$.

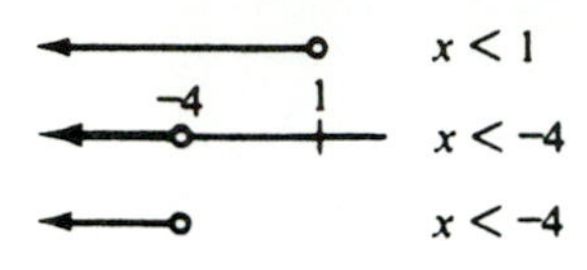

FIGURE 11-2

Since either Case 1 or Case 2 can occur, the original inequality is true when $x > 1$ or $x < -4$. We can write the solution as

$$x < -4 \quad \text{or} \quad x > 1.$$

We can relate solving our inequality $x^2 + 3x - 4 > 0$ to solving the corresponding *equation* $x^2 + 3x - 4 = 0$. The roots of $x^2 + 3x - 4 = 0$ are -4 and 1 [since $x^2 + 3x - 4 = (x + 4)(x - 1) = 0$]. In Fig. 11-3(a), the roots are marked on

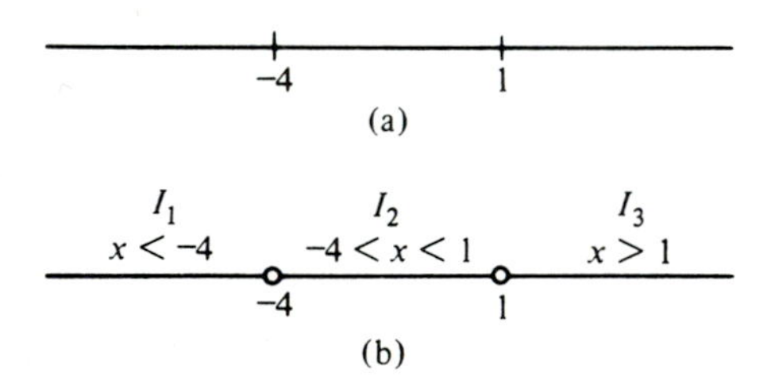

FIGURE 11-3

a number line. These two roots determine three sets of numbers, called *intervals*, which are labeled I_1, I_2, and I_3 in Fig. 11-3(b):

I_1: all values of x such that $x < -4$.

I_2: all values of x such that $x > -4$ and $x < 1$.

I_3: all values of x such that $x > 1$.

The conditions on x in I_2 are usually written $-4 < x < 1$, which means that $x > -4$ and $x < 1$ *simultaneously*.

As we found before, $x^2 + 3x - 4$ is positive on I_1 and I_3. It can be shown that $x^2 + 3x - 4$ is negative on I_2. Thus, on each of the intervals determined by the roots of $x^2 + 3x - 4 = 0$, the polynomial $x^2 + 3x - 4$ is either strictly positive or strictly negative.

In general, for any polynomial P, the roots of the equation $P = 0$ give rise to certain intervals.* On each of these intervals, P is always positive or always negative. Thus if P is positive (or negative) at a point on one such interval, then P must be positive (or negative) on that *entire* interval. This fact lets us solve inequalities without the bother of going into different cases. Example 1 will show you how.

EXAMPLE 1

Solve $x^2 - 4x - 12 < 0$.

First we factor the left side.

$$x^2 - 4x - 12 < 0,$$
$$(x + 2)(x - 6) < 0.$$

Next we find the roots of the equation $(x + 2)(x - 6) = 0$.

$$\text{Roots of } (x + 2)(x - 6) = 0 \text{ are } -2, 6.$$

These two roots give rise to three intervals (see Fig. 11-4). On each interval the

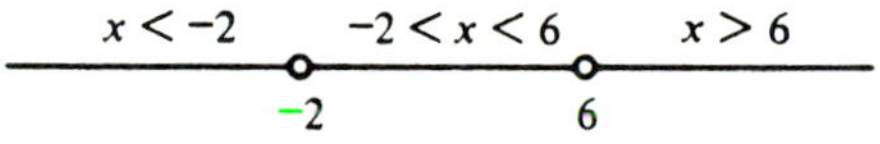

$x < -2$ $\quad -2 < x < 6$ $\quad x > 6$

-2 $\quad 6$

FIGURE 11-4

product $(x + 2)(x - 6)$ must be always positive or always negative. Let's find the sign of the product for $x < -2$. To do this we choose *any* value of x in the interval and see if $(x + 2)(x - 6)$ is positive or negative for that value. We choose $x = -8$. Then

$$\begin{aligned}(x + 2)(x - 6) &= (-8 + 2)(-8 - 6)\\ &= (-)(-) = (+).\end{aligned}$$

Note how we conveniently found the sign of the product by using the signs of its factors. Since the product $(x + 2)(x - 6)$ is positive (+) for $x = -8$, it is positive on the entire interval where $x < -2$. To find the sign of the product on the other intervals, we choose and test a value of x in each of them.

For $-2 < x < 6$, let's choose $x = 0$.

$$(x + 2)(x - 6) = (0 + 2)(0 - 6) = (+)(-) = (-).$$

---

* We are concerned only with those roots that are real numbers.

Thus

$$(x+2)(x-6)<0 \quad \text{for} \quad -2<x<6.$$

For $x>6$, let's choose $x = 10$.

$$(x+2)(x-6) = (10+2)(10-6) = (+)(+) = (+).$$

Thus

$$(x+2)(x-6)>0 \quad \text{for} \quad x>6.$$

Figure 11-5 gives a summary of the signs. With the aid of this sign chart, we can

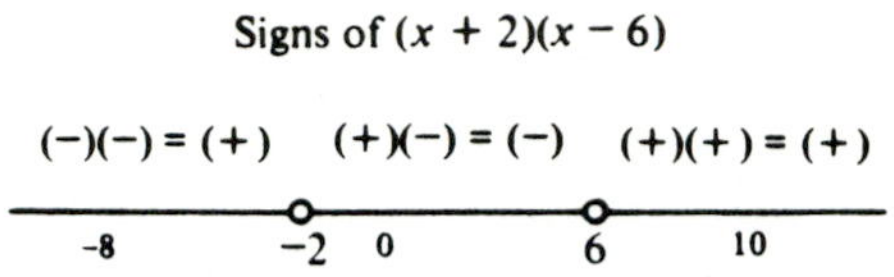

Signs of $(x+2)(x-6)$

$(-)(-)=(+)$ $(+)(-)=(-)$ $(+)(+)=(+)$

-8 −2 0 6 10

FIGURE 11-5

conclude that the solution of $(x+2)(x-6)<0$ is

$$\boxed{-2<x<6.}$$

EXAMPLE 2

Solve $x^2 - 3x \geq 6x - 20$.

First we write the inequality so that one side is 0.

$$x^2 - 9x + 20 \geq 0.$$

Factoring the left side gives

$$(x-4)(x-5) \geq 0.$$

$$\text{Roots of } (x-4)(x-5) = 0 \text{ are } 4, 5.$$

These roots give us the intervals in Fig. 11-6. We now choose a point in each inter-

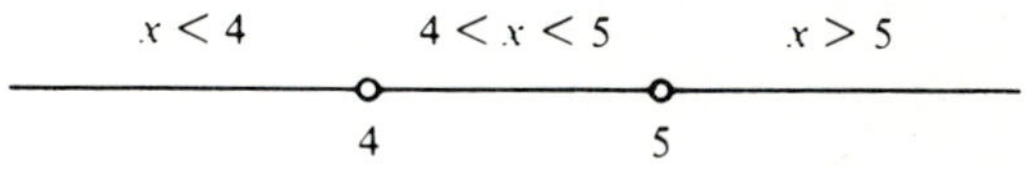

$x<4$ $4<x<5$ $x>5$

4 5

FIGURE 11-6

val and find the sign of the product $(x-4)(x-5)$ at those points, as in Fig. 11-7. Notice that the product is positive when $x=3$ and when $x=6$. Therefore,

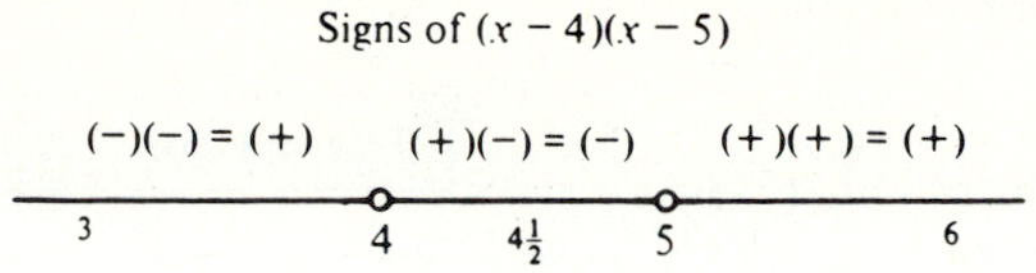

Signs of $(x-4)(x-5)$

$(-)(-) = (+)$ $(+)(-) = (-)$ $(+)(+) = (+)$

3 4 $4\frac{1}{2}$ 5 6

FIGURE 11-7

$(x-4)(x-5) > 0$ for $x < 4$ or for $x > 5$. But we want the solution of the inequality $(x-4)(x-5) \geq 0$. Since $(x-4)(x-5) = 0$ when $x = 4$ or $x = 5$, the solution must include these numbers and is

$$\boxed{x \leq 4 \quad \text{or} \quad x \geq 5.}$$

*In Example 2, **don't** think that you can write the answer as $4 \geq x \geq 5$. These symbols imply that $4 \geq 5$, which is false.*

EXAMPLE 3

Solve $t^3 - 8t^2 + 16t > 0$.

Factoring the left side gives

$$t(t^2 - 8t + 16) > 0,$$

$$t(t-4)^2 > 0.$$

$$\text{Roots of } t(t-4)^2 = 0 \text{ are } 0, 4.$$

These roots determine the intervals in Fig. 11-8. At some point in each interval we

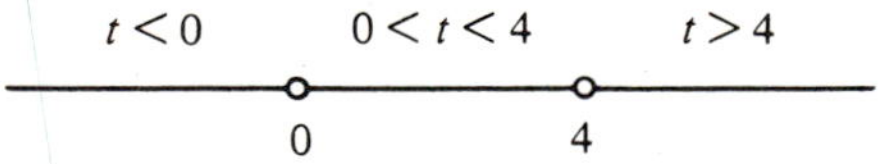

$t < 0$ $0 < t < 4$ $t > 4$

0 4

FIGURE 11-8

find the sign of $t(t-4)^2$. From Fig. 11-9, we see that $t(t-4)^2 > 0$ for those values

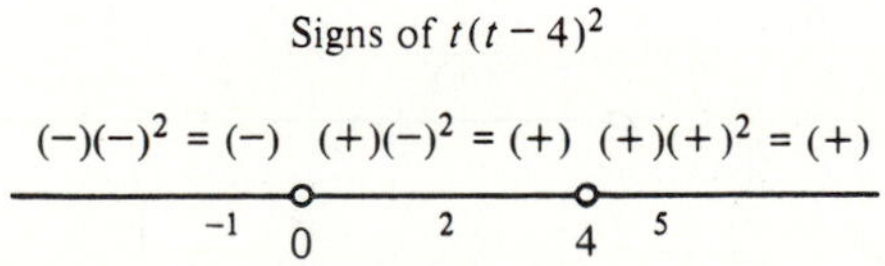

Signs of $t(t-4)^2$

$(-)(-)^2 = (-)$ $(+)(-)^2 = (+)$ $(+)(+)^2 = (+)$

-1 0 2 4 5

FIGURE 11-9

of t such that

$$\boxed{0 < t < 4 \quad \text{or} \quad t > 4.}$$

The next example involves solving an inequality of the form $P/Q > 0$, where P and Q are polynomials. Here we'll find the sign of the quotient P/Q on the intervals determined by the roots of *two* equations. We get these equations by setting the numerator and denominator equal to 0. The same method is used if the inequality symbol $>$ is replaced by $\geq$, $<$, or $\leq$.

EXAMPLE 4

Solve $\dfrac{x^2 - 4x + 3}{x + 2} > 0.$

Factoring the numerator gives

$$\frac{(x-1)(x-3)}{x+2} > 0.$$

Setting the numerator and denominator equal to 0 gives the roots, 1, 3, and -2.

$(x-1)(x-3) = 0.$	$x + 2 = 0.$
Roots are $1, 3.$	Root is $-2.$

These three roots give rise to four intervals. See Fig. 11-10.

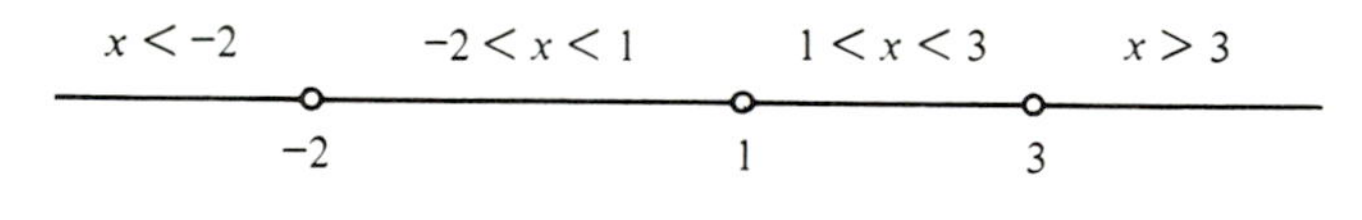

FIGURE 11-10

Now we choose a point in each of the intervals (see Fig. 11-11) and find the sign of $\dfrac{(x-1)(x-3)}{x+2}$ at each point. Since this quotient is positive for $x = 0$ and $x = 4$, then

Signs of $\dfrac{(x-1)(x-3)}{x+2}$

$\dfrac{(-)(-)}{(-)} = (-)$ $\quad \dfrac{(-)(-)}{(+)} = (+)$ $\quad \dfrac{(+)(-)}{(+)} = (-)$ $\quad \dfrac{(+)(+)}{(+)} = (+)$

−3 −2 0 1 2 3 4

FIGURE 11-11

$$\frac{(x-1)(x-3)}{x+2} > 0 \text{ for}$$

$$\boxed{-2 < x < 1 \quad \text{or} \quad x > 3.}$$

To solve the inequality $\frac{3x}{x-2} \leq 4$, you may be tempted to first clear of fractions by multiplying both sides by $x - 2$:

$$\frac{3x}{x-2}(x-2) \leq 4(x-2),$$

$$3x \leq 4(x-2).$$

This is **incorrect**, because $x - 2$ is negative for $x < 2$ and positive for $x > 2$. This means that for some values of x you have multiplied both sides by a negative number, so the direction of the inequality symbol must change. For other values of x, the direction does not change. In general, you should avoid multiplying (or dividing) both sides of an inequality by an expression containing a variable. To solve $\frac{3x}{x-2} \leq 4$, you should first rewrite it in the form $P/Q \leq 0$ and then solve, as Example 5 shows.

EXAMPLE 5

Solve $\frac{3x}{x-2} \leq 4$.

We first rewrite the inequality so that one side is a single fraction and the other side is 0.

$$\frac{3x}{x-2} \leq 4,$$

$$\frac{3x}{x-2} - 4 \leq 0,$$

$$\frac{3x - 4(x-2)}{x-2} \leq 0 \quad \text{[combining into one fraction]},$$

$$\frac{8-x}{x-2} \leq 0.$$

Note that this inequality has the form $P/Q \leq 0$. Setting both numerator and denominator equal to 0, we get

$8 - x = 0.$	$x - 2 = 0.$
Root is 8.	Root is 2.

These two roots give rise to the three intervals in Fig. 11-12. From the sign chart in

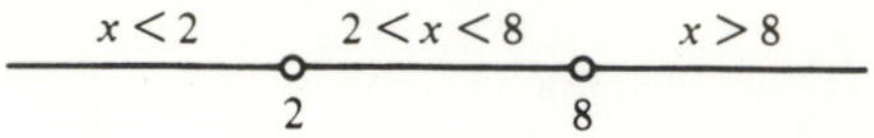

$x<2$ $\quad 2<x<8 \quad$ $x>8$

2 8

FIGURE 11-12

Fig. 11-13, we see $\frac{8-x}{x-2}<0$ for $x<2$ or $x>8$. Now, $\frac{8-x}{x-2}$ is 0 only if the numerator

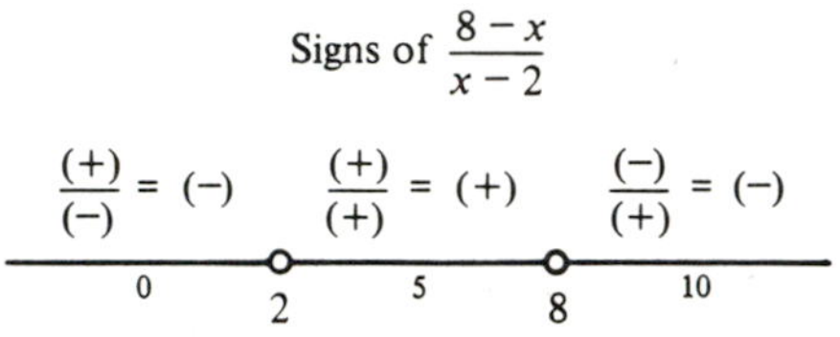

Signs of $\frac{8-x}{x-2}$

$\frac{(+)}{(-)}=(-)$ $\quad \frac{(+)}{(+)}=(+)$ $\quad \frac{(-)}{(+)}=(-)$

0 2 5 8 10

FIGURE 11-13

is 0, that is, when $x=8$. The solution of $\frac{8-x}{x-2}\leq 0$ must include 8, so it is

$$x<2 \quad \text{or} \quad x\geq 8.$$

Exercise 11.1

Solve each inequality.

1. $(x-1)(x-5)<0.$
2. $(x+3)(x-8)>0.$
3. $(x+1)(x-3)>0.$
4. $(x+4)(x+5)<0.$
5. $x^2-1<0.$
6. $x^2>9.$
7. $x^2-x-6>0.$
8. $x^2-2x-3\leq 0.$
9. $5s-s^2\leq 0.$
10. $t^2+9t+18\geq 0.$
11. $x^2+5x<-6.$
12. $x^2-12>x.$
13. $x^2+4x-5\geq 3x+15.$
14. $x^2+9x+9\leq 2-x^2.$
15. $2z^2-5z-12<0.$
16. $5t^2-1>4t.$
17. $x^2+2x+1>0.$
18. $x^2+9\leq 6x.$
19. $(x+2)(x-1)(x-4)>0.$
20. $x(x+3)(5-x)<0.$
21. $(x+5)(x-3)^2<0.$
22. $(x+1)^2(x-4)^2>0.$
23. $y^3-y\leq 0.$
24. $x^3+8x^2+15x\geq 0.$
25. $x^3-2x^2\geq 0.$
26. $p^4-2p^3-3p^2\leq 0.$

27. $x^4 - 1 > 0.$

28. $x^4 - 2x^2 + 1 \leq 0.$

29. $\dfrac{x - 4}{x + 8} > 0.$

30. $\dfrac{x + 3}{x + 5} < 0.$

31. $\dfrac{t}{3 - t} \leq 0.$

32. $\dfrac{s - 1}{2s - 1} > 0.$

33. $\dfrac{x + 3}{x^2 - 1} < 0.$

34. $\dfrac{x^2 - x - 6}{x + 4} > 0.$

35. $\dfrac{x^2 - 5x + 4}{x^2 + 5x + 4} \geq 0.$

36. $\dfrac{x^2 - 4}{x^2 + 2x + 1} \leq 0.$

37. $\dfrac{5}{x + 2} > 1.$

38. $\dfrac{3}{9(x - 8)} < \dfrac{1}{18}.$

39. $\dfrac{x + 1}{x + 4} > 3.$

40. $\dfrac{x}{1 - 2x} < 4.$

41. $\dfrac{x + 3}{x - 1} \leq x.$

42. $\dfrac{2x + 3}{x} \geq -\dfrac{9}{x - 6}.$

11.2 INEQUALITIES INVOLVING ABSOLUTE VALUE

In calculus and other areas of mathematics, it is useful to work with inequalities involving absolute value. Recall that the absolute value of a number is the distance of that number from 0.

For example, if $|x| < 3$, then x is less than 3 units from 0. Thus x must lie between -3 and 3. That is, $-3 < x < 3$ [Fig. 11-14(a)]. However, if $|x| > 3$, then x must be more than 3 units from 0. Thus, one of two things must be true: either $x > 3$ *or* $x < -3$ [Fig. 11-14(b)]. We can extend these ideas. If $|x| \leq 3$, then we have $-3 \leq x \leq 3$. If $|x| \geq 3$, then $x \geq 3$ or $x \leq -3$.

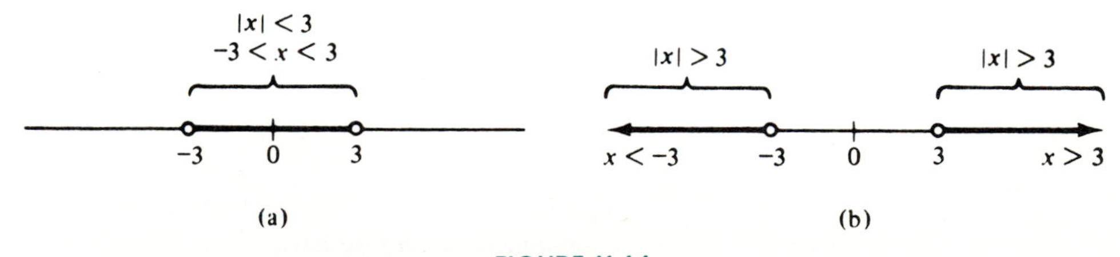

FIGURE 11-14

In general, the solution of $|x| < d$ or $|x| \leq d$, where d is a positive number, consists of one interval. However, when $|x| > d$ or $|x| \geq d$, there are two intervals in the solution. More precisely, we have the following rules.

	If $d > 0$, then
Rule 1.	$\lvert x\rvert < d$ if and only if $-d < x < d$;
	$\lvert x\rvert \le d$ if and only if $-d \le x \le d$.
Rule 2.	$\lvert x\rvert > d$ if and only if $x < -d$ or $x > d$;
	$\lvert x\rvert \ge d$ if and only if $x \le -d$ or $x \ge d$.

EXAMPLE 1

Solve the following absolute value inequalities.

a. $|x - 2| < 4$.

The number $x - 2$ must be less than 4 units from 0. By Rule 1, this means that $x - 2$ lies between -4 and 4. Thus $-4 < x - 2 < 4$. We solve this as follows.

$$-4 < x - 2 < 4,$$

$$-4 + 2 < x < 4 + 2 \qquad \text{[adding 2 to each member]},$$

$$\boxed{-2 < x < 6.}$$

b. $|3 - 2x| \le 5$.

$$-5 \le 3 - 2x \le 5 \qquad \text{[Rule 1]},$$

$$-5 - 3 \le -2x \le 5 - 3 \qquad \text{[subtracting 3 from each member]},$$

$$-8 \le -2x \le 2,$$

$$4 \ge x \ge -1 \qquad \text{[dividing each member by } -2 \text{ and changing the direction of inequalities]},$$

$$\boxed{-1 \le x \le 4} \qquad \text{[rewriting]}.$$

EXAMPLE 2

Solve the following inequalities.

a. $|x + 5| \ge 7$.

The number $x + 5$ must be *at least* 7 units from 0. By Rule 2, either $x + 5 \ge 7$ *or* $x + 5 \le -7$. Solving these inequalities, we have

$$\boxed{x \ge 2 \quad \text{or} \quad x \le -12.}$$

b. $|3x - 4| > 1$.

By Rule 2, either $3x - 4 > 1$ or $3x - 4 < -1$. Thus either $3x > 5$ or $3x < 3$. The solution is

$$x > \frac{5}{3} \quad \text{or} \quad x < 1.$$

The numbers 5 and 9 are 4 units apart. Also,

$$|9 - 5| = |4| = 4,$$
$$|5 - 9| = |-4| = 4.$$

In general, you may think of $|a - b|$ or $|b - a|$ as the distance between a and b.

EXAMPLE 3

Use absolute value notation to express each fact.

a. *x is less than 3 units from 5.*

Since the distance between x and 5 is less than 3, we must have

$$|x - 5| < 3.$$

b. *x differs from 6 by at least 7.*

$$|x - 6| \geq 7.$$

c. *x is more than 2 units from -3.*

$$|x - (-3)| > 2,$$
$$|x + 3| > 2.$$

d. *y is less than ϵ (a Greek letter read "epsilon") units from L.*

$$|y - L| < \epsilon.$$

e. *$x < 9$ and $x > -9$ simultaneously.*

$$|x| < 9.$$

The following are three basic properties of absolute value:

BASIC PROPERTIES OF INEQUALITIES

1. $|ab| = |a| \cdot |b|$.

2. $\left|\frac{a}{b}\right| = \frac{|a|}{|b|}$.

3. $|a - b| = |b - a|$.

EXAMPLE 4

a. $|(-7)\cdot 3| = |-7|\cdot|3| = 21$; $|(-7)(-3)| = |-7|\cdot|-3| = 21$.

b. $|4-2| = |2-4| = 2$.

c. $|7-x| = |x-7|$.

d. $\left|\frac{-7}{3}\right| = \frac{|-7|}{|3|} = \frac{7}{3}$; $\left|\frac{-7}{-3}\right| = \frac{|-7|}{|-3|} = \frac{7}{3}$.

e. $\left|\frac{x-3}{-5}\right| = \frac{|x-3|}{|-5|} = \frac{|x-3|}{5}$.

Exercise 11.2

In Problems **1–20**, *solve each inequality.*

1. $|x| < 3$.

2. $|x| < 10$.

3. $|x| > 6$.

4. $|x| > 3$.

5. $|2x| \leq 2$.

6. $|4x| \leq 3$.

7. $|3x| \geq 9$.

8. $\left|\frac{x}{2}\right| \geq 3$.

9. $|x-4| < 16$.

10. $|y+5| \leq 6$.

11. $|y+1| \geq 6$.

12. $|x-2| \geq 4$.

13. $|3x-5| \leq 1$.

14. $\left|\frac{x}{3} - 5\right| < 4$.

15. $|1-3x| < 2$.

16. $|4x-1| > 7$.

17. $|\frac{1}{2} - t| > \frac{1}{2}$.

18. $|5-2x| > 1$.

19. $\left|\frac{3x-8}{2}\right| \geq 4$.

20. $\left|\frac{x-8}{4}\right| \leq 2$.

21. Why does $|x-5| < -12$ have no solution?

22. Why does every value of x satisfy $|2x+1| > -1$?

23. Use absolute value notation to express each fact.

a. x is less than 3 units from 7.

b. x differs from 2 by less than 3.

c. x is no more than 5 units from -7.

d. The distance between 7 and x is 4.

e. $x+4$ is less than 2 units from 0.

f. x is between 3 and -3, but is not equal to 3 or -3.

g. $x < -6$ or $x > 6$.

h. $x - 6 \geq 4$ or $x - 6 \leq -4$.

i. The number, x, of hours that a machine will operate efficiently differs from 105 by less than 3.

j. The average monthly income x (in dollars) of a family differs from 1000 by less than 100.

24. Use absolute value notation to indicate that the prices p and q of two products may differ by no more than 2 (dollars).

25. Show that if $|y - L| < \epsilon$, then $L - \epsilon < y < L + \epsilon$.

26. Show that if $|x - \mu| \leq 2\sigma$, (μ and σ are the Greek letters mu and sigma, respectively), then $-2\sigma + \mu \leq x \leq 2\sigma + \mu$.

27. In the manufacture of widgets, the average length of a part is .01 cm. Use absolute value notation to express the fact that an individual measurement x of a part does not differ from the average by more than .005 cm.

11.3 LINEAR INEQUALITIES IN TWO VARIABLES

As you have seen, solutions of inequalities in one variable can be represented by intervals on the real number line. But you will see that the solution of an inequality in *two* variables, such as $2x + y < 5$, is represented by a *region* in the coordinate plane.

We call $2x + y < 5$ a *linear* inequality in x and y. In general, any inequality that can be put in the form $ax + by + c < 0$ (or $\leq$, $>$, $\geq$) is called a **linear inequality** in the variables x and y.* Geometrically, the solution to an inequality like this consists of all points in the plane whose coordinates satisfy it. These points make up the **graph** of the inequality.

Now, consider the nonvertical line $y = mx + b$ in Fig. 11-15. The following can be shown.

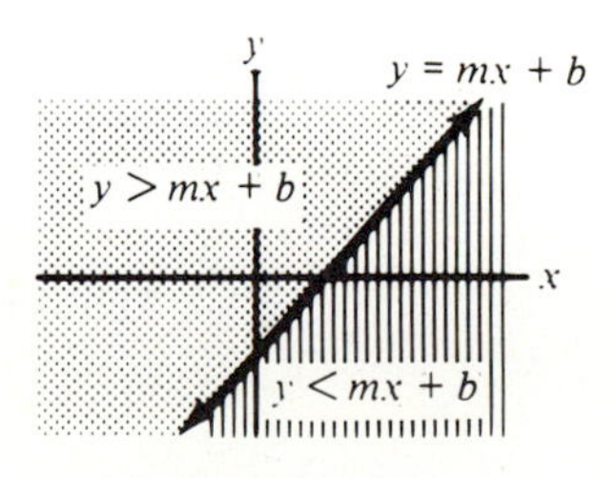

FIGURE 11-15

* We assume that a and b are not both zero.

1. The region above the line is the graph of the inequality $y > mx + b$.
2. The region below the line is the graph of the inequality $y < mx + b$.
3. The line itself is the graph of the equation $y = mx + b$.

For a vertical line $x = a$ (see Fig. 11-16), the region to the *right* of the line is the graph of $x > a$, while the region to the *left* is the graph of $x < a$.

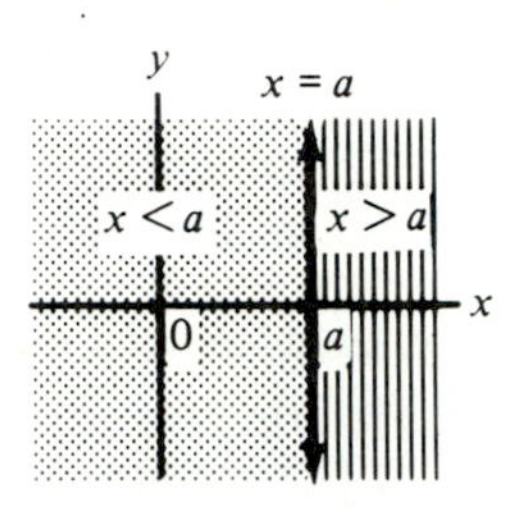

FIGURE 11-16

We can use these facts to graph $2x + y < 5$. First we rewrite the inequality in one of the forms $y < mx + b$ or $y > mx + b$. In our case we have

$$y < -2x + 5.$$

Next we indicate the graph of the corresponding *equation* $y = -2x + 5$ by drawing a broken line in a coordinate plane, as in Fig. 11-17. From Statement 2, the graph of $y < -2x + 5$ is all points *below* the line. Part of this region is shaded in Fig. 11-17.

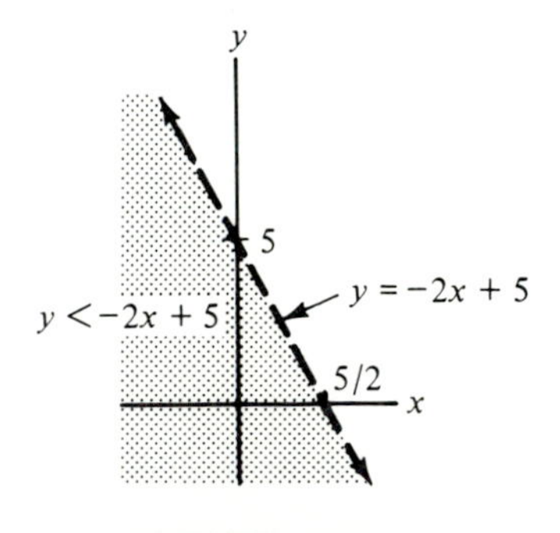

FIGURE 11-17

If (x_0, y_0) is *any* point in this region, its coordinates satisfy $y < -2x + 5$. That is, its second coordinate y_0 is less than the number $-2x_0 + 5$. See Fig. 11-18. For example, the point $(-2, -1)$ is in the region and $-1 < -2(-2) + 5$.

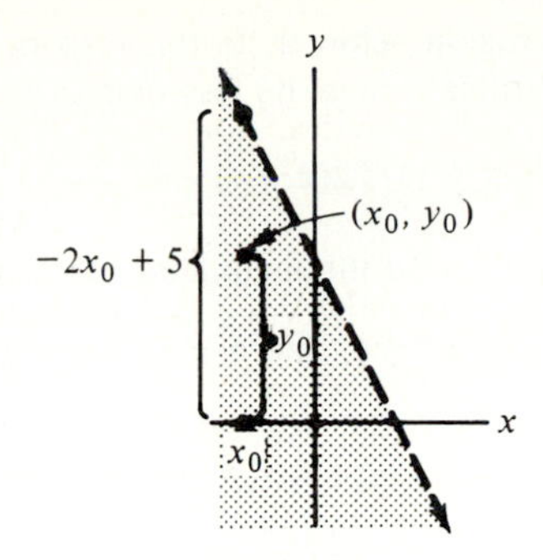

FIGURE 11-18

If we wanted to find the graph of $y \leq -2x + 5$, it would not only be the region below the line $y = -2x + 5$, but it would also include the line itself. This is indicated by the solid line in Fig. 11-19. It will be our custom that **a solid line *is* included in a graph, but a broken line *is not*.**

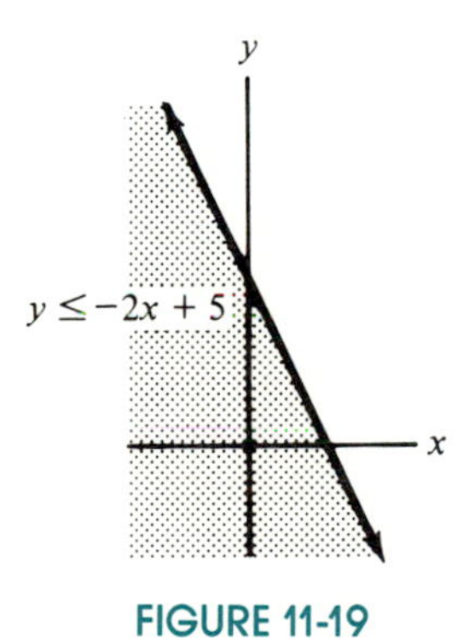

FIGURE 11-19

EXAMPLE 1

a. *Graph* $y \leq 5$.

The inequality has the form $y \leq mx + b$, where $m = 0$ and $b = 5$. The graph of $y = 5$ is a horizontal line (see Fig. 11-20). Thus the graph of $y \leq 5$ *is the line* $y = 5$

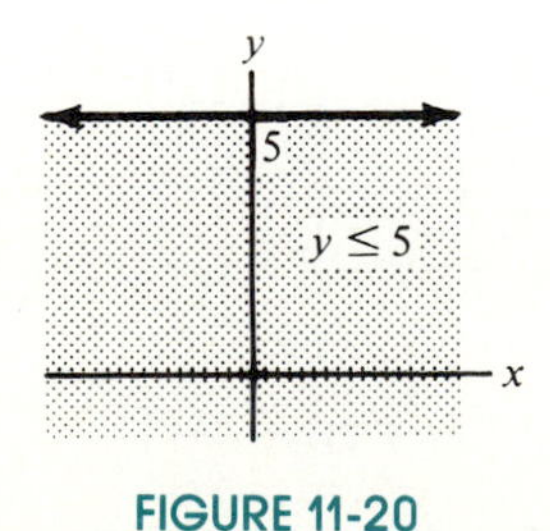

FIGURE 11-20

and the region below it. In this region, a point can have any x-coordinate, but its y-coordinate must be less than or equal to 5.

b. *Graph* $2(2x - y) < 2(x + y) - 4$.

Let's try to write the inequality in the form $y < mx + b$ or $y > mx + b$.

$$2(2x - y) < 2(x + y) - 4,$$
$$4x - 2y < 2x + 2y - 4,$$
$$-4y < -2x - 4,$$
$$y > \frac{1}{2}x + 1 \qquad \text{[dividing both sides by } -4 \text{ and changing direction of inequality].}$$

The inequality has the form $y > mx + b$. We now indicate the graph of $y = \frac{1}{2}x + 1$ by drawing a broken line. See Fig. 11-21. Then we shade the region *above* it.

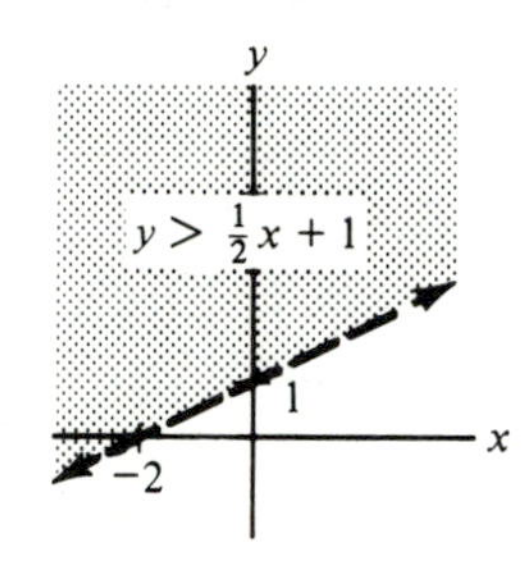

FIGURE 11-21

Exercise 11.3

Graph each linear inequality.

1. $y > 2$.

2. $2y \leq 6$.

3. $y \geq x + 4$.

4. $y < 4x - 12$.

5. $y \leq 2x - 6$.

6. $y > 6 - 3x$.

7. $x > 2$.

8. $3x - 1 \leq x$.

9. $2x + 3y > 6$.

10. $3x - 2y \geq 12$.

11. $x + 2y \leq 7$.

12. $y > 6 - 2x$.

13. $-x \leq 2y - 4$.

14. $2x + y \geq 10$.

15. $3x + y < 0$.

16. $x + 5y < -5$.

17. $3 - (6x + 2y) > y$.

18. $4(x - 3) \leq 3(y + 2)$.

19. $4(x - 2) + y \geq 2 - (x + y)$.

20. $3(x - 2y) + 2y > 2(1 - 2y) + 3$.

11.4 REVIEW

IMPORTANT TERMS AND SYMBOLS

interval *(11.1)*

$a < x < b$ *(11.1)*

linear inequality in two variables *(11.3)*

REVIEW PROBLEMS

In Problems **1–28**, *solve the inequalities.*

1. $x^2 + 4x - 12 < 0$.

2. $x^2 + 11x + 28 > 0$.

3. $y^2 > 6y$.

4. $z^2 + 6z < -5$.

5. $2x^2 + 5x \geq x^2 - 4x - 20$.

6. $3x(x - 4) \leq 2x^2 - 27$.

7. $x(3x + 2) < 1$.

8. $6x(x - 1) > 1 - 5x$.

9. $(x + 4)(x - 5)(x - 9) > 0$.

10. $(x + 2)(x + 4)^2 < 0$.

11. $p^3 - 8p^2 \leq 0$.

12. $r^3 - 9r \geq 0$.

13. $\frac{x + 9}{x + 2} \geq 0$.

14. $\frac{x + 3}{x^2 - 3x + 2} \leq 0$.

15. $\frac{x^2 - 6x + 9}{x^2 + 7x + 10} > 0$.

16. $\frac{x^2 + 2x - 8}{x^2 - 25} < 0$.

17. $\frac{2x - 1}{x + 1} > 1$.

18. $\frac{1 - x}{1 - 2x} \leq -1$.

19. $\frac{x^2}{x + 4} \leq 2$.

20. $\frac{6}{x} \leq x + 1$.

21. $|3x| > 6$.

22. $|x + 4| \leq 6$.

23. $|4x - 1| < 1$.

24. $\left|\frac{5x - 8}{12}\right| \geq 1$.

25. $|4 - 2x| \geq 4$.

26. $|-1 - x| < 1$.

27. $|x + \frac{1}{2}| \leq \frac{3}{2}$.

28. $|\frac{2}{3}x - 5| > 4$.

In Problems **29–32**, *graph each linear inequality in two variables.*

29. $y \geq 3x - 9$.

30. $5x < 7x$.

31. $x - (3y - 2x) < x - y$.

32. $2(5x + 1) - y \geq 4(y - 2)$.

12 Functions

12.1 FUNCTIONS

In this chapter we look at one of the most basic concepts in all of mathematics—the notion of a *function*.

Briefly, a function is a special type of input-output relation that expresses how one quantity (the *output*) depends on another quantity (the *input*). For example, when money is invested at some interest rate, the interest I (output) depends on the length of time t (input) that the money is invested. To express this dependence, we say that "I is a function of t." Functional relations like this are usually specified by a formula that shows what must be done to the input to find the output.

To illustrate, suppose \$100 earns simple interest at an annual rate of 6%. Then it can be shown that interest and time are related by the formula

$$I = 100(.06)t, \tag{1}$$

where I is in dollars and t is in years. For example,

$$\text{if } t = \tfrac{1}{2}, \quad \text{then} \quad I = 100(.06)(\tfrac{1}{2}) = 3. \tag{2}$$

Thus Formula (1) assigns to the input $\frac{1}{2}$ the output 3. We can think of Formula (1) as defining a *rule*: Multiply t by 100(.06). The rule assigns to each input number t exactly one output number I, which we symbolize by the following arrow notation:

$$t \to I \qquad \text{or} \qquad t \to 100(.06)t.$$

This rule is an example of a *function* in the following sense:

> A **function** is a rule that assigns to each input number exactly one output number. The set of all input numbers to which the rule applies is called the **domain** of the function. The set of all output numbers is called the **range.**

For the interest function defined by Formula (1), the input number t cannot be negative because negative time makes no sense. Thus the domain consists of all nonnegative numbers; that is, all $t \geq 0$. From (2) we see that when the input is $\frac{1}{2}$, the output is 3. Thus 3 is in the range.

We have been using the term *function* in a restricted sense because, in general, the inputs or outputs do not have to be numbers. For example, a table of cities and their populations assigns to each city (which is not a number) its population (exactly one output). Thus a function is implied. However, we shall consider only functions whose domain and range consist of real numbers.

A variable that represents input numbers for a function is called an **independent variable.** One that represents output numbers is called a **dependent variable** because its value *depends* on the value of the independent variable. We say that the dependent variable is a *function of* the independent variable. That is, output is a function of input. Thus for the interest formula $I = 100(.06)t$, the independent variable is t, the dependent variable is I, and I is a function of t.

As another example, the equation (or formula)

$$y = x + 2 \tag{3}$$

defines y as a function of x. It gives the rule: Add 2 to x. This rule assigns to each input x exactly one output $x + 2$, which is y. Thus if $x = 1$, then $y = 3$; if $x = -4$, then $y = -2$. The independent variable is x and the dependent variable is y.

Not all equations in x and y define y as a function of x, as Example 1 shows.

EXAMPLE 1

Let $y^2 = x$.

a. Suppose x is a positive input number such as 9. Then $y^2 = 9$, and so $y = \pm 3$. Thus with the input number 9 there are assigned not one but *two* output numbers, 3 and -3. Consequently, y is **not** a function of x.

b. If y is an input number, then we square y and get exactly one output number, x. For example, if $y = 3$, then $x = y^2 = 3^2 = 9$.

$$y \to x \qquad (\text{where } x = y^2).$$

Thus x is a function of y. The independent variable is y and the dependent variable is x. Since y can be any real number, the domain is all real numbers.

Some equations in x and y define either variable as a function of the other variable, as Example 2 shows.

EXAMPLE 2

Let $x + y - 1 = 0$.

To determine if y is a function of x, we treat x as input and see if there is exactly one y-value for output. Thus we solve the equation for y.

$$y = 1 - x.$$

Clearly, with each input x there is exactly one output number, $1 - x$. Thus y is a function of x.

Now, let's see if x is a function of y. Here y is input and x is output. Solving $x + y - 1 = 0$ for x, we have

$$x = 1 - y.$$

Clearly, for each y there is exactly one output number, $1 - y$. Thus x is a function of y.

We have therefore shown that the equation $x + y - 1 = 0$ defines y as a function of x, and x as a function of y.

In general, if we merely say that one variable is a function of another variable, this does not indicate what rule is involved. Usually the letters f, g, h, F, G, and so on are used to represent function rules. For example, Eq. (3) above ($y = x + 2$) defines y as a function of x, where the rule is add 2 to the input. Suppose we let f represent this rule. Then we say that f is the function. To indicate that f assigns to the input 1 the output 3, we write $f(1) = 3$, which is read "f of 1 equals 3." Similarly, $f(-4) = -2$. More generally, if x is any input, we have the following notation:

> $f(x)$, which is read "f of x," means the output number in the range of the function f that corresponds to the input number x in the domain.

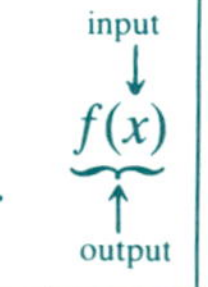

Thus the output $f(x)$ is the same as y. But since $y = x + 2$, we may write $y = f(x) = x + 2$ or simply

$$f(x) = x + 2.$$

For example, to find $f(3)$, which is the output corresponding to the input 3, we replace each x in $f(x) = x + 2$ by 3:

$$f(3) = 3 + 2 = 5.$$

Likewise,

$$f(8) = 8 + 2 = 10,$$
$$f(-5) = -5 + 2 = -3.$$

Output numbers such as $f(-5)$, are called **function values** (or functional values). Keep in mind that they are in the range of f.

*$f(x)$ **does not** mean f times x.*

Functions are often defined by "functional notation." For example, the equation $g(x) = x^3 + x^2$ defines the function g that assigns to an input number x the output number $x^3 + x^2$.

$$g: \quad x \rightarrow x^3 + x^2.$$

Thus g adds the cube and the square of an input number. Some function values are

$$g(2) = 2^3 + 2^2 = 12,$$
$$g(-1) = (-1)^3 + (-1)^2 = -1 + 1 = 0,$$
$$g(t) = t^3 + t^2,$$
$$g(x+1) = (x+1)^3 + (x+1)^2.$$

Note that $g(x+1)$ was found by replacing each x in $x^3 + x^2$ by the input $x + 1$.

When we refer to the function g defined by $g(x) = x^3 + x^2$, we shall feel free to call the equation itself a function. Thus we speak of the "function $g(x) = x^3 + x^2$" and the "function $y = x + 2$."

Let's be specific about the domain of a function that is given by an equation. Unless otherwise stated, the domain consists of all real numbers for which that equation makes sense and gives function values that are real numbers. For example, suppose

$$h(x) = \frac{1}{x-6}.$$

Here any real number can be used for x except 6, because the denominator is 0 when x is 6 (we cannot divide by zero). Thus the domain of h is understood to be all real numbers except 6.

EXAMPLE 3

Domains of functions.

a. Let $f(x) = \dfrac{x}{x^2 - x - 2}$.

We cannot divide by zero, so we must find any values of x that make the denominator 0. These *cannot* be input numbers. Thus we set the denominator equal to 0 and solve for x.

$$x^2 - x - 2 = 0 \qquad \text{[quadratic equation]},$$
$$(x-2)(x+1) = 0 \qquad \text{[factoring]},$$
$$x = 2, -1.$$

Therefore the domain of f is all real numbers *except* 2 and -1.

b. Let $g(t) = \sqrt{2t - 1}$.

We cannot have function values that involve imaginary numbers. To avoid square roots of negative numbers, $2t - 1$ must be greater than or equal to 0.

$$2t - 1 \geq 0,$$
$$2t \geq 1 \qquad \text{[adding 1 to both sides]},$$
$$t \geq \frac{1}{2} \qquad \text{[dividing both sides by 2]}.$$

Thus the domain is all real numbers t such that $t \geq \frac{1}{2}$.

EXAMPLE 4

Domains and function values.

a. Let $f(x) = 2x + 3$.

Here we don't have to rule out any values of x, so the domain of f is all real numbers. Let's find some function values.

$$f(x) = 2x + 3.$$

Find $f(4)$: $\quad f(4) = 2(4) + 3 = 11.$

Find $f(x+1)$: $\quad f(x+1) = 2(x+1) + 3 = 2x + 5.$

b. Let $g(x) = 3x^2 - x + 5$.

Any real number can be used for x, so the domain of g is all real numbers.

$$g(x) = 3x^2 - x + 5.$$

Find $g(z)$: $\quad g(z) = 3(z)^2 - z + 5 = 3z^2 - z + 5.$

Find $g(r^2)$: $\quad g(r^2) = 3(r^2)^2 - r^2 + 5 = 3r^4 - r^2 + 5.$

Find $g(x+h)$: $\quad g(x+h) = 3(x+h)^2 - (x+h) + 5$

$$= 3(x^2 + 2hx + h^2) - x - h + 5$$
$$= 3x^2 + 6hx + 3h^2 - x - h + 5.$$

Don't be confused by notation. In Example 4b *we found $g(x+h)$ by replacing each x in $g(x)=3x^2-x+5$ by $x+h$.* ***Don't*** *write the function and then add h. That is, $g(x+h) \neq g(x)+h$.*

$$g(x+h) \neq 3x^2-x+5+h.$$

Also, ***don't*** *use the distributive law on $g(x+h)$. It does* ***not*** *stand for multiplication.*

$$g(x+h) \neq g(x)+g(h).$$

EXAMPLE 5

Let $h(x)=2$.

The domain of h is all real numbers. All function values are 2. For example,

$$h(10)=2, \qquad h(-387)=2, \qquad h(x+3)=2.$$

We call h a *constant function*. More generally, we have this definition:

> *Any function of the form $h(x)=c$, where c is a constant, is called a* ***constant function****.*

Sometimes more than one equation is needed to define a function, as Example 6 shows.

EXAMPLE 6

Let

$$F(s)=\begin{cases} s^2+1, & \text{if } s>0, \\ -3, & \text{if } s=0, \\ 2-s, & \text{if } s<0. \end{cases}$$

This is called a **compound function** because it is defined by more than one equation. The value of an input number s determines which equation to use. We see that s can be any real number.

Find $F(2)$: Since $2>0$, we substitute 2 for s in $F(s)=s^2+1$.

$$F(2)=2^2+1=5.$$

Find $F(0)$: Since $s=0$, we have $F(0)=-3$.

Find $F(-1)$: Since $-1<0$, we substitute -1 for s in $F(s)=2-s$.

$$F(-1)=2-(-1)=3.$$

Exercise 12.1

In Problems **1–22**, *give the domain of each function.*

1. $f(x) = \frac{3}{x}$.

2. $g(x) = \frac{4}{x^2}$.

3. $g(x) = \frac{x}{3}$.

4. $f(x) = \frac{x+2}{5}$.

5. $h(x) = \sqrt{x}$.

6. $f(r) = 7r - 2$.

7. $H(z) = 10$.

8. $h(t) = (2t + 1)^2$.

9. $F(t) = 3t^2 + 5$.

10. $G(s) = \frac{s+1}{s}$.

11. $H(x) = \frac{x}{x+2}$.

12. $f(y) = \frac{8-y}{4-7y}$.

13. $f(x) = \frac{3x-1}{2x+5}$.

14. $f(x) = \sqrt{x-3}$.

15. $g(x) = \sqrt{4x+3}$.

16. $h(x) = \frac{x^3-1}{x^2-4x+4}$.

17. $G(y) = \frac{4}{y^2-y}$.

18. $H(z) = \frac{-3}{z^2+2z-8}$.

19. $f(x) = \frac{x+1}{x^2+6x+5}$.

20. $f(x) = \frac{4}{x} + \frac{x}{x-3}$.

21. $h(s) = \frac{4-s^2}{2s^2-7s-4}$.

22. $G(r) = \frac{2}{r^2+1}$.

In Problems **23–40**, *find the function values for each function.*

23. $f(x) = 5x$; $f(0)$, $f(3)$, $f(-4)$.

24. $g(x) = 2x - 5$; $g(-1)$, $g(4)$, $g\left(-\frac{1}{2}\right)$.

25. $h(t) = 4 - 3t$; $h(1)$, $h\left(-\frac{2}{3}\right)$, $h\left(\frac{1}{2}\right)$.

26. $H(s) = s^2 - 3$; $H(4)$, $H(\sqrt{2})$, $H\left(\frac{2}{3}\right)$.

27. $f(x) = 7x$; $f(s)$, $f(t+1)$, $f(x+3)$.

28. $G(x) = 2 - x^2$; $G(-8)$, $G(u)$, $G(u^2)$.

29. $g(u) = u^2 + u$; $g(-2)$, $g(2v)$, $g(x^2)$.

30. $h(v) = \frac{1}{\sqrt{v}}$; $h(16)$, $h\left(\frac{1}{4}\right)$, $h(1-x)$.

31. $f(x) = 12$; $f(2)$, $f(t+8)$, $f(-\sqrt{17})$.

32. $H(x) = (x+4)^2$; $H(0)$, $H(2)$, $H(t-4)$.

33. $f(x) = x^2 + 2x + 1;\quad f(1),\quad f(-1),\quad f(x+h).$

34. $f(x) = \dfrac{x^2}{x+3};\quad f(0),\quad f(t^3),\quad f(xy).$

35. $g(x) = \dfrac{x-5}{x^2+4};\quad g(5),\quad g(3x),\quad g(x+h).$

36. $g(x) = |x-3|;\quad g(10),\quad g(3),\quad g(-3).$

37. $F(t) = \begin{cases} 1, & \text{if } t > 0 \\ 0, & \text{if } t = 0; \\ -1, & \text{if } t < 0 \end{cases}\quad F(10),\quad F(-\sqrt{3}),\quad F(0),\quad F\left(-\frac{18}{5}\right).$

38. $f(x) = \begin{cases} 4, & \text{if } x \geq 0 \\ 3, & \text{if } x < 0 \end{cases};\quad f(3),\quad f(-4),\quad f(0).$

39. $G(x) = \begin{cases} x, & \text{if } x \geq 3 \\ 2-x, & \text{if } x < 3 \end{cases};\quad G(8),\quad G(3),\quad G(-1),\quad G(1).$

40. $h(r) = \begin{cases} 3r-1, & \text{if } r > 2 \\ r^2-4r+7, & \text{if } r < 2 \end{cases};\quad h(3),\quad h(-3),\quad h(2).$

In Problems **41–44**, *find (a)* $f(x+h)$, *and (b)* $\dfrac{f(x+h)-f(x)}{h}$; *simplify your answers.*

41. $f(x) = 3x - 4.$

42. $f(x) = 2x + 1.$

43. $f(x) = 2.$

44. $f(x) = x^2.$

In Problems **45–48**, *is y a function of x? Is x a function of y?*

45. $y - 3x - 4 = 0.$

46. $x^2 + y = 0.$

47. $y = 7x^2.$

48. $x^2 + y^2 = 1.$

49. In a lab experiment, different weights were attached to a spring, causing it to stretch. Table 12-1 gives the recorded data.

TABLE 12-1

WEIGHT (lb) w	AMOUNT OF STRETCH (in.) s
1	$\frac{1}{2}$
4	2
6	3

Since for each value of w there is exactly one value of s, we can think of s as a function of w. Here w's are inputs and s's are outputs. Let's say that $s = F(w)$. Find $F(4)$, $F(1)$, and $F(6)$.

50. The distance s (in feet) that an object falls as a function of elapsed time t (in seconds) is given by $s = f(t) = 16t^2$.
a. Find $f(0)$, $f(1)$, and $f(2)$.
b. From a practical standpoint, what would you define the domain of f to be?

51. Suppose that consumers will buy q units of a certain product at a price of p dollars per unit. Then the function f given by $p = f(q) = 80 - 2q$ is called a *demand function* for the product. We say that price per unit, p, is a function of quantity demanded, q. Find $f(30)$. What does this number represent? Also find $f(25)$, $f(10)$, and $f(1)$.

52. The formula for the area A of a circle of radius r is $A = \pi r^2$. Is the area a function of the radius?

53. In the study of circular motion, the formula

$$a = \frac{v^2}{20}$$

occurs. Is a a function of v?

54. The formula for the circumference C of a circle of diameter d is $C = \pi d$. Is C a function of d? Is d a function of C?

55. Suppose that a ball is thrown up in the air and the equation $s = 48t - 16t^2$ gives the height s (in feet) of the ball after t seconds. Find the heights when $t = 1$ and when $t = 2$. Is t a function of s? Is s a function of t?

56. Suppose $f(b) = ab^2 + a^2b$. (a) Find $f(a)$. (b) Find $f(ab)$.

In Problems **57–60,** *determine the domain of each function.*

57. $f(x) = \sqrt{x^2 + 4x - 12}$.

58. $g(x) = \sqrt{x^4 - x^2}$.

59. $g(t) = \sqrt{\dfrac{t-1}{t+5}}$.

60. $F(p) = \sqrt{\dfrac{p^2 - 4}{p+1}}$.

12.2 GRAPHS OF FUNCTIONS

Often it is useful to see what a function "looks like." This is done by graphing. For example, to graph the *absolute value function* $f(x) = |x|$, we let $y = f(x)$ and graph this equation. That is, we graph

$$y = f(x) = |x|,$$

or

$$y = |x|.$$

We choose various values of the independent variable x (input numbers) and find corresponding values of y (output numbers). See the table in Fig. 12-1. Plotting

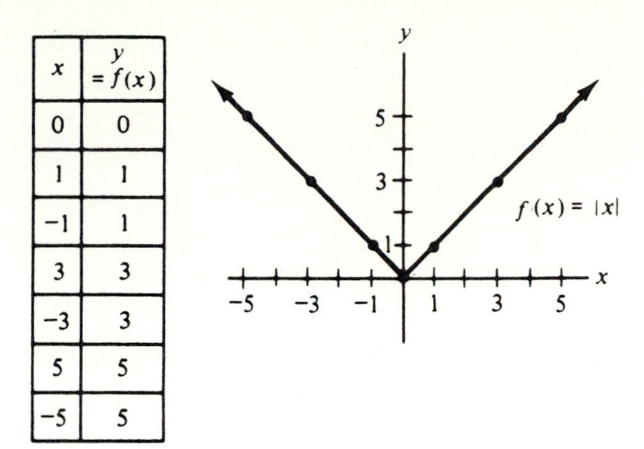

x	$y = f(x)$
0	0
1	1
−1	1
3	3
−3	3
5	5
−5	5

FIGURE 12-1

the points given by the table and connecting them, we get the graph of f (see Fig. 12-1). *We always label the horizontal axis with the independent variable.* The vertical axis can be labeled either y or $f(x)$ and can be called the **function-value axis.**

In Example 1, we graph a function of the form $f(x) = mx + b$. If $y = f(x)$, then $y = mx + b$, which has a straight line as its graph. We call the function $f(x) = mx + b$ a **linear function.** (Note that $mx + b$ is a polynomial of degree 1.) Thus

the graph of a linear function is a straight line.

EXAMPLE 1

Graph $f(x) = 2x - 1$.

Since f is a linear function, its graph is a *straight line.* Thus we need to plot only two points to draw the line, as in Fig. 12-2. The function-value axis is labeled $f(x)$.

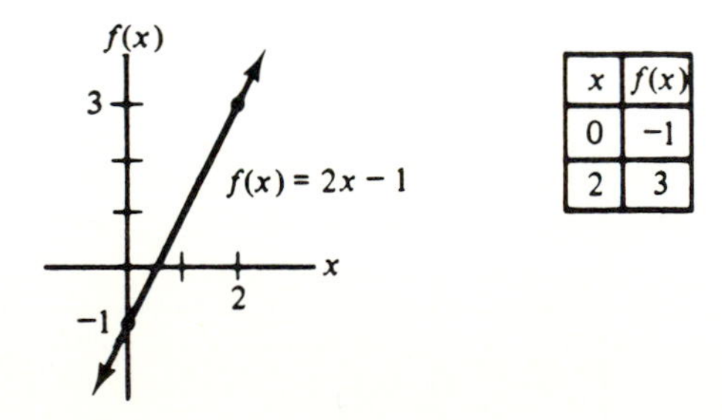

x	$f(x)$
0	−1
2	3

FIGURE 12-2

Figure 12-3 shows the graph of some function $y = f(x)$. Corresponding to the input number x on the horizontal axis is the output number $f(x)$ on the vertical

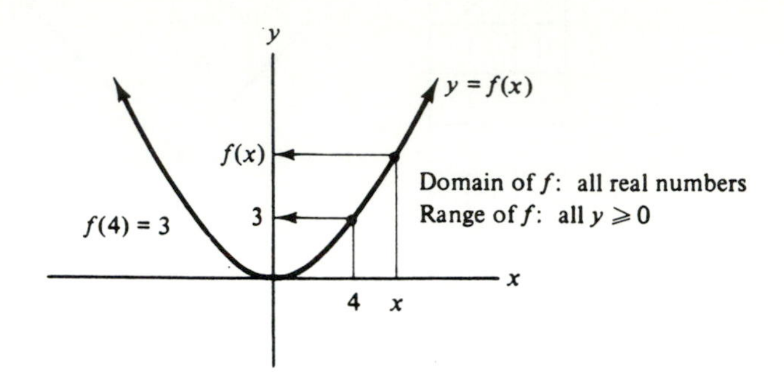

FIGURE 12-3

axis. For example, corresponding to the input 4 is the output 3, so $f(4) = 3$. From the graph, it seems reasonable to assume that there is an output number for any value of x, so the domain of f is all real numbers. Notice that the y-coordinates of all points on the graph are nonnegative, and for any $y \geq 0$ there is at least one x such that $y = f(x)$. Thus the range of f is all $y \geq 0$. This shows that we can make an "educated" guess about the domain and range of a function by looking at its graph. In general, the domain consists of all x-values that are included in the graph, and the range is all y-values that are included in the graph. For example, from Fig. 12-1 we see that the domain of $f(x) = |x|$ is all real numbers, and its range is all nonnegative numbers ($y \geq 0$).

EXAMPLE 2

Graph $G(u) = \sqrt{u}$.

Here u is the independent variable, so the horizontal axis is labeled u. The function-value axis is labeled $G(u)$. See Fig. 12-4. Recall that $\sqrt{u}$ denotes the *principal*

u	$G(u)$
0	0
$\frac{1}{4}$	$\frac{1}{2}$
1	1
4	2
9	3

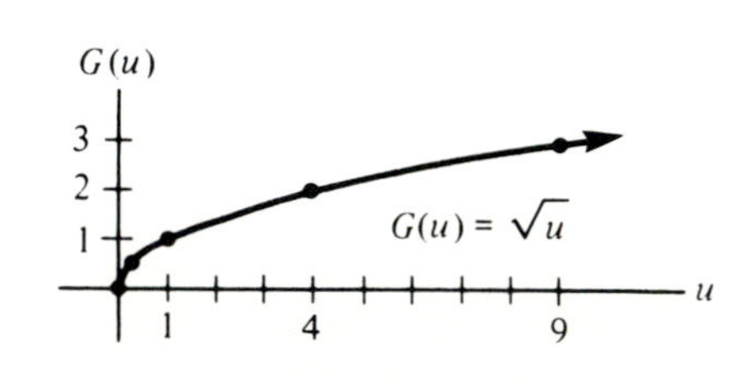

FIGURE 12-4

square root of u. Thus $G(9) = \sqrt{9} = 3$, not ± 3. Also, we cannot choose negative values for u because we don't want imaginary numbers for $\sqrt{u}$. That is, we must have $u \geq 0$, so the domain of G is all nonnegative numbers. From the graph, the range of G is clearly all nonnegative numbers.

EXAMPLE 3

Graph $f(x) = \frac{1}{x}$.

The domain of f is all numbers except 0, so the graph has *no* point corresponding to $x = 0$ (see Fig. 12-5). From the graph we see that the range of f is all real numbers except 0. In the table we chose some values of x between -1 and 1 so that we could see what happens to the graph for values of x *near* 0.

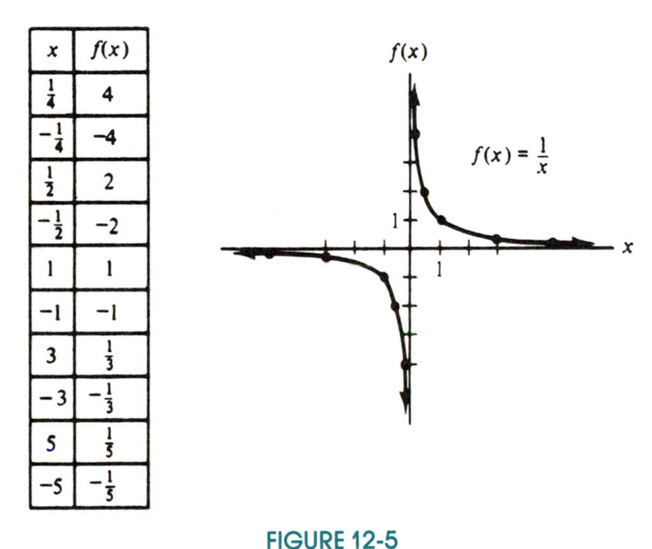

x	$f(x)$
$\frac{1}{4}$	4
$-\frac{1}{4}$	-4
$\frac{1}{2}$	2
$-\frac{1}{2}$	-2
1	1
-1	-1
3	$\frac{1}{3}$
-3	$-\frac{1}{3}$
5	$\frac{1}{5}$
-5	$-\frac{1}{5}$

FIGURE 12-5

EXAMPLE 4

Sketch the graph of the compound function

$$f(x) = \begin{cases} x, & \text{if } 0 \leq x < 3, \\ x - 1, & \text{if } 3 \leq x \leq 5, \\ 4, & \text{if } 5 < x \leq 7. \end{cases}$$

The domain of f is $0 \leq x \leq 7$. The graph is given in Fig. 12-6, where the *open dot* means that the point is *not* included in the graph. Notice that the range of f is all real numbers y such that $0 \leq y \leq 4$.

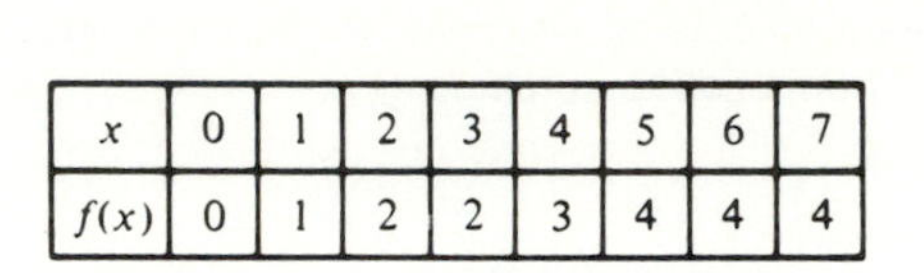

x	0	1	2	3	4	5	6	7
$f(x)$	0	1	2	2	3	4	4	4

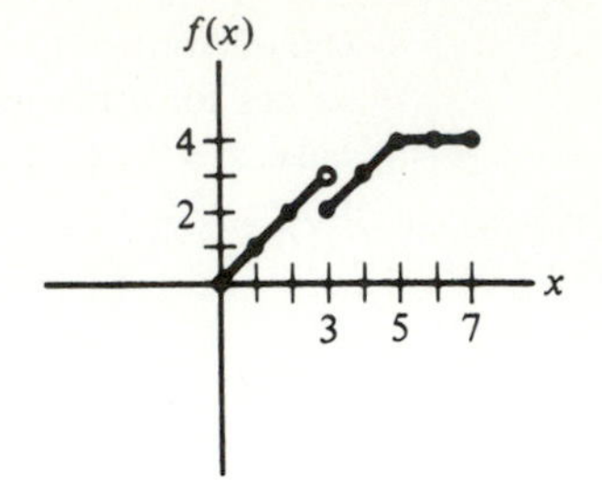

FIGURE 12-6

There is an easy way to tell whether or not a graph represents a function. Consider the leftmost diagram in Fig. 12-7(a). Notice that with the given input number x, there are *two* output numbers, y_1 and y_2. Thus the graph is *not* that of a function of x.

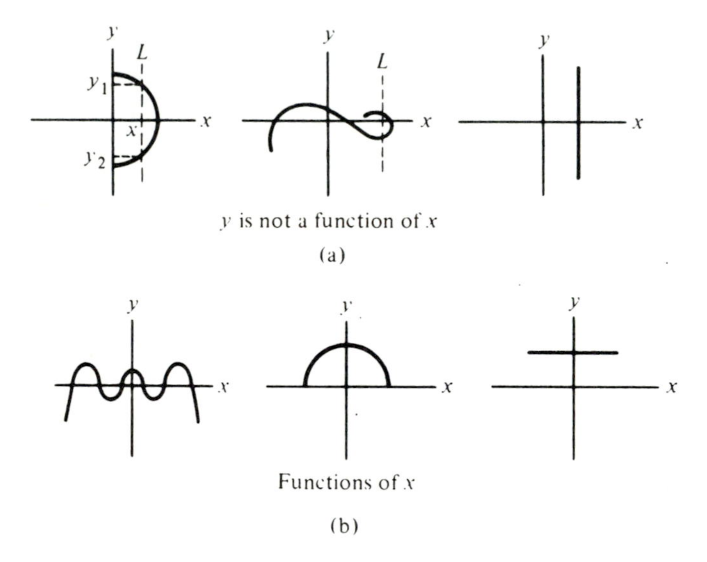

FIGURE 12-7

In general, if you can draw a *vertical* line L that intersects a graph in at least two points, then you *do not* have a function of x. When you can't draw such a line, then you do have a function of x. Thus the graphs in Fig. 12-7(a) do not represent functions of x, but those in Fig. 12-7(b) do.

Given a function $y = f(x)$, any value of x for which $f(x)$ is 0 is called a **zero** of f. For example, 3 is a zero of $f(x) = 2x - 6$, since $f(3) = 2(3) - 6 = 0$. We call 3 a *real* zero because it is a real number.

There is a connection between the real zeros and the graph of a function $y = f(x)$. The real zeros are the x-coordinates of the points where the graph touches the x-axis. To illustrate, Fig. 12-8 shows graphically that the real zeros of $f(x) = x^2 - 2x - 3 = (x - 3)(x + 1)$ are 3 and -1.

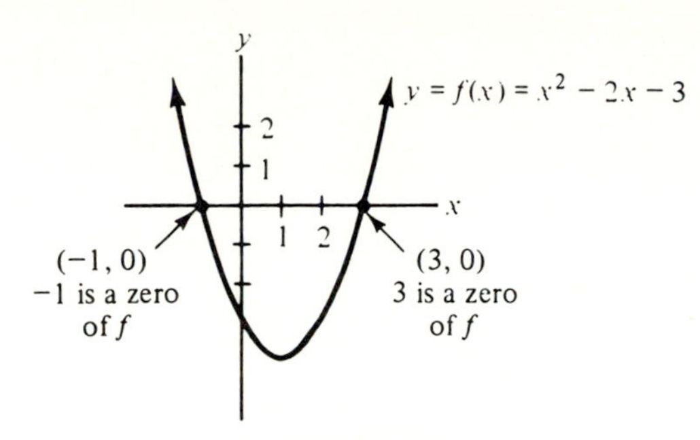

FIGURE 12-8

Exercise 12.2

In Problems **1–16**, *graph each function and give the domain and range.*

1. $f(x) = 2x + 2.$

2. $g(x) = 4 - x.$

3. $g(x) = |x - 2|.$

4. $f(x) = |x| - 2.$

5. $G(t) = 2\sqrt{t}.$

6. $F(t) = \sqrt{4t}.$

7. $f(x) = 4.$

8. $f(x) = 0.$

9. $h(x) = 4 - x^2.$

10. $h(x) = (x + 1)^2.$

11. $F(z) = \dfrac{3}{z}.$

12. $F(w) = -\dfrac{4}{w}.$

13. $f(x) = \dfrac{2}{x - 4}.$

14. $f(x) = \dfrac{1}{x + 2}.$

15. $f(x) = \begin{cases} x, & \text{if } 0 < x \leq 3, \\ 4, & \text{if } 3 < x \leq 5, \\ -1, & \text{if } x > 5. \end{cases}$

16. $g(x) = \begin{cases} x + 6, & \text{if } x \geq 3, \\ x^2, & \text{if } x < 3. \end{cases}$

17. In Fig. 12-9 is the graph of $y = f(x)$. (a) Estimate $f(0), f(2), f(4)$, and $f(-2)$. (b) What is the domain of f? (c) What is the range of f? (d) What is a real zero of f?

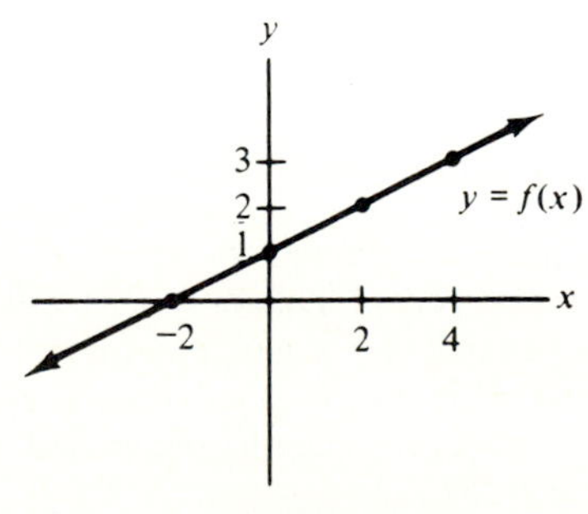

FIGURE 12-9

18. In Fig. 12-10 is the graph of $y = f(x)$. (a) Estimate $f(0)$ and $f(2)$. (b) What is the domain of f? (c) What is the range of f? (d) What is a real zero of f?

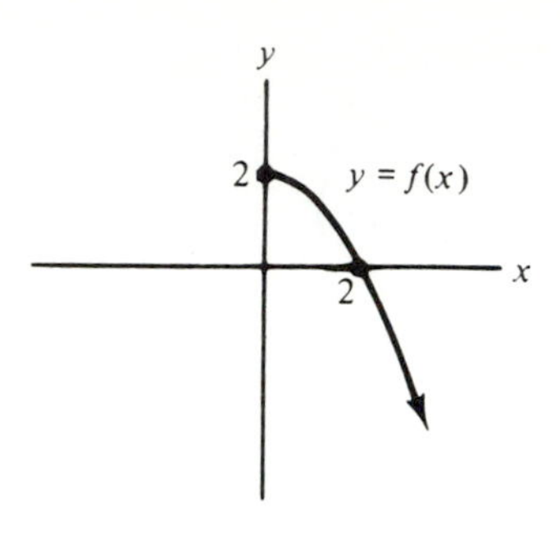

FIGURE 12-10

19. In Fig. 12-11 is the graph of $y = f(x)$. (a) Estimate $f(0)$, $f(1)$, and $f(-1)$. (b) What is the domain of f? (c) What is the range of f? (d) What is a real zero of f?

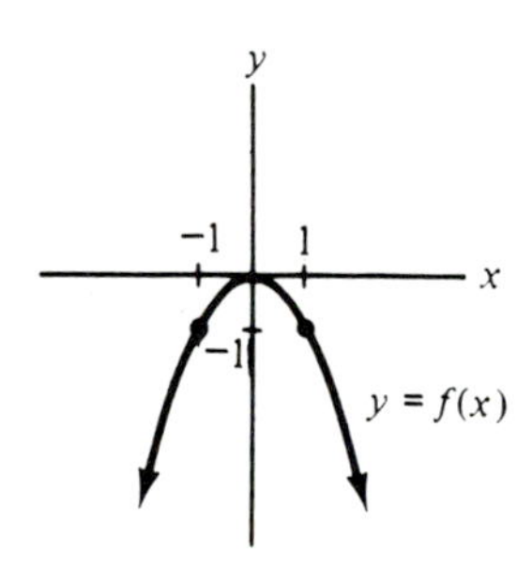

FIGURE 12-11

20. The electric potential V (in volts) as a function of its distance r (in meters) from an object having a certain electric charge is given by

$$V = \frac{100}{r}.$$

Here r must be positive. Sketch the graph of this function.

21. Table 12-2 is called a *demand schedule*. It gives a correspondence between the quantity q of a certain product that consumers will buy (demand) at the price p per unit.

(a) Suppose we think of q as an input number and p as an output number. With each input number there corresponds exactly one output number. Thus, price per unit, p, is a function of the quantity demanded, q. Call this function f. Find $f(5)$, $f(10)$, $f(20)$, and $f(25)$.

TABLE 12-2 Demand Schedule

QUANTITY DEMANDED q	PRICE PER UNIT p
5	20
10	10
20	5
25	4

(b) Draw a rectangular coordinate plane and label the horizontal axis q and the vertical axis p. Plot each quantity-price pair [for example, plot (5, 20), (10, 10), and so on]. Connect the points by a smooth curve. In this way we can approximate quantity-price pairs in between the given data. This curve is called a *demand curve.*

22. The dividends paid per share (in cents) by a corporation during the last 10 years are as follows.

year	1	2	3	4	5	6	7	8	9	10
dividend	5	5	5	6	$11\frac{1}{2}$	15	20	22	$27\frac{1}{4}$	30

These data give rise to a function f—namely, the one defined by thinking of the year y as input and the corresponding dividend d as output. Thus, $d = f(y)$. For example, $f(1) = 5, f(2) = 5$, and $f(8) = 22$. Plot the year-dividend pairs (1, 5), (2, 5), (8, 22), and so on, and connect the points by a smooth curve.

23. In Fig. 12-12, which graphs represent functions of x?

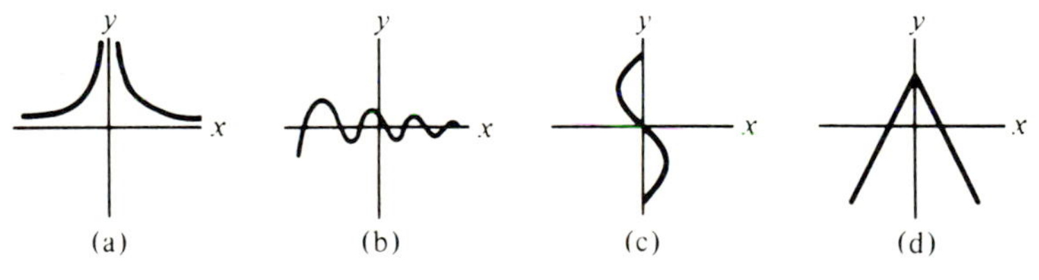

FIGURE 12-12

24. Sketch the graph of

$$y = f(x) = \begin{cases} -100x + 600, & \text{if } 0 \leq x < 5, \\ -100x + 1100, & \text{if } 5 \leq x < 10, \\ -100x + 1600, & \text{if } 10 \leq x < 15. \end{cases}$$

A function such as this might describe the inventory y of a company at time x.

25. In a psychological experiment on visual information, a subject briefly viewed an array of letters and was then asked to recall as many letters from the array as possible.* The procedure was repeated several times. Suppose y is the average number of letters recalled from arrays with x letters. The graph of the results is approximately the graph of

$$y = f(x) = \begin{cases} x, & \text{if } 0 \le x \le 4, \\ \frac{1}{2}x + 2, & \text{if } 4 < x \le 5, \\ 4.5, & \text{if } 5 < x \le 12. \end{cases}$$

Graph this function.

12.3 GRAPHS OF QUADRATIC FUNCTIONS

We now consider the graph of the function

$$y = f(x) = ax^2 + bx + c \qquad (a \neq 0),$$

where a, b, and c are constants. This is called a **quadratic function** because $ax^2 + bx + c$ is a polynomial of degree two. For example, the functions $f(x) = x^2 - 3x + 2$ and $y = -3t^2$ are quadratic, but $g(x) = \frac{1}{x^2}$ is not.

The graph of the quadratic function $y = f(x) = ax^2 + bx + c$ is called a **parabola** and has a shape like the curves in Fig. 12-13. If $a > 0$, the parabola *opens upward* [Fig. 12-13(a)]. If $a < 0$, then the parabola *opens downward* [Fig. 12-13(b)].

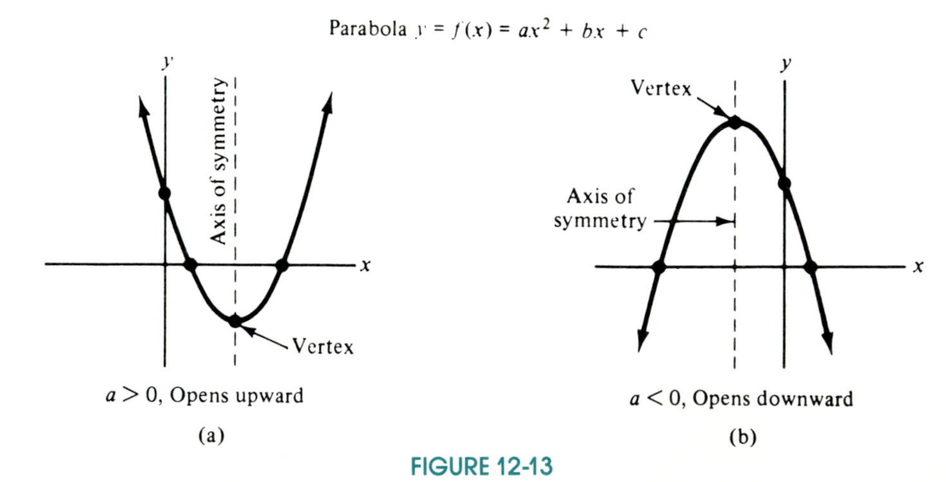

FIGURE 12-13

* Adapted from G. R. Loftus and E. F. Loftus, *Human Memory: The Processing of Information*. (New York: L. Erlbaum Associates, distributed by the Halsted Press Division of John Wiley, 1976).

Each parabola in Fig. 12-13 is *symmetric* about a vertical line, called the **axis of symmetry** of the parabola. That is, if the page were folded on one of these lines, then the two halves of the corresponding parabola would coincide. The axis (of symmetry) is *not* part of the parabola.

Figure 12-13 also shows points labeled **vertex.** At such a point, the axis of symmetry cuts the parabola. If $a > 0$, the vertex is the *lowest* point on the parabola. This means that y has a minimum value at this point. Similarly, if $a < 0$, the vertex is the *highest* point on the parabola. Here y has its maximum value.

Figure 12-14 shows a parabola $y = f(x) = ax^2 + bx + c$. We can find a formula for the coordinates of its vertex. The point where the parabola intersects the

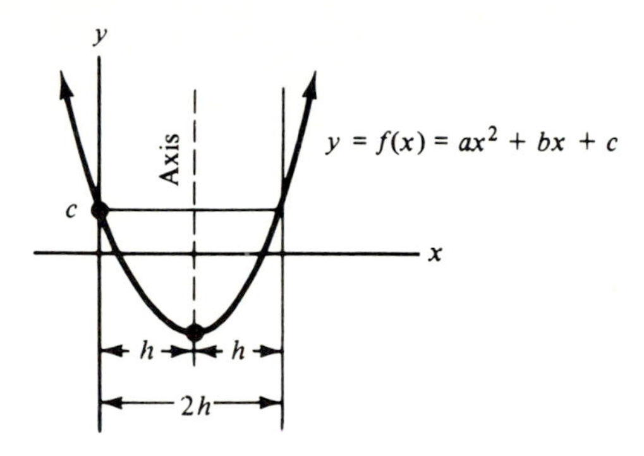

FIGURE 12-14

y-axis (that is, the *y-intercept*) occurs when $x = 0$. The y-coordinate of the point is $y = a(0)^2 + b(0) + c = c$, so the y-intercept is $(0, c)$. More simply, we say that the y-intercept is c. Suppose the x-coordinate of the vertex is h. Then, because of symmetry, when x is $2h$ the y-coordinate should also be c. That is, the point $(2h, c)$ must satisfy $ax^2 + bx + c = y$. This means

$$a(2h)^2 + b(2h) + c = c,$$

$$4ah^2 + 2bh = 0,$$

$$2h(2ah + b) = 0.$$

Setting each factor equal to 0 and solving for h gives $h = 0$ or $h = -\frac{b}{2a}$. It can be shown that $h = -\frac{b}{2a}$ gives the x-coordinate of the vertex. Since the y-coordinate of the vertex is $f\left(-\frac{b}{2a}\right)$, we have

$$\text{vertex} = \left(-\frac{b}{2a}, f\left(-\frac{b}{2a}\right)\right).$$

In summary, we have the following.

> The quadratic function $y = ax^2 + bx + c$ has a parabola for its graph.
>
> 1. If $a > 0$, the parabola opens upward.
>   If $a < 0$, the parabola opens downward.
> 2. The vertex is $\left(-\frac{b}{2a}, f\left(-\frac{b}{2a}\right)\right)$.
> 3. The y-intercept is c.

We can quickly sketch the graph of a quadratic function by first finding the vertex. Passing a (broken) vertical line through the vertex gives the axis of symmetry. We then plot some points to one side of the axis and, by using symmetry, obtain corresponding points on the other side. The y-intercept is a convenient point to plot. Once these points are found, it is easy to draw a parabola through them.

EXAMPLE 1

Graph the quadratic function $y = f(x) = -x^2 - 4x + 12$.

Here $a = -1$, $b = -4$, and $c = 12$. Because $a < 0$, the parabola opens downward. The x-coordinate of the vertex is

$$-\frac{b}{2a} = -\frac{-4}{2(-1)} = -2.$$

The y-coordinate is $f(-2) = -(-2)^2 - 4(-2) + 12 = 16$. Thus the vertex (highest point) is $(-2, 16)$. After plotting the vertex, we draw the axis by passing a vertical line through it (see Fig. 12-15). Now let's plot two points to the right of the axis. One can be the y-intercept, which is 12 since $c = 12$. For another point, let $x = 2$. Then $y = f(2) = -(2)^2 - 4(2) + 12 = 0$, so $(2, 0)$ lies on the graph. By symmetry we can get

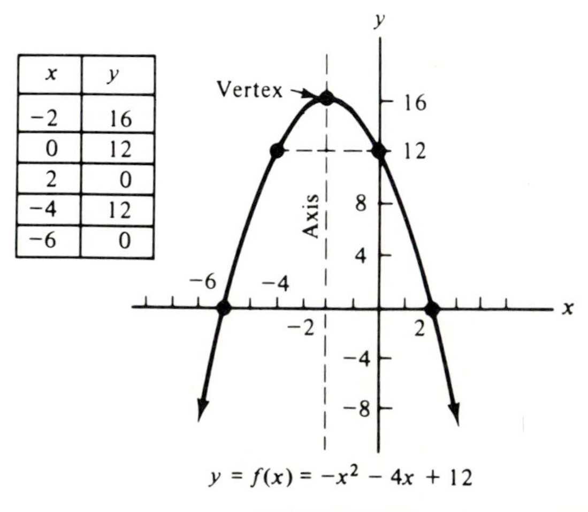

x	y
−2	16
0	12
2	0
−4	12
−6	0

FIGURE 12-15

two more points to the *left* of the axis. For example, since $(0, 12)$ is *two* units to the *right* of the axis, there is a corresponding point *two* units to the *left* of the axis with the same y-coordinate. Thus we get the point $(-4, 12)$. Similarly, to the point $(2, 0)$—which is four units to the right of the axis—there corresponds the point $(-6, 0)$. Thus we get the table and graph in Fig. 12-15.

In Example 1, two of the plotted points were $(2, 0)$ and $(-6, 0)$. At these points notice that the parabola intersects the x-axis; that is, these points are the ***x*-intercepts**. Sometimes it is convenient to use the x-intercepts when you are sketching a parabola. They are obtained by setting $y = 0$ and solving for x. To illustrate, for the parabola in Example 1 we have

$$0 = -x^2 - 4x + 12,$$

$$0 = x^2 + 4x - 12,$$

$$0 = (x + 6)(x - 2).$$

Thus $x = -6$ or $x = 2$, so the x-intercepts are $(-6, 0)$ and $(2, 0)$. More simply, we say that the x-intercepts are -6 and 2.

EXAMPLE 2

Graph the following quadratic functions.

a. $f(x) = 2x^2 + 2x + 3$.

Here $a = 2$, $b = 2$, and $c = 3$. Since $a > 0$, the parabola opens upward. The x-coordinate of the vertex is

$$-\frac{b}{2a} = -\frac{2}{2(2)} = -\frac{1}{2}.$$

Since $f(-\frac{1}{2}) = 2(-\frac{1}{2})^2 + 2(-\frac{1}{2}) + 3 = \frac{5}{2}$, the vertex is $(-\frac{1}{2}, \frac{5}{2})$. The y-intercept is $c = 3$. If $x = 1$, then $f(x) = f(1) = 7$. This gives the point $(1, 7)$. Using symmetry with the points $(0, 3)$ and $(1, 7)$, we obtain the points $(-1, 3)$ and $(-2, 7)$. See Fig. 12-16.

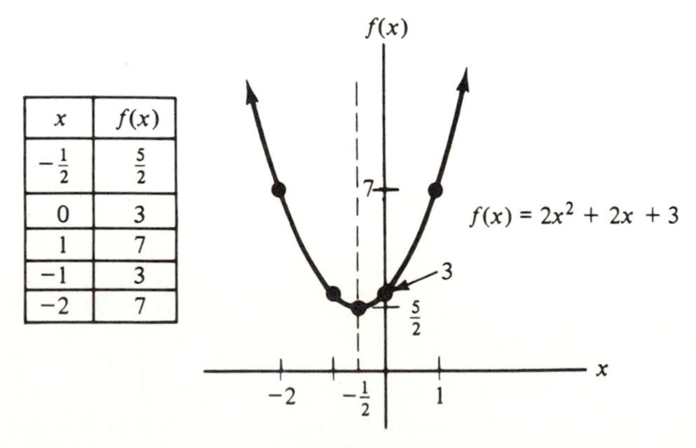

x	$f(x)$
$-\frac{1}{2}$	$\frac{5}{2}$
0	3
1	7
−1	3
−2	7

FIGURE 12-16

b. $p = 2q^2$.

Here p is a quadratic function of q, where $a = 2$, $b = 0$, and $c = 0$. Since $a > 0$, the parabola opens upward. The q-coordinate of the vertex is

$$-\frac{b}{2a} = -\frac{0}{2(2)} = 0,$$

and the p-coordinate is $p = 2(0)^2 = 0$. Thus the vertex is $(0, 0)$. In this case the p-axis is the axis of symmetry. If $q = 2$, then $p = 8$. This gives the point $(2, 8)$ and, by symmetry, the point $(-2, 8)$. See Fig. 12-17.

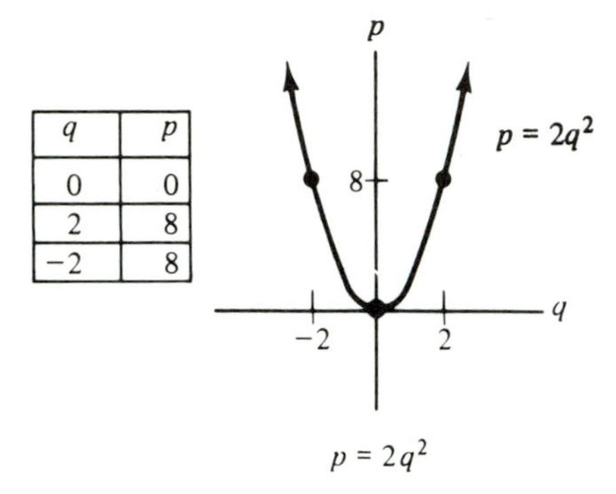

q	p
0	0
2	8
−2	8

FIGURE 12-17

The following examples show some applications of quadratic functions.

EXAMPLE 3

The height s of a ball thrown vertically upward from the ground is given by

$$s = 19.6t - 4.9t^2,$$

where s is in meters and t is elapsed time in seconds. When does the ball reach its greatest height and what is that height?

Note that s is a quadratic function of t, where $a = -4.9$, $b = 19.6$, and $c = 0$. Because $a < 0$ (the parabola opens downward), s has its maximum value at the vertex, where

$$t = -\frac{b}{2a} = -\frac{19.6}{2(-4.9)} = 2$$

and $s = 19.6(2) - 4.9(2)^2 = 19.6$. Thus the maximum height is 19.6 m, which occurs after 2 s. Figure 12-18 shows the graph of the given function. Only the part for which $t \geq 0$ and $s \geq 0$ is drawn, because the elapsed time and height cannot be negative.

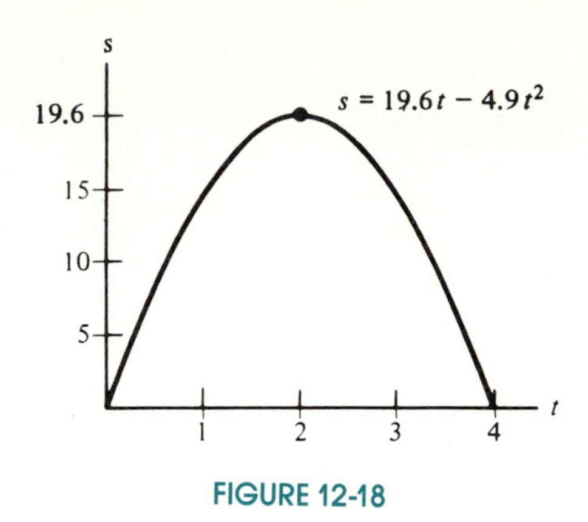

FIGURE 12-18

EXAMPLE 4

A rectangle is to have a perimeter of 20 m. *What should the length and width be so that the rectangle has the maximum area?*

Let x be the length (in meters) of one side of the rectangle (see Fig. 12-19). Then the opposite side also has length x. This leaves $20 - 2x$ to be divided equally between

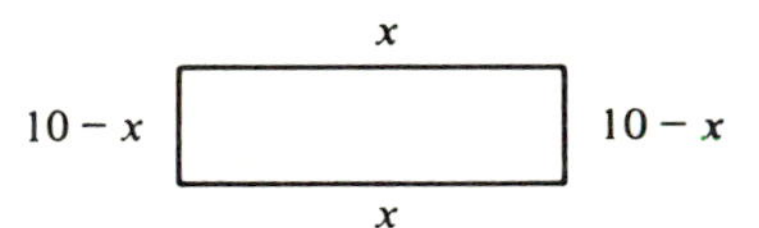

FIGURE 12-19

the other two sides. Thus each of these sides has length $(20 - 2x)/2$, which simplifies to $10 - x$. Let A be the area of the rectangle. Then

$$\text{area} = (\text{length})(\text{width}).$$

$$A = x(10 - x),$$

$$A = 10x - x^2.$$

Here A is a quadratic function of x, where $a = -1$ and $b = 10$. Because $a < 0$, A has a maximum value when

$$x = -\frac{b}{2a} = -\frac{10}{2(-1)} = 5.$$

Therefore, to get maximum area, the length should be 5 m and the width should be $10 - x = 10 - 5 = 5$ m. In this case the rectangle is a square of area 25 m^2.

Exercise 12.3

In Problems **1–8**, *state whether or not the function is quadratic.*

1. $f(x) = 26 - 3x$.

2. $g(x) = (7 - x)^2$.

3. $g(x) = 4x^2$.

4. $h(s) = 6(4s + 1)$.

5. $h(q) = \frac{1}{2q - 4}$.

6. $f(t) = 2t(3 - t) + 4t$.

7. $f(s) = \frac{s^2 - 4}{2}$.

8. $g(t) = (t^2 - 1)^2$.

In Problems **9–12**, *do not include a graph.*

9. For the parabola $y = f(x) = -4x^2 + 8x + 7$, (a) find the vertex. (b) Does the vertex correspond to the highest point, or the lowest point, on the graph?

10. Repeat Problem 9 if $y = f(x) = 8x^2 + 4x - 1$.

11. For the parabola $y = f(x) = x^2 + 2x - 8$, find (a) the y-intercept, (b) the x-intercepts, and (c) the vertex.

12. Repeat Problem 11 if $y = f(x) = 3 + x - 2x^2$.

In Problems **13–22**, *graph each function. Give the vertex.*

13. $f(x) = x^2 - 6x + 5$.

14. $f(x) = -3x^2$.

15. $y = -2x^2 - 6x$.

16. $y = x^2 - 1$.

17. $s = h(t) = t^2 + 2t + 1$.

18. $s = h(t) = 2t^2 + 3t - 2$.

19. $y = -9 + 8x - 2x^2$.

20. $y = 1 - x - x^2$.

21. $t = f(s) = s^2 - 8s + 13$.

22. $t = f(s) = s^2 + 6s + 11$.

In Problems **23–26**, *graph each function. Give the vertex and intercepts.*

23. $y = -x^2 + x + 6$.

24. $y = x^2 + 3x$.

25. $y = x^2 - 6x + 7$.

26. $y = 1 + 2x - x^2$.

In Problems **27–30**, *state whether $f(x)$ has a maximum value or a minimum value, and find that value.*

27. $f(x) = 100x^2 - 20x + 25$.

28. $f(x) = -2x^2 - 16x + 3$.

29. $f(x) = 4x - 50 - .1x^2$.

30. $f(x) = x(x + 3) - 12$.

31. The height s of a ball thrown vertically upward from the ground is given by

$$s = -4.9t^2 + 58.8t,$$

where s is in meters and t is elapsed time in seconds. After how many seconds will the ball reach its maximum height? What is the maximum height?

32. Biologists studied the nutritional effects on rats that were fed a diet containing 10% protein*. The protein consisted of yeast and corn flour. By varying the percentage P of yeast in the protein mix, the biologists estimated that the average weight gain (in grams) of a rat over a period of time was given by

$$f(P) = -\frac{1}{50}P^2 + 2P + 20, \qquad 0 \leq P \leq 100.$$

Find the maximum weight gain.

33. The total weekly revenue r (in dollars) when q units of a manufacturer's product are sold is given by $r = 1000q - 2q^2$. Find the value of q that maximizes total revenue and determine this revenue.

34. Express the area of the rectangle shown in Fig. 12-20 as a quadratic function of x. For what value of x will the area be a maximum?

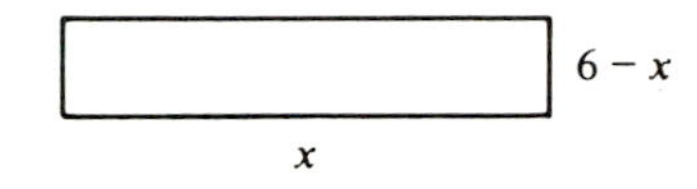

FIGURE 12-20

35. The owner of a garden center wants to fence a rectangular plot adjacent to a building by using the building as one side of the enclosed area. If the owner has 200 yd of fence, what should be the dimensions of the enclosed plot if the area is to be a maximum?

36. Find two numbers whose sum is 40 and whose product is a maximum.

37. A real estate firm owns 80 garden-type apartments. At \$300 per month each apartment can be rented. However, for each \$10 per month increase, there will be two vacancies with no possibility of filling them. What rent per apartment will maximize monthly revenue?

38. A television cable company has 2000 subscribers who are paying \$10 per month. It can get 200 more subscribers for each \$.20 decrease in the monthly rate. What rate will yield maximum revenue, and what will this revenue be?

39. A company has set aside \$3000 to fence a rectangular portion of land adjacent to a stream by using the stream for one side of the enclosed area. The cost of the fencing parallel to the stream is \$5 per foot installed, and the fencing for the remaining two sides is \$3 per foot installed. Find the dimensions of the maximum enclosed area. (*Hint*: If the side opposite the stream has length x, first show that each of the other two sides has length $(3000 - 5x)/6$.)

---

* Adapted from R. Bressani, "The Use of Yeast in Human Foods," ed. Mateles and Tannenbaum, *Single-Cell Protein* (Cambridge, Mass.: M.I.T. Press, 1968).

12.4 NEW FUNCTIONS FROM OLD ONES

There are different ways of combining two functions to get a new function. Suppose f and g are the functions

$$f(x) = x^2 \quad \text{and} \quad g(x) = 3x.$$

We can define a new function called the **sum** of f and g, denoted by $\boldsymbol{f+g}$. Its function value at x is $f(x) + g(x)$. That is,

$$(f + g)(x) = f(x) + g(x) = x^2 + 3x.$$

For example,

$$(f + g)(2) = 2^2 + 3(2) = 10.$$

Similarly, we define the **difference** $\boldsymbol{f-g}$, **product** $\boldsymbol{fg}$, and **quotient** $\boldsymbol{\frac{f}{g}}$ as follows*:

$$(f - g)(x) = f(x) - g(x) = x^2 - 3x,$$
$$(fg)(x) = f(x) \cdot g(x) = x^2(3x) = 3x^3,$$
$$\left(\frac{f}{g}\right)(x) = \frac{f(x)}{g(x)} = \frac{x^2}{3x} = \frac{x}{3}, \qquad \text{for } x \neq 0.$$

EXAMPLE 1

Let $f(x) = 3x^2 + 4$ and $g(x) = x^2 - x + 5$. Find $(f - g)(x)$ and use the result to find $(f - g)(3)$.

$$(f - g)(x) = f(x) - g(x) = (3x^2 + 4) - (x^2 - x + 5)$$
$$= 2x^2 + x - 1 \qquad \text{[simplifying]}.$$
$$(f - g)(3) = 2(3)^2 + 3 - 1 = 20.$$

EXAMPLE 2

Let $f(x) = x + 1$ and $g(x) = x + 2$. Find $(fg)(x)$ and $(fg)(-3)$.

$$(fg)(x) = f(x) \cdot g(x) = (x + 1)(x + 2)$$
$$= x^2 + 3x + 2.$$
$$(fg)(-3) = (-3)^2 + 3(-3) + 2 = 2.$$

---

* In each of the four combinations, we assume that x is in the domains of both f and g. In the quotient we also do not allow any value of x for which $g(x)$ is 0.

EXAMPLE 3

Let $f(x) = x$ *and* $g(x) = x - 5$. *Find* $\left(\frac{f}{g}\right)(x)$ *and* $\left(\frac{g}{f}\right)(x)$.

$$\left(\frac{f}{g}\right)(x) = \frac{f(x)}{g(x)} = \frac{x}{x-5}.$$

$$\left(\frac{g}{f}\right)(x) = \frac{g(x)}{f(x)} = \frac{x-5}{x}.$$

We can also combine two functions by first applying one function to a number and then applying the other function to the result. For example, suppose $f(x) = x^2$, $g(x) = 3x$, and $x = 2$. Then $g(2) = 3(2) = 6$. Thus g sends the input 2 into the output 6.

$$2 \xrightarrow{g} 6.$$

Let's see what f does to 6. That is, the output 6 becomes input to f.

$$f(6) = 6^2 = 36,$$

so f sends 6 into 36.

$$6 \xrightarrow{f} 36.$$

By first applying g and then f, we get a new function h, which sends 2 into 36.

$$\underbrace{2 \xrightarrow{g} 6 \xrightarrow{f} 36}_{h}.$$

Thus $h(2) = 36$. In the previous diagram, let's replace the 6 and the 36 by functional notation. Since

$$6 = g(2) \quad \text{and} \quad 36 = f(6) = f(g(2)),$$

where $f(g(2))$ is read "f of g of 2," our diagram becomes

$$\underbrace{2 \xrightarrow{g} g(2) \xrightarrow{f} f(g(2))}_{h},$$

Consequently,

$$h(2) = f(g(2)).$$

To be more general, let's replace the 2 by the variable x. Then we have $h(x) = f(g(x))$. We can think of $f(g(x))$ as a function of a function. This means that g is applied to x, and the output $g(x)$ serves as the input to f. The function h is called the *composition of* f *with* g and is denoted by $f \circ g$.

> If f and g are functions, then the **composition function of f with g**, denoted by $f \circ g$, is given by*
>
> $$(f \circ g)(x) = f(g(x)).$$

For $f(x) = x^2$ and $g(x) = 3x$, we can get a simple form for $f \circ g$:

$$(f \circ g)(x) = f(g(x)) = f(3x) = (3x)^2 = 9x^2.$$

For example, $(f \circ g)(2) = 9(2)^2 = 36$, as we saw before.

EXAMPLE 4

Let $f(x) = 2x + 3$ and $g(x) = x - 4$. Find $(f \circ g)(x)$ and $(f \circ g)(4)$.

$$\begin{aligned}(f \circ g)(x) &= f(g(x)) = f(x - 4)\\ &= 2(x - 4) + 3\\ &= 2x - 5.\\ (f \circ g)(4) &= 2(4) - 5 = 3.\end{aligned}$$

EXAMPLE 5

Let $f(x) = 2x^2$ and $g(x) = 3x + 1$. Find $(f \circ g)(x)$ and $(g \circ f)(x)$.

$$(f \circ g)(x) = f(g(x)) = f(3x + 1) = 2(3x + 1)^2.$$

$$(g \circ f)(x) = g(f(x)) = g(2x^2) = 3(2x^2) + 1 = 6x^2 + 1.$$

Generally, $f \circ g \neq g \circ f$. In Example 5, notice that $(f \circ g)(x) = 2(3x + 1)^2$, while $(g \circ f)(x) = 6x^2 + 1$. Also, do not confuse the product $f(x)g(x)$ with $f(g(x))$.

Sometimes we can think of a particular function in terms of a composition, as Example 6 shows.

EXAMPLE 6

The function $h(x) = (2x - 1)^3$ can be considered to be a composition. We note that $h(x)$ is obtained by finding $2x - 1$ and cubing the result. Suppose we let $g(x) = 2x - 1$ and $f(x) = x^3$. Then

---

* The domain of $f \circ g$ consists of all x in the domain of g such that $g(x)$ is in the domain of f.

$$h(x) = (2x-1)^3 = [g(x)]^3 = f(g(x)) = (f \circ g)(x),$$

which gives h as a composition of two functions.

Exercise 12.4

1. If $f(x) = x + 1$ and $g(x) = x + 4$, find the following.

a. $(f+g)(x)$. b. $(f+g)(0)$. c. $(f-g)(x)$.

d. $(fg)(x)$. e. $(fg)(-2)$. f. $\left(\frac{f}{g}\right)(x)$.

g. $(f \circ g)(x)$. h. $(f \circ g)(3)$. i. $(g \circ f)(x)$.

2. If $f(x) = 8x$ and $g(x) = 8 + x$, find the following.

a. $(f+g)(x)$. b. $(f-g)(x)$. c. $(f-g)(4)$.

d. $(fg)(x)$. e. $\left(\frac{f}{g}\right)(x)$. f. $\left(\frac{f}{g}\right)(2)$.

g. $(f \circ g)(x)$. h. $(g \circ f)(x)$. i. $(g \circ f)(2)$.

3. If $f(x) = x^2$ and $g(x) = x^2 + x$, find the following.

a. $(f+g)(x)$. b. $(f-g)(x)$. c. $(f-g)(-\frac{1}{2})$.

d. $(fg)(x)$. e. $\left(\frac{f}{g}\right)(x)$. f. $\left(\frac{f}{g}\right)(-\frac{1}{2})$.

g. $(f \circ g)(x)$. h. $(g \circ f)(x)$. i. $(g \circ f)(-3)$.

4. If $f(x) = x^2 - 1$ and $g(x) = 4$, find the following.

a. $(f+g)(x)$. b. $(f+g)(\frac{1}{2})$. c. $(f-g)(x)$.

d. $(fg)(x)$. e. $(fg)(4)$. f. $\left(\frac{f}{g}\right)(x)$.

g. $(f \circ g)(x)$. h. $(f \circ g)(100)$. i. $(g \circ f)(x)$.

In Problems **5–8**, *find* $(f \circ g)(x)$ *and* $(g \circ f)(x)$.

5. $f(x) = \sqrt{x}$, $g(x) = x - 1$.

6. $f(x) = \frac{1}{x}$, $g(x) = 3$.

7. $f(x) = \frac{1}{x}$, $g(x) = x^2 - 1$.

8. $f(x) = x^2 - 1$, $g(x) = x^2 + 1$.

In Problems **9–14**, *find functions f and g such that* $h(x) = (f \circ g)(x)$.

9. $h(x) = (x^2 + 2)^2$.

10. $h(x) = \sqrt{x - 2}$.

11. $h(x) = (x^2 - 1)^2 + 2(x^2 - 1)$.

12. $h(x) = (3x^3 - 2x)^3 - (3x^3 - 2x)^2 + 7$.

13. $h(x) = \sqrt[5]{\frac{x+1}{3}}$.

14. $h(x) = \frac{x+1}{(x+1)^2+2}$.

15. A manufacturer determines that the total number of units of output per day, q, is a function of the number of employees, m, where $q = f(m) = (40m - m^2)/4$. The total revenue, r, that is received for selling q units is given by the function g, where $r = g(q) = 40q$. Determine $(g \circ f)(m)$. What does this composition function describe?

12.5 REVIEW

IMPORTANT TERMS AND SYMBOLS

function *(12.1)*
domain *(12.1)*
range *(12.1)*
independent variable *(12.1)*
dependent variable *(12.1)*
$f(x)$ *(12.1)*
function value *(12.1)*
constant function *(12.1)*
function-value axis *(12.2)*
linear function *(12.2)*
zero of function *(12.2)*
quadratic function *(12.3)*
parabola *(12.3)*
axis of symmetry *(12.3)*
vertex *(12.3)*
$f+g, f-g, fg, f/g$ *(12.4)*
composition function *(12.4)*
$f \circ g$ *(12.4)*

REVIEW PROBLEMS

In Problems **1–6**, *give the domain of each function.*

1. $f(x) = \frac{x}{x^2 - 3x + 2}$.

2. $g(x) = x^2 + 3x$.

3. $F(t) = 7t + 4t^2$.

4. $G(x) = 18$.

5. $h(x) = \frac{\sqrt{x}}{x-1}$.

6. $H(s) = \frac{\sqrt{s-5}}{4}$.

In Problems **7–14**, *find the function values for the given function.*

7. $f(x) = 3x^2 - 4x + 7$; $f(0)$, $f(-3)$, $f(5)$, $f(t)$.

8. $g(x) = 4$; $g(4)$, $g(\frac{1}{100})$, $g(-156)$, $g(x+4)$.

9. $G(x) = \sqrt{x-1}$; $G(1)$, $G(10)$, $G(t+1)$, $G(x^2)$.

10. $F(x) = \frac{x-3}{x+4}$; $F(-1)$, $F(0)$, $F(5)$, $F(x+3)$.

11. $h(u) = \frac{\sqrt{u+4}}{u}$; $h(5)$, $h(-4)$, $h(x)$, $h(u-4)$.

12. $H(s) = \frac{(s-4)^2}{3}$; $H(-2)$, $H(7)$, $H(\frac{1}{2})$, $H(x^2)$.

13. $f(x) = \begin{cases} 4, & \text{if } x < 2 \\ 8 - x^2, & \text{if } x > 2 \end{cases}$; $f(4)$, $f(-2)$, $f(0)$, $f(10)$.

14. $h(q) = \begin{cases} q, & \text{if } -1 \le q < 0 \\ 3 - q, & \text{if } 0 \le q < 3 \\ 2q^2, & \text{if } 3 \le q \le 5 \end{cases}$; $h(0)$, $h(4)$, $h(-\frac{1}{2})$, $h(\frac{1}{2})$.

In Problems **15** *and* **16**, *for the given function find (a)* $f(x+h)$, *and (b)* $\frac{f(x+h)-f(x)}{h}$, *and simplify your answers.*

15. $f(x) = 3 - 7x$.

16. $f(x) = x^2 + 4$.

In Problems **17–28**, *graph the given function. For those that are quadratic, give the vertex.*

17. $y = f(x) = 5x - 4$.

18. $f(x) = 5$.

19. $g(x) = 9 - x^2$.

20. $G(u) = \sqrt{u+4}$.

21. $F(t) = \frac{6}{t}$.

22. $f(x) = |x| + 1$.

23. $y = f(x) = \begin{cases} 1 - x, & \text{if } x \le 0, \\ 1, & \text{if } x > 0. \end{cases}$

24. $f(t) = t^2 + 2t$.

25. $h(t) = t^2 - 4t - 5$.

26. $s = g(t) = 8 - 2t - t^2$.

27. $y = F(x) = -x^2 - 2x - 3$.

28. $y = F(x) = (2x - 1)^2$.

29. If $f(x) = 3x - 1$ and $g(x) = 2x + 3$, find the following.

a. $(f+g)(x)$.
b. $(f+g)(4)$.
c. $(f-g)(x)$.
d. $(fg)(x)$.
e. $(fg)(1)$.
f. $\left(\frac{f}{g}\right)(x)$.
g. $(f \circ g)(x)$.
h. $(f \circ g)(5)$.
i. $(g \circ f)(x)$.

30. If $f(x) = x^2$ and $g(x) = 2x + 1$, find the following.

a. $(f+g)(x)$.
b. $(f-g)(x)$.
c. $(f-g)(-3)$.
d. $(fg)(x)$.
e. $\left(\frac{f}{g}\right)(x)$.
f. $\left(\frac{f}{g}\right)(2)$.
g. $(f \circ g)(x)$.
h. $(g \circ f)(x)$.
i. $(g \circ f)(-4)$.

In Problems **31–34**, *find* $(f \circ g)(x)$ *and* $(g \circ f)(x)$.

31. $f(x) = \frac{1}{x}$, $g(x) = x - 1$.

32. $f(x) = \frac{x+1}{4}$, $g(x) = \sqrt{x}$.

33. $f(x) = x + 2$, $g(x) = x^3$.

34. $f(x) = 2$, $g(x) = 3$.

35. Suppose that manufacturers will supply q units of a product at a price of p dollars per unit. Then the function f given by $p = f(q) = .5q + 10$ is called a *supply function* for the product. We say that price per unit p is a function of the quantity supplied q. Find $f(10)$. What does this number represent? Also, find $f(20)$, $f(100)$, and $f(500)$.

36. Table 12-3 is called a *supply schedule*. It gives the quantity q of a certain product that manufacturers will supply at the price p per unit.

(a) Suppose we think of q as an input number and p as an output number. With each input number there corresponds exactly one output number. Thus price per unit, p, is a function of quantity supplied, q. Call this function f. Find $f(30)$, $f(100)$, $f(150)$, $f(190)$, and $f(210)$.

TABLE 12-3 Supply Schedule

QUANTITY SUPPLIED q	PRICE PER UNIT p
30	10
100	20
150	30
190	40
210	50

(b) Draw a rectangular coordinate plane and label the horizontal axis q and the vertical axis p. Plot each quantity-price pair [plot (30, 10), (100, 20), and so on]. Connect these points by a smooth curve. In this way we can approximate the quantity-price pairs in between the given data. This curve is called a *supply curve*.

37. Figure 12-21 gives the graph of $y = f(x)$. (a) Find $f(0)$, $f(1)$, $f(10)$, and $f(-30)$. (b) What is the domain of f? (c) What is the range of f?

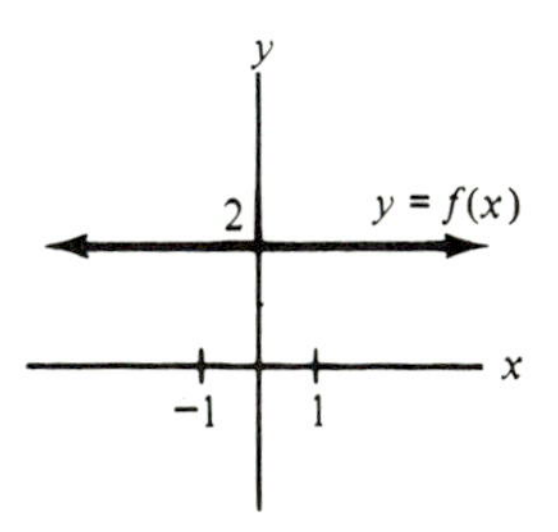

FIGURE 12-21

38. In Fig. 12-22, which graphs represent functions of x?

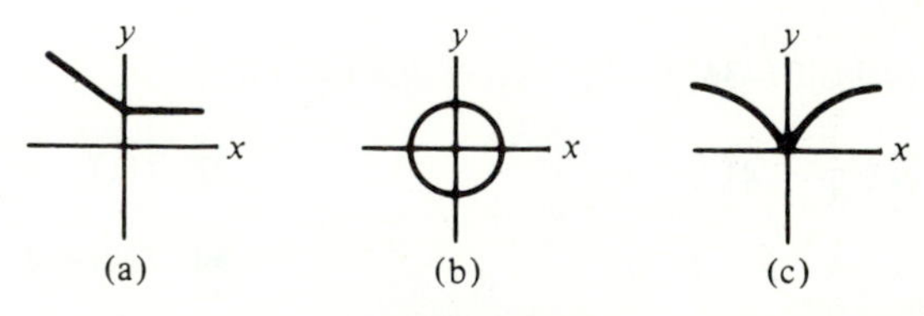

FIGURE 12-22

39. Find the maximum value of y if $y = -3x^2 + 6x - 4$.

40. Find the minimum value of y if $y = 4x^2 + 16x - 6$.

41. A marketing company estimates that n months after the introduction of a client's new product, $f(n)$ thousand households will use it, where

$$f(n) = \frac{40}{3}n - \frac{10}{9}n^2, \qquad 0 \le n \le 12.$$

Estimate the maximum number of households that will use the product.

42. A rectangular field is to be enclosed by a fence and divided equally into three parts by two fences parallel to one pair of the sides. If a total of 800 ft of fencing is to be used, find the dimensions of the field if its area is to be maximized.

43. The Vista TV Cable Company currently has 2000 subscribers who are paying a monthly rate of \$5. A survey reveals that there will be 50 more subscribers for each \$0.10 decrease in the rate. At what rate will maximum revenue be obtained, and how many subscribers will there be at this rate?

13 Exponential and Logarithmic Functions

13.1 EXPONENTIAL FUNCTIONS

In this chapter we'll look at two special types of functions. The first involves a positive constant raised to a variable power, such as $f(x) = 2^x$. We call this an *exponential function*. Notice that the base is a fixed number (2), and the exponent is a variable (x). Thus, if $x = 3$, then $f(3) = 2^3 = 8$. Another exponential function is $g(x) = 5^x$. In general we have the following definition.

> A function of the form
>
> $$y = f(x) = b^x,$$
>
> where $b > 0$, $b \neq 1$,* and x is any real number, is called an **exponential function** with base b.

Do not confuse the exponential function $y = 2^x$ with the *power function* $y = x^2$, which has a variable base and a constant exponent.

Since the exponent in b^x can be any real number, you may wonder how we assign a value to something like $4^{\sqrt{2}}$. Stated simply, we use approximations. First, $4^{\sqrt{2}}$ is approximately $4^{1.4} = 4^{7/5} = \sqrt[5]{4^7}$, which *is* defined. Better approximations

---

*If $b = 1$, then $f(x) = 1^x = 1$. This function is so uninteresting that we do not call it an exponential function.

are $4^{1.41} = 4^{141/100} = \sqrt[100]{4^{141}}$, and so on. In this way the meaning of $4^{\sqrt{2}}$ becomes clear.

When you work with exponential functions, it may be necessary to apply rules for exponents. You may assume that the usual rules for exponents hold for all real-number exponents. If m and n are real numbers and a and b are positive, these rules are as follows.

1. $a^m a^n = a^{m+n}$.	**2.** $\dfrac{a^m}{a^n} = a^{m-n}$.
3. $(a^m)^n = a^{mn}$.	**4.** $(ab)^n = a^n b^n$.
5. $\left(\dfrac{a}{b}\right)^n = \dfrac{a^n}{b^n}$.	**6.** $a^1 = a$.
7. $a^0 = 1$.	**8.** $a^{-n} = \dfrac{1}{a^n}$.

EXAMPLE 1

a. *Express a^{kx}, where k is a constant, in the form b^x.*

We have

$$a^{kx} = (a^k)^x = b^x, \qquad \text{where } b = a^k.$$

b. *Express a^{bx+c} in the form $k_1(k_2)^x$,* where *k_1 and k_2 are constants.*

We have

$$a^{bx+c} = a^{bx} a^c = a^c(a^b)^x = k_1(k_2)^x,$$

where $k_1 = a^c$ and $k_2 = a^b$. Thus we can think of a^{bx+c} as a "multiple" of the exponential function $(k_2)^x$.

Figure 13-1 shows the graphs of the exponential functions $y = 2^x$, $y = 3^x$, and $y = (\frac{1}{2})^x = 2^{-x}$. Since $b^0 = 1$ for every base b, all the graphs have y-intercept 1. They have no x-intercepts. The graphs can get very close to the x-axis, but they never touch it.

There are two basic shapes for the graphs of $y = b^x$. They depend on whether $b > 1$ or $0 < b < 1$. (Look at Fig. 13-1 again.)

> If $b > 1$, the graph of $y = b^x$ *rises* from left to right. In Quadrant I, the greater the value of b, the more quickly the graph rises. (Compare the graphs of $y = 2^x$ and $y = 3^x$.)
>
> If $0 < b < 1$, the graph of $y = b^x$ *falls* from left to right. [See the graph of $y = (\frac{1}{2})^x$.]

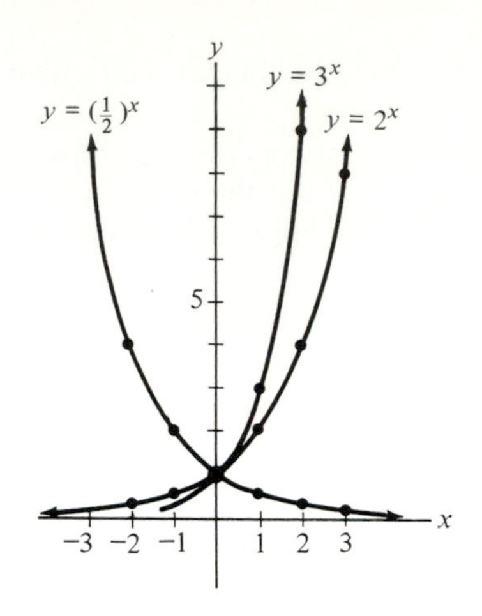

x	2^x	3^x	$(\frac{1}{2})^x$
−2	$\frac{1}{4}$	$\frac{1}{9}$	4
−1	$\frac{1}{2}$	$\frac{1}{3}$	2
0	1	1	1
1	2	3	$\frac{1}{2}$
2	4	9	$\frac{1}{4}$
3	8	27	$\frac{1}{8}$

FIGURE 13-1

Finally, we note the following.

1. The domain of an exponential function is all real numbers.
2. The range is all positive numbers.

An irrational number called e is often used as a base in $y = b^x$.

e is approximately 2.71828 . . . .*

The number e frequently occurs in studies of growth (populations) and decay (radioactive decay). Appendix B gives a table of (approximate) values of e^x and e^{-x}. The graph of $y = e^x$ is shown in Fig. 13-2.

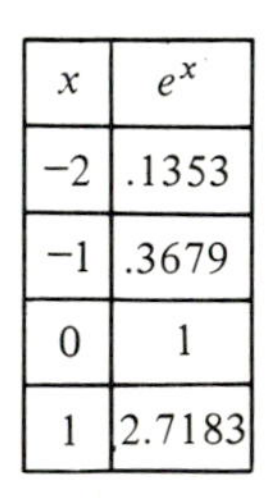

x	e^x
−2	.1353
−1	.3679
0	1
1	2.7183

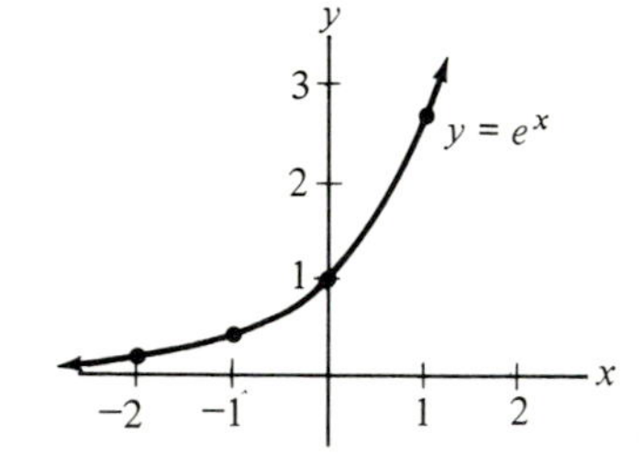

FIGURE 13-2

* We use the letter e in honor of the Swiss mathematician Leonhard Euler (1707–1783).

EXAMPLE 2

The projected population P of a certain city is given by

$$P = 100{,}000e^{.05t},$$

where t is the number of years after 1985. *Predict the population for* 2005.

The number of years from 1985 to 2005 is 20, so let $t = 20$. Then

$$P = 1000{,}000e^{.05t} = 100{,}000e^{.05(20)}$$
$$= 100{,}000e^{1} = 100{,}000e.$$

Since $e \approx 2.71828$ ($\approx$ means "is approximately"),

$$P \approx 100{,}000(2.71828) = 271{,}828.$$

Many economic forecasts are based on population studies.

EXAMPLE 3

A mail-order company advertises in a national magazine. The company finds that of all small towns, the percentage P (given as a decimal) of those in which exactly x people respond to an ad is given (approximately) by the formula

$$P = \frac{e^{-.5}(.5)^x}{1 \cdot 2 \cdot 3 \cdots x}.$$

From what percentage of small towns can the company expect exactly two people to respond?

We want to find P when $x = 2$. The denominator $1 \cdot 2 \cdot 3 \cdots x$ is the product of the integers from 1 to x. Thus

$$P = \frac{e^{-.5}(.5)^x}{1 \cdot 2 \cdot 3 \cdots x} = \frac{e^{-.5}(.5)^2}{1 \cdot 2}.$$

From the table in Appendix B, $e^{-.5} = .6065$. Therefore

$$P = \frac{.6065(.25)}{2} \approx .0758.$$

Thus in approximately 7.58% of all small towns, exactly two people will respond to the ad.

Exercise 13.1

In Problems **1–4**, *graph each function.*

1. $y = 4^x$. **2.** $y = (\frac{1}{3})^x$. **3.** $y = (\frac{1}{4})^x$. **4.** $y = 3^{x/2}$.

In Problems **5–8**, *use the table in Appendix* B *to find the value of each number.*

5. e^3. **6.** $e^{.5}$. **7.** $e^{-1.5}$. **8.** $e^{-.3}$.

In Problems **9–14**, *find the given function values.*

9. $f(x) = 9^x$; $f(2)$, $f(-2)$, $f(\frac{1}{2})$.

10. $g(x) = 4(3)^x$; $g(0)$, $g(2)$, $g(-1)$.

11. $h(t) = 3(16)^{t/2}$; $h(1)$, $h(\frac{1}{2})$, $h(-\frac{1}{2})$.

12. $f(t) = 5 - (\frac{1}{2})^t$; $f(2)$, $f(3)$, $f(-3)$.

13. $g(x) = 1 + 2(\frac{1}{8})^{1-x}$; $g(1)$, $g(-1)$, $g(\frac{1}{3})$.

14. $h(x) = \dfrac{6(.25)^{(4-x)/2}}{5}$; $h(0)$, $h(2)$, $h(3)$.

15. The projected population P of a city is given by $P = 125{,}000(1.12)^{t/20}$, where t is the number of years after 1984. Predict the population in 2004.

16. For a certain city the population P grows at the rate of 2% per year. The formula $P = 1{,}000{,}000(1.02)^t$ gives the population t years after 1984. Find the population in (a) 1984, (b) 1987.

17. The probability P that a telephone operator will receive exactly x calls during a certain time period is given by

$$P = \frac{e^{-3}3^x}{1 \cdot 2 \cdot 3 \cdots x}.$$

Find the probability that the operator receives exactly two calls.

18. Express e^{kt} in the form b^t.

19. At a certain time there are 100 mg of a radioactive substance. It decays so that after t years, the number of milligrams present, A, is given by $A = 100e^{-.035t}$. How many milligrams are present after 20 years? Give your answer to the nearest milligram.

20. A function used in statistics is

$$y = f(x) = \frac{1}{\sqrt{2\pi}} e^{-x^2/2}.$$

Find $f(0)$, $f(-1)$, and $f(1)$. Assume that $\dfrac{1}{\sqrt{2\pi}} = .399$.

Problems **21** *and* **22** *refer to the following: We say that interest is* ***compounded continuously*** *when it is compounded at every instant of time. The formula* $\mathbf{A = Pe^{rn}}$ *gives the amount* **A** *of a principal* **P** *after* **n** *years when the principal is compounded continuously at an annual rate* **r** *(given as a decimal).*

21. Find the amount that a principal of \$1000 will become after 5 years with interest compounded continuously at an annual rate of 8 percent.

22. Find the amount that \$100 will become at the end of one year if interest is compounded continuously at an annual rate of 10%.

23. For a manufacturer of a product, the daily output y on the tth day after the start of a production run is given approximately by $y = 500(1 - e^{-.2t})$. Such an equation is called a *learning equation*. It indicates that as time goes on, output per day will increase until a certain level is reached. This may be due to the gain in the workers' skills at their jobs. Find, to the nearest complete unit, the output on (a) the first day, and (b) the tenth day after the start of a production run.

24. The formula $A = Pe^{-rn}$ gives the amount A at the end of n years of a principal P that depreciates at the rate of r per year compounded continuously. Here r is given as a decimal. What is the value at the end of ten years of \$6000 of machinery that depreciates at an annual rate of 8%, compounded continuously? Give the answer to the nearest dollar.

25. *Calculator problem.* If $f(x) = 4(1.3)^x$, find $f(5.6)$. Give your answer to four decimal places.

26. *Calculator problem.* If $g(t) = (.35)^{-t/5} + (.35)^{t/5}$, find $g(-2)$. Give your answer to four decimal places.

13.2 LOGARITHMIC FUNCTIONS

The second function of interest to us in this chapter is related to an exponential function. Figure 13-3 shows the graph of the exponential function $s = f(t) = 2^t$. The function f sends an input number t into a *positive* output number s.

$$f : t \rightarrow s, \qquad \text{where } s = 2^t.$$

For example, f sends 3 into 8:

$$f : 3 \rightarrow 8.$$

Now look at the same curve in Fig. 13-4. The direction of the arrows indicates that with each positive number s we can associate exactly one value of t. Notice

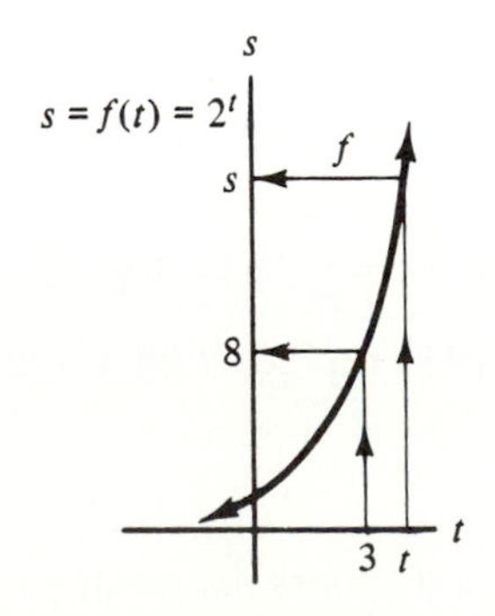

FIGURE 13-3

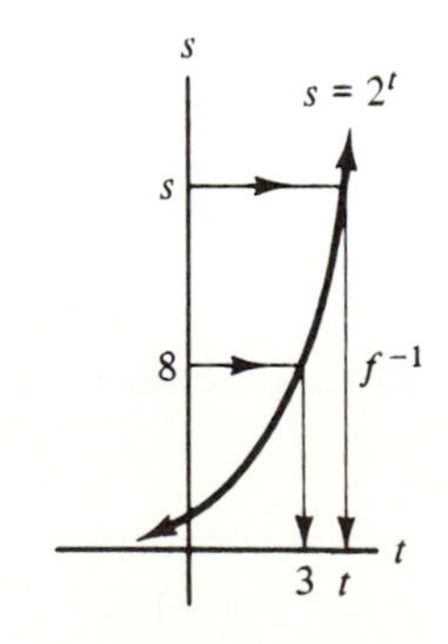

FIGURE 13-4

that with $s = 8$ we associate $t = 3$. Let's think here of s as an input and t as an output. Then we have a function that sends s's into t's. We'll denote this function by the symbol f^{-1} (read "f inverse").*

$$f^{-1}: s \to t, \qquad \text{where } s = 2^t.$$

Thus $f^{-1}(s) = t$.

The functions f and f^{-1} are related. In Fig. 13-5 you can see that f^{-1} *reverses*

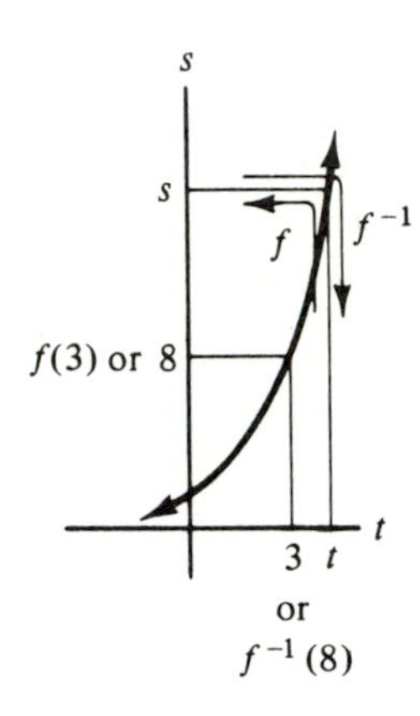

FIGURE 13-5

the action of f, and vice versa. For example,

$$f \text{ sends 3 into 8, and } f^{-1} \text{ sends 8 into 3.}$$

That is,

$$f(3) = 8 \quad \text{and} \quad f^{-1}(8) = 3.$$

Notice that the domain of f^{-1} is the range of f (all positive numbers) and the range of f^{-1} is the domain of f (all real numbers).

We give a special name to f^{-1}. It is called the **logarithmic** (or log) **function with base 2**. Usually we write f^{-1} as $\log_2$ (read "log base 2"). Thus $f^{-1}(8) = \log_2 8 = 3$, where $\log_2$ is just a symbol for a certain function. Putting everything together, we have the following.

$$\text{If } s = f(t) = 2^t, \text{ then } t = f^{-1}(s) = \log_2 s. \tag{1}$$

Let's generalize to other bases and in (1) replace s by x and t by y.

* The -1 in f^{-1} is not an exponent, so f^{-1} does *not* mean $\dfrac{1}{f}$.

The **logarithmic function** with base b, where $b > 0$ and $b \neq 1$, is denoted by $\log_b$ and

$$y = \log_b x \quad \text{if and only if} \quad b^y = x.$$

The domain of $\log_b$ is all positive numbers, and the range is all real numbers.

The logarithmic function reverses the action of the exponential function, and vice versa. Because of this we say that the logarithmic function is the *inverse* of the exponential function, and the exponential function is the inverse of the logarithmic function.

Always remember: When we say that the log base b of x is y, we mean that b raised to the y power is x.

$$\log_b x = y \quad \text{means} \quad b^y = x.$$

In this sense, *a logarithm of a number is an exponent.* It is the power to which we must raise the base to get the number. For example,

$$\log_2 8 = 3 \quad \text{because} \quad 2^3 = 8.$$

We say that $\log_2 8 = 3$ is the **logarithmic form** of the **exponential form** $2^3 = 8$.

EXAMPLE 1

	Exponential form		*Logarithmic form*
a. Since	$5^2 = 25$,	then	$\log_5 25 = 2$.
b. Since	$3^4 = 81$,	then	$\log_3 81 = 4$.
c. Since	$10^0 = 1$,	then	$\log_{10} 1 = 0$.

EXAMPLE 2

Logarithmic form		*Exponential form*
a. $\log_{10} 1000 = 3$	means	$10^3 = 1000$.
b. $\log_{64} 8 = \frac{1}{2}$	means	$64^{1/2} = 8$.
c. $\log_2 \frac{1}{16} = -4$	means	$2^{-4} = \frac{1}{16}$.

EXAMPLE 3

Graph the function $y = \log_2 x$.

Substituting values of x and then finding corresponding values of y can be awkward. For example, if $x = 3$, then $y = \log_2 3$, which is not easily determined. An easier way to plot points is to use the equivalent exponential form $x = 2^y$. We choose values of y and find corresponding values of x. If $y = 0$, then $x = 1$. Thus $(1, 0)$ lies on the graph. In Fig. 13-6 are other points. From the graph you can see that the domain is all positive numbers and the range is all real numbers.

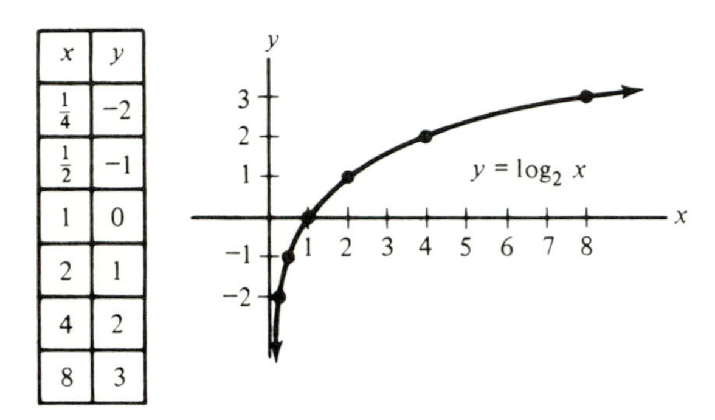

x	y
$\frac{1}{4}$	-2
$\frac{1}{2}$	-1
1	0
2	1
4	2
8	3

FIGURE 13-6

It is obvious from Fig. 13-6 that if two numbers are different, then their logarithms (base 2) are different. More generally, this gives the following property.

$$\text{If } \log_b m = \log_b n, \text{ then } m = n.$$

There is a similar property for the exponential function.

$$\text{If } b^m = b^n, \text{ then } m = n.$$

Logarithms with base 10 are called **common logarithms**. They were often used to simplify computations before the calculator age. We usually omit the subscript 10 when writing common logs. Thus

$$\log x \quad \text{means} \quad \log_{10} x.$$

Logarithms with base e are called **natural** (or Napierian) **logarithms**.* We use the symbol ln for natural logs. Thus

---

* After the Scottish mathematician John Napier (1550–1617), the inventor of logarithms.

$\ln x$ means $\log_e x$.

The symbol $\ln x$ is read "ell-en of x." Appendix C gives a table of natural logarithms (accurate to five decimal places) and instructions on how to use the table. For example, you can see that $\ln 2 = .69315$. This means that $e^{.69315} = 2$. Figure 13-7 shows the graph of $y = \ln x$. Notice that it has the same shape as does Fig. 13-6.

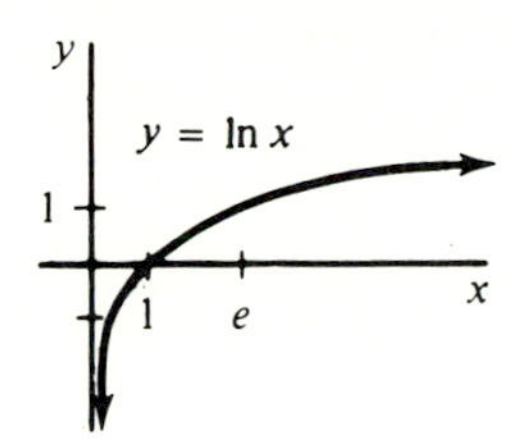

FIGURE 13-7

EXAMPLE 4

Find each of the following.

a. $\log 100$.

Here the base is 10. Thus $\log 100$ is the power to which we must raise 10 to get 100. Since $10^2 = 100$, $\log 100 = 2$.

b. $\ln 1$.

Here the base is e. Because $e^0 = 1$, $\ln 1 = 0$.

c. $\log .1$.

Since $.1 = \frac{1}{10} = 10^{-1}$, $\log .1 = -1$.

d. $\ln e$.

Since $e^1 = e$, $\ln e = 1$.

e. $\log_{36} 6$.

Because $6 = \sqrt{36} = 36^{1/2}$, $\log_{36} 6 = \frac{1}{2}$.

EXAMPLE 5

Solve for x.

a. $\log_2 x = 4$.

$$\log_2 x = 4,$$

$$2^4 = x \qquad \text{[converting to exponential form]},$$

$$\boxed{16 = x.}$$

b. $x = \log_5 125$.

$$x = \log_5 125,$$

$$5^x = 125 \qquad \text{[converting to exponential form]}.$$

Clearly

$$\boxed{x = 3.}$$

c. $x + 1 = \log_4 16$.

$$x + 1 = \log_4 16,$$

$$4^{x+1} = 16 \qquad \text{[converting to exponential form]}.$$

From inspection, the exponent $x + 1$ must be 2. Thus $x + 1 = 2$, or

$$\boxed{x = 1}$$

In the original equation, you might have noticed that $\log_4 16$ is 2. Thus we have $x + 1 = 2$, from which $x = 1$.

d. $\log_x 49 = 2$.

$$\log_x 49 = 2,$$

$$x^2 = 49,$$

$$x = \pm 7.$$

A negative number cannot be a base in a logarithm, so we disregard $x = -7$ and choose

$$\boxed{x = 7.}$$

Exercise 13.2

In Problems **1–16**, *write each exponential form logarithmically and each logarithmic form exponentially.*

1. $4^3 = 64$.

2. $2 = \log_{12} 144$.

3. $10^5 = 100{,}000$.

4. $\log_9 3 = \frac{1}{2}$.

5. $\log_2 64 = 6$.

6. $4^4 = 256$.

7. $\log_8 2 = \frac{1}{3}$.

8. $8^{2/3} = 4$.

9. $6^0 = 1$.

10. $\log_{1/2} 4 = -2$.

11. $\log_2 x = 14$.

12. $10^{.48302} = 3.041$.

13. $e^2 = 7.3891$.

14. $e^{.33647} = 1.4$.

15. $\ln 3 = 1.0986$.

16. $\log 5 = .6990$.

In Problems **17** *and* **18**, *graph the function.*

17. $y = \log_3 x$.

18. $y = \log_4 x$.

In Problems **19–22**, *use Appendix* C *to find the value.*

19. $\ln 5$.

20. $\ln 3.12$.

21. $\ln 7.39$.

22. $\ln 9.98$.

In Problems **23–36**, *find the value.*

23. $\log_6 36$.

24. $\log_2 32$.

25. $\log_3 27$.

26. $\log_{27} 3$.

27. $\log_{16} 4$.

28. $\log_7 7$.

29. $\log 10$.

30. $\log 10{,}000$.

31. $\log .01$.

32. $\ln(e^4)$.

33. $\log_5 1$.

34. $\log_2 \sqrt{2}$.

35. $\log_2 \frac{1}{8}$.

36. $\log_5 \frac{1}{25}$.

In Problems **37–62**, *find x.*

37. $\log_2 x = 4$.

38. $\log_3 x = 2$.

39. $\log_5 x = 3$.

40. $\log_4 x = 0$.

41. $\log x = -1$.

42. $\ln x = 1$.

43. $\ln x = 2$.

44. $\log_x 100 = 2$.

45. $\log_x 8 = 3$.

46. $\log_x 3 = \frac{1}{2}$.

47. $\log_x \frac{1}{6} = -1$.

48. $\log_x y = 1$.

49. $\log_4 16 = x$.

50. $\log_3 1 = x$.

51. $\log 10{,}000 = x$.

52. $\log_2 \frac{1}{16} = x$.

53. $\log_{25} 5 = x$.

54. $\log_9 9 = x$.

55. $\log_3 x = -4$.

56. $\log_x(2x - 3) = 1$.

57. $\log_x 81 = 2$.

58. $\log_{1/2} x = -2$.

59. $\log_8 64 = x - 1$.

60. $\log_x(6 - x) = 2$.

61. $\log_3(x - 2) = 2$.

62. $\log_2 4 = 3x - 1$.

63. For a firm, the cost c (in dollars) of producing q units of a product is given by the cost equation $c = (2q \ln q) + 20$. Use Appendix C to evaluate the cost when $q = 6$. (Give your answer to two decimal places.)

64. *Calculator problem.* Evaluate $\log 87 + 3 \log .834$. Give your answer to four decimal places.

65. *Calculator problem.* Evaluate $\ln 54 - (\ln 17.2)^2$. Give your answer to four decimal places.

13.3 PROPERTIES OF LOGARITHMS

Logarithms have many properties. One involves the log of a product, like $\log_b(mn)$. If we let $x = \log_b m$ and $y = \log_b n$, then $b^x = m$ and $b^y = n$. Therefore

$$mn = b^x b^y = b^{x+y}.$$

Thus $mn = b^{x+y}$. Converting this to logarithmic form gives $\log_b(mn) = x + y$. Since x is $\log_b m$ and y is $\log_b n$, we have our first property.

Property 1. $\log_b(mn) = \log_b m + \log_b n.$

The log of a product is a sum of logs.

In some of the examples and exercises, we'll use Table 13-1. It gives the values of a few common logs to four decimal places. For example, $\log 4 = .6021$. This means $10^{.6021} = 4$.

TABLE 13-1 Common Logarithms

x	$\log x$	x	$\log x$
2	.3010	7	.8451
3	.4771	8	.9031
4	.6021	9	.9542
5	.6990	10	1.0000
6	.7782	e	.4343

EXAMPLE 1

a. *Find* log 15.

Log 15 is not in Table 13-1. But we can write 15 as the product $3 \cdot 5$.

$$\begin{aligned} \log 15 &= \log(3 \cdot 5) \\ &= \log 3 + \log 5 \qquad \text{[Property 1]} \\ &= .4771 + .6990 \qquad \text{[Table 13-1]}. \\ \log 15 &= 1.1761. \end{aligned}$$

b. *Find* log 56.

$$\log 56 = \log(8 \cdot 7) = \log 8 + \log 7 = .9031 + .8451 = 1.7482.$$

The next two properties can be proven in the same basic way as was Property 1.

Property 2. $\log_b \frac{m}{n} = \log_b m - \log_b n.$

The log of a quotient is a difference of logs.

Property 3. $\log_b(m^n) = n \log_b m.$

The log of a power of a number is the power times a log.

EXAMPLE 2

a. *Find* $\log \frac{9}{2}$.

$$\begin{aligned} \log \frac{9}{2} &= \log 9 - \log 2 && \text{[Property 2]} \\ &= .9542 - .3010 && \text{[Table 13-1].} \\ \log \frac{9}{2} &= .6532. \end{aligned}$$

b. *Find* $\log 64$.

$$\begin{aligned} \log 64 &= \log(8^2) \\ &= 2 \log 8 && \text{[Property 3]} \\ &= 2(.9031) && \text{[Table 13-1].} \\ \log 64 &= 1.8062. \end{aligned}$$

c. *Find* $\log \sqrt{5}$.

$$\log \sqrt{5} = \log(5^{1/2}) = \frac{1}{2} \log 5 = \frac{1}{2}(.6990) = .3495.$$

d. *Find* $\log \frac{16}{21}$.

$$\begin{aligned} \log \frac{16}{21} &= \log 16 - \log 21 = \log(4^2) - \log(3 \cdot 7) \\ &= 2 \log 4 - [\log 3 + \log 7] \\ &= 2(.6021) - [.4771 + .8451] = -.1180. \end{aligned}$$

You should note the use of brackets in the second line. It is wrong to write $2 \log 4 - \log 3 + \log 7$.

Make sure that you clearly understand Properties 1–3. They do not apply to the log of a sum $[\log_b(m+n)]$, *log of a difference* $[\log_b(m-n)]$, *or quotient of logs* $\left[\dfrac{\log_b m}{\log_b n}\right]$. *For example,*

$$\log_b(m+n) \neq \log_b m + \log_b n,$$

$$\log_b(m-n) \neq \log_b m - \log_b n,$$

$$\frac{\log_b m}{\log_b n} \neq \log_b(m-n),$$

and

$$\frac{\log_b m}{\log_b n} \neq \log_b\left(\frac{m}{n}\right).$$

EXAMPLE 3

a. *Write* $\log \dfrac{1}{x^2}$ *in terms of* $\log x$.

$$\log \frac{1}{x^2} = \log x^{-2} = -2 \log x \qquad \text{[Property 3].}$$

b. *Write* $\log x - \log(x+3)$ *as a single logarithm.*

$$\log x - \log(x+3) = \log \frac{x}{x+3} \qquad \text{[Property 2].}$$

c. *Write* $3 \log_2 10 + \log_2 15$ *as a single logarithm.*

$$\begin{aligned} 3 \log_2 10 + \log_2 15 &= \log_2(10^3) + \log_2 15 && \text{[Property 3]} \\ &= \log_2[(10^3)15] && \text{[Property 1]} \\ &= \log_2 15{,}000. \end{aligned}$$

d. *Write* $\ln 3 + \ln 7 - \ln 2 - 2 \ln 4$ *as a single logarithm.*

$$\begin{aligned} &\ln 3 + \ln 7 - \ln 2 - 2 \ln 4 \\ &= \ln 3 + \ln 7 - \ln 2 - \ln(4^2) \\ &= \ln 3 + \ln 7 - [\ln 2 + \ln(4^2)] \\ &= \ln(3 \cdot 7) - \ln(2 \cdot 4^2) \\ &= \ln 21 - \ln 32 \\ &= \ln \frac{21}{32}. \end{aligned}$$

EXAMPLE 4

a. *Write* $\ln \dfrac{x^5 y}{zw}$ *in terms of* $\ln x$, $\ln y$, $\ln z$, *and* $\ln w$.

$$\ln \frac{x^5 y}{zw} = \ln(x^5 y) - \ln(zw)$$
$$= \ln(x^5) + \ln y - [\ln z + \ln w]$$
$$= 5 \ln x + \ln y - \ln z - \ln w.$$

b. *Write* $\log \sqrt[3]{\frac{x^2}{x-4}}$ *in terms of* $\log x$ *and* $\log(x-4)$.

$$\log \sqrt[3]{\frac{x^2}{x-4}} = \log\left(\frac{x^2}{x-4}\right)^{1/3} = \frac{1}{3} \log \frac{x^2}{x-4}$$
$$= \frac{1}{3}[\log(x^2) - \log(x-4)]$$
$$= \frac{1}{3}[2 \log x - \log(x-4)].$$

Do not confuse $\ln x^2$ with $(\ln x)^2$. We have

$$\ln x^2 = \ln(x \cdot x),$$
$$\text{but} \qquad (\ln x)^2 = (\ln x)(\ln x),$$

which can be written as $\ln^2 x$. Thus in $\ln x^2$ we square x; in $(\ln x)^2$, or $\ln^2 x$, we square $\ln x$.

Since $b^0 = 1$ and $b^1 = b$, by converting to logarithmic forms we get two more properties of logs.

Property 4. $\log_b 1 = 0.$

Property 5. $\log_b b = 1.$

EXAMPLE 5

a. *Find* $\log 1$.

$$\log 1 = \log_{10} 1 = 0 \qquad \text{[Property 4]}.$$

b. *Find* $\ln e$.

$$\ln e = \log_e e = 1 \qquad \text{[Property 5]}.$$

c. *Find* $\log 1000$.

$$\log 1000 = \log_{10} 10^3$$
$$= 3 \log_{10} 10$$
$$= 3 \cdot 1 \qquad \text{[Property 5]}$$
$$= 3.$$

d. *Find* $\log(10^n)$.

$$\log(10^n) = n \log_{10} 10 = n \cdot 1 = n.$$

EXAMPLE 6

a. *Find* $\log_7 \sqrt[9]{7^8}$.

$$\log_7 \sqrt[9]{7^8} = \log_7(7^{8/9}) = \frac{8}{9} \log_7 7$$

$$= \frac{8}{9} \cdot 1 = \frac{8}{9} \qquad \text{[Property 5]}.$$

b. *Find* $\ln(e^2) + \log \frac{1}{10}$.

$$\ln(e^2) + \log \frac{1}{10} = 2 \ln e + \log 10^{-1}$$

$$= 2 \log_e e + (-1) \log_{10} 10$$

$$= 2(1) + (-1)(1) = 1.$$

c. *Find* $\log \frac{200}{21}$.

$$\log \frac{200}{21} = \log 200 - \log 21 = \log(2 \cdot 10^2) - \log(3 \cdot 7)$$

$$= \log 2 + \log(10^2) - [\log 3 + \log 7]$$

$$= \log 2 + 2 \log 10 - \log 3 - \log 7$$

$$= .3010 + 2(1) - .4771 - .8451 \qquad \text{[Table 13-1]}$$

$$= .9788.$$

Exercise 13.3

In Problems **1–48**, *use Table* 13-1 *and properties of logs to find each value.*

1. $\log 5$.
2. $\log 3$.
3. $\log 21$.
4. $\log 14$.
5. $\log 35$.
6. $\log 12$.
7. $\log 25$.
8. $\log 49$.
9. $\log \frac{9}{4}$.
10. $\log \frac{7}{3}$.
11. $\log \frac{1}{2}$.
12. $\log \frac{7}{9}$.
13. $\log(8^5)$.
14. $\log(4^{-3})$.
15. $\log \frac{1}{5^4}$.
16. $\log 2^{.01}$.
17. $\log \sqrt{2}$.
18. $\log \sqrt[3]{4}$.

19. $\log \sqrt[3]{6^2}$.

20. $\log \dfrac{1}{\sqrt[3]{5}}$.

21. $\log \frac{14}{5}$.

22. $\log \frac{3}{35}$.

23. $\log \frac{15}{28}$.

24. $\log \frac{81}{25}$.

25. $\log 10{,}000$.

26. $\log 10^{10}$.

27. $\log .01$.

28. $\log(.1)^3$.

29. $\log 300$.

30. $\log \sqrt[3]{400}$.

31. $\log \frac{100}{9}$.

32. $\log \frac{1}{600}$.

33. $\log \frac{27}{5000}$.

34. $\log \sqrt{\frac{8}{3}}$.

35. $\log_7 7^{48}$.

36. $\log_4 \sqrt[5]{4}$.

37. $\log_5 \sqrt[4]{5^3}$.

38. $\log_2 \dfrac{1}{\sqrt{2}}$.

39. $\ln e^4$.

40. $\ln \dfrac{1}{e}$.

41. $\log_2 \dfrac{2^6}{2^{10}}$.

42. $\log_3(3^5 \cdot 3^4)^6$.

43. $\log[(10\sqrt{10})^3]$.

44. $\log_5 \dfrac{25}{\sqrt{5}}$.

45. $\log_7 \dfrac{\sqrt[3]{49}}{7}$.

46. $\log_8(64\sqrt[4]{8})$.

47. $\log 10 + \ln(e^2)$.

48. $(\log 100)(\ln \sqrt{e})$.

In Problems **49–62**, *write each expression as a single logarithm.*

49. $\log 7 + \log 4$.

50. $\log_3 10 - \log_3 5$.

51. $\log_2(x + 2) - \log_2(x + 1)$.

52. $\log(x^2) + \log x$.

53. $2 \ln 5 - 3 \ln 4$.

54. $3 \log_7 2 + 2 \log_7 3$.

55. $\frac{1}{2} \log_4 2 + 3 \log_4 3$.

56. $\frac{1}{3} \ln x - 5 \ln(x - 2)$.

57. $\log_3 x + \log_3 y - \log_3 z$.

58. $\ln x + \ln(x - 1) - 2 \ln y$.

59. $\log_2 5 + 2 \log_2 3 - \log_2 7$.

60. $\frac{1}{2}(\log_3 x + \log_3 y)$.

61. $2 \ln x + 3 \ln y - 4 \ln z - 2 \ln w$.

62. $\log_6 5 - 2 \log_6 \frac{1}{5} - 2 \log_6 4$.

In Problems **63–68**, *write each expression in terms of* $\log x$ *and* $\log(x + 1)$.

63. $\log[x(x + 1)^2]$.

64. $\log \dfrac{\sqrt{x}}{x + 1}$.

65. $\log \dfrac{x^2}{(x + 1)^3}$.

66. $\log[x(x + 1)]^3$.

67. $\log\left(\dfrac{x}{x + 1}\right)^3$.

68. $\log \sqrt{x(x + 1)}$.

In Problems **69–74**, *write each expression in terms of* $\ln x$, $\ln(x + 1)$, *and* $\ln(x + 2)$.

69. $\ln \dfrac{x}{(x + 1)(x + 2)}$.

70. $\ln \dfrac{x^2(x + 1)}{x + 2}$.

71. $\ln \dfrac{\sqrt{x}}{(x + 1)^2(x + 2)^3}$.

72. $\ln \dfrac{1}{x(x + 1)(x + 2)}$.

73. $\ln\left[\dfrac{1}{x+2}\sqrt[5]{\dfrac{x^2}{x+1}}\right].$

74. $\ln\sqrt{\dfrac{x^4(x+1)^3}{x+2}}.$

A chemist can determine the acidity or basicity of an aqueous solution at room temperature by finding the pH *of the solution. To do this, the chemist can first find the hydrogen-ion concentration (in moles per liter). The symbol* $[H^+]$ *stands for this concentration. The* pH *is then given by*

$$\text{pH} = -\log[H^+].$$

If $\text{pH} < 7$, *the solution is acidic. If* $\text{pH} > 7$, *it is basic. If* $\text{pH} = 7$, *we say that the solution is neutral. Use this information in Problems* **75** *and* **76**.

75. What is the pH of vinegar with $[H^+]$ equal to 3×10^{-4}?

76. A cleaning solution has a pH of 8. What is the $[H^+]$ of this solution?

For an aqueous solution at room temperature, the product of the hydrogen-ion concentration, $[H^+]$, *and the hydroxide-ion concentration,* $[OH^-]$, *is* 10^{-14} *(where the concentrations are in moles per liter).*

$$[H^+][OH^-] = 10^{-14}.$$

In Problems **77** *and* **78**, *find the* pH *of a solution (see explanation above Problem* 75) *with the given* $[OH^-]$.

77. $[OH^-] = 10^{-4}$.

78. $[OH^-] = 3 \times 10^{-2}$.

79. The intensity level β (beta) of a sound wave of intensity I is given by

$$\beta = 10 \log \frac{I}{I_0},$$

where I_0 is a reference intensity taken to be 10^{-12} (watts per square meter), which approximately corresponds to the faintest sound that you can hear. The intensity level is measured in decibels (db). For example, the intensity level of an ordinary conversation is 40 db, and that for a subway train is 100 db. Find the intensity level of the sound of rustling leaves, which has an intensity of 10^{-11}.

80. According to Richter,* the magnitude M of an earthquake occurring 100 km from a certain type of seismometer is given by $M = 3 + \log A$, where A is the recorded trace amplitude (in millimeters) of the quake. (a) Find the magnitude of an earthquake that records a trace amplitude of 1 mm. (b) If a particular earthquake has amplitude A_1 and magnitude M_1, determine the magnitude of a quake with amplitude $100A_1$. Express your answer to (b) in terms of M_1.

In Problems **81–84**, *use Appendix* C *to find each value.*

81. ln 250.

82. ln .0412.

83. ln .000105.

84. ln 100,000.

---

* C. F. Richter, *Elementary Seismology*. (San Francisco: W. H. Freeman, 1958).

13.4 EXPONENTIAL AND LOGARITHMIC EQUATIONS

An equation in which an unknown appears in an exponent is called an **exponential equation**. It can generally be solved by first taking logarithms of both sides, as the following examples show.

EXAMPLE 1

Solve $e^{3t} = \frac{4}{3}$.

Taking natural logarithms of both sides, we have

$$\begin{aligned} \ln e^{3t} &= \ln \frac{4}{3}, \\ 3t \ln e &= \ln 4 - \ln 3 \qquad \text{[Properties 2 and 3]}, \\ 3t &= \ln 4 - \ln 3 \qquad \text{[Property 5]}, \\ t &= \frac{\ln 4 - \ln 3}{3}. \end{aligned}$$

From Appendix C, we have

$$t = \frac{1.38629 - 1.09861}{3} = .096.$$

EXAMPLE 2

Solve $5^{2x+1} = 36$.

Taking natural logarithms of both sides gives

$$\begin{aligned} \ln 5^{2x+1} &= \ln 36, \\ (2x+1)(\ln 5) &= \ln 36, \\ 2x + 1 &= \frac{\ln 36}{\ln 5}, \\ 2x &= \frac{\ln 36}{\ln 5} - 1, \\ x &= \frac{\ln 36}{2 \ln 5} - \frac{1}{2}. \end{aligned}$$

Using Appendix C, we have

$$x = \frac{3.58352}{2(1.60944)} - \frac{1}{2} = .613.$$

In Example 2 we could have used logarithms to any base, such as 5. We chose base e because a table of natural logarithms is available to us.

Another type of equation is a **logarithmic equation**. It involves a logarithm of an expression containing an unknown. A logarithmic equation can generally be solved by rewriting it in exponential form, as the next example shows.

EXAMPLE 3

Solve $\log_2 x = 5 - \log_2(x + 4)$.

We first get all logarithms on one side, so that we can combine them.

$$\log_2 x + \log_2(x + 4) = 5,$$

$$\log_2[x(x + 4)] = 5 \qquad \text{[Property 1]}.$$

In exponential form we have

$$x(x + 4) = 2^5,$$

$$x^2 + 4x = 32,$$

$$x^2 + 4x - 32 = 0 \qquad \text{[quadratic equation]},$$

$$(x - 4)(x + 8) = 0,$$

$$x = 4 \quad \text{or} \quad x = -8.$$

Since the original equation is not defined for a negative value of x, -8 is not a solution. However, 4 satisfies the original equation. Check this! Thus only 4 is a solution.

Exercise 13.4

In Problems **1–20**, *solve the equation.*

1. $e^{2t} = 5$.

2. $e^{4t} = \frac{3}{4}$.

3. $e^{3t+1} = 15$.

4. $e^{1-t} = 25$.

5. $3^x = 7$.

6. $2^{8x} = 9$.

7. $5^{2x-5} = 9$.

8. $3^{x-2} = 14$.

9. $4^{t/2} = 20$.

10. $2.5^x = 16$.

11. $e^{-x/10} = \frac{3}{2}$.

12. $2^{-2t/3} = \frac{4}{5}$.

13. $\log(x - 3) = 3$.

14. $\log_2(x + 1) = 4$.

15. $\log_4(3x - 4) = 2$.

16. $\log x + \log(x - 15) = 2$.

17. $\log(3x - 1) - \log(x - 3) = 2$.

18. $\log_3(2x + 3) = 4 - \log_3(x + 6)$.

19. $\log_2\left(\frac{2}{x}\right) = 3 + \log_2 x$.

20. $\log(2x + 1) = \log(x + 6)$.

21. In one city the population P grows at the rate of 2% per year. The equation

$P = 1{,}000{,}000(1.02)^t$ gives the population t years after 1983. Find the value of t for which the population is 1,500,000.

22. The number, Q, of milligrams of a radioactive substance present after t years is given by

$$Q = 100e^{-.035t}.$$

(a) How many milligrams are present after 0 years? (b) After how many years will there be 50 mg present?

23. A marketing-research company needs to determine how people adapt to the taste of a new cough drop. In one experiment, a person was given a cough drop and was asked periodically to assign a number, on a scale from 0 to 10, to the perceived taste. This number was called the *response magnitude*. The number 10 was assigned to the initial taste. After conducting the experiment several times, the company estimated that the response magnitude, R, is given by

$$R = 10e^{-t/40},$$

where t is the number of seconds after the person is given the cough drop. (a) Find the response magnitude after 20 s. Give your answer to the nearest integer. (b) After how many seconds does a person have a response magnitude of 5? Give your answer to the nearest second.

24. When the price of a product is p dollars per unit, consumers will buy q thousand units of the product, where $q = 80 - 2^p$. Find p when $q = 60$.

25. The equation $A = P(1.1)^t$ gives the value, A, at the end of t years of an investment of P dollars compounded annually at an annual interest rate of 10%. How many years will it take for an investment to double? Give your answer to the nearest year.

26. A translucent material has the property that although light passes through the material, its intensity is reduced. A particular translucent plastic has the property that a sheet 1 mm thick reduces the intensity of light by 10%. How many such sheets are necessary to reduce the intensity of a beam of light to about 50% of its original value?

27. The water in a midwestern lake contains sediment, and the presence of the sediment reduces the transmission of light through the water. Experiments indicate that the intensity of light is reduced by 10% by passage through 20 cm of water. Suppose that the lake is uniform with respect to the amount of sediment contained by the water. A measuring instrument can detect light at the intensity of .17% of full sunlight. This measuring instrument is lowered into the lake. At what depth will it first cease to record the presence of light?

13.5 CHANGE-OF-BASE FORMULAS

Suppose you need to find $\log_5 2$. This would be easy if you had a table of logs in base 5. However, you can find $\log_5 2$ by using the table of natural logs in Appendix C. Here's how.

Let $x = \log_5 2$. Then $5^x = 2$. Now take the natural log of each side.

$$\ln(5^x) = \ln 2,$$
$$x \ln 5 = \ln 2,$$
$$x = \frac{\ln 2}{\ln 5} = \frac{.69315}{1.60944} = .4307.$$

Thus $\log_5 2 = .4307$.

If we generalize this procedure, we get a *change of base formula.*

A CHANGE OF BASE FORMULA

$$\log_b N = \frac{\ln N}{\ln b}.$$

This formula lets us find logs in base b by using natural logs.

EXAMPLE 1

a. *Find* $\log_3 7$.

Using the change of base formula above with $b = 3$ and $N = 7$ gives

$$\log_3 7 = \frac{\ln 7}{\ln 3}$$
$$= \frac{1.94591}{1.09861} = 1.771 \qquad \text{[Appendix C]}.$$

b. *Find* $\log 7$ *by using natural logs.*

Here $b = 10$ and $N = 7$. Thus

$$\log 7 = \frac{\ln 7}{\ln 10} = \frac{1.94591}{2.30259} = .8451.$$

We can write a more general form of the previous change of base formula.

GENERAL CHANGE OF BASE FORMULA

$$\log_b N = \frac{\log_a N}{\log_a b}.$$

This formula lets us find logs in base b by using logs in base a.

Exercise 13.5

In Problems **1–8,** *find the value by using Appendix* C.

1. $\log_5 3$. **2.** $\log_7 4$. **3.** $\log_4 9$. **4.** $\log_2 3$.

5. $\log_8 10$. **6.** $\log_3 20$. **7.** $\log 2$. **8.** $\log 9$.

9. Suppose $\log 59 = 1.7709$ and $\log 15 = 1.1761$. Find $\log_{15} 59$.

10. Suppose $\log_4 13 = 1.8502$ and $\log_4 7 = 1.4037$. Find $\log_7 13$.

11. Suppose $\log_3 11 = 2.1826$. Find $\log_9 11$.

13.6 REVIEW

IMPORTANT TERMS AND SYMBOLS

exponential function *(13.1)*
$y = b^x$ *(13.1)*
e *(13.1)*
logarithmic function *(13.2)*
$y = \log_b x$ *(13.2)*
exponential form *(13.2)*
logarithmic form *(13.2)*
common logarithm *(13.2)*
$\log x$ *(13.2)*
natural logarithm *(13.2)*
$\ln x$ *(13.2)*

REVIEW PROBLEMS

In Problems **1–4**, *find the given function values.*

1. $f(x) = 4 + 3^{2x}$; $f(0)$, $f(1)$, $f(-\frac{1}{2})$.

2. $g(t) = 4(9)^{t/2}$; $g(1)$, $g(-1)$, $g(3)$.

3. $h(s) = 100e^{(s+3)/2}$; $h(-3)$, $h(5)$, $h(-6)$.

4. $F(x) = 6 - 3(\frac{1}{2})^{x+4}$; $F(-3)$, $F(-1)$, $F(-6)$.

In Problems **5** *and* **6**, *graph the equation.*

5. $y = \log_5 x$. **6.** $y = 5^x$.

In Problems **7–12**, *write each exponential form logarithmically and each logarithmic form exponentially.*

7. $3^5 = 243$. **8.** $\log_7 343 = 3$. **9.** $\log_{16} 2 = \frac{1}{4}$.

10. $10^5 = 100{,}000$. **11.** $e^4 = 54.598$. **12.** $\log_9 9 = 1$.

In Problems **13–18**, *find the value.*

13. $\log_5 125$.
14. $\log_4 16$.
15. $\log_2 \frac{1}{16}$.
16. $\log_{1/3} \frac{1}{9}$.
17. $\log_{1/3} 9$.
18. $\log_4 2$.

In Problems **19–26**, *find x.*

19. $\log_5 \frac{1}{25} = x$.
20. $\log_x 1000 = 3$.
21. $\log x = -2$.
22. $\log_8 64 = x$.
23. $\log_x 81 = 2$.
24. $\log_3 x = \frac{1}{2}$.
25. $\log_x(4x - 9) = 1$.
26. $\log_4(x - 6) = 2$.

In Problems **27–44**, *use Table* 13-1 *of Sec.* 13-3 *and properties of logs to find the value.*

27. $\log 16$.
28. $\log(12^3)$.
29. $\log \frac{8}{3}$.
30. $\log \frac{10}{7}$.
31. $\log \sqrt[4]{8}$.
32. $\log \sqrt[3]{2^2}$.
33. $\log 500$.
34. $\log \frac{1}{200}$.
35. $\log \frac{24}{35}$.
36. $\log \frac{800}{21}$.
37. $\log \sqrt{\frac{5}{2}}$.
38. $\log(10 \sqrt[4]{10})$.
39. $\log_9 1$.
40. $\log_3 3^{10}$.
41. $\ln \sqrt[3]{e}$.
42. $\log_5 \sqrt[7]{5^4}$.
43. $\log_7(49\sqrt{7})$.
44. $\log_3 \dfrac{\sqrt{3}}{27}$.

In Problems **45–50**, *write each expression as a single logarithm.*

45. $2 \log 5 - 3 \log 3$.
46. $6 \ln x + 4 \ln y$.
47. $2 \ln x + \ln y - 3 \ln z$.
48. $\log_6 2 - \log_6 4 - 2 \log_6 3$.
49. $\frac{1}{2} \log_2 x + 2 \log_2(x^2) - 3 \log_2(x + 1) - 4 \log_2(x + 2)$.
50. $3 \log x + \log y - 2(\log z + \log w)$.

In Problems **51–56**, *write the expression in terms of* $\ln x$, $\ln y$, *and* $\ln z$.

51. $\ln \dfrac{x^2 y}{z^3}$.
52. $\ln \dfrac{\sqrt{x}}{(yz)^2}$.
53. $\ln \sqrt[3]{xyz}$.
54. $\ln\left[\dfrac{xy^3}{z^2}\right]^4$.
55. $\ln\left[\dfrac{1}{x}\sqrt{\dfrac{y}{z}}\right]$.
56. $\ln\left[\left(\dfrac{x}{y}\right)^2\left(\dfrac{x}{z}\right)^3\right]$.

57. Because of poor advertising, a company finds that its annual revenues have been cut sharply. The annual revenue R (in dollars) at the end of t years of business is given by the equation $R = 200{,}000e^{-.2t}$. Find the annual revenue at the end of (a) 2 years and (b) 3 years.

In Problems **58–60**, *use the change of base formulas in Sec.* 13-5 *to find the given value.*

58. $\log_6 9$.

59. $\log_3 5$.

60. Suppose that $\log_2 19 = 4.2479$ and $\log_2 5 = 2.3219$. Find $\log_5 19$.

In Problems **61–64**, *solve the equation.*

61. $e^{3x} = 2$.

62. $4^{x+3} = 7$.

63. $\log x + \log(10x) = 3$.

64. $\log_3(x + 1) = \log_3(x - 1) + 1$.

65. On the surface of a glass slide is a grid that divides the surface into 225 equal squares. Suppose a blood sample containing N red cells is spread on the slide and the cells are randomly distributed. Then the number of squares containing no cells is (approximately) given by $225e^{-N/225}$. If 100 of the squares contain no cells, estimate the number of cells the blood sample contained. Round your answer to the nearest 10 cells.

14 Sequences and Series

14.1 SEQUENCES AND SERIES

You probably have heard, or have even used, the word *sequence*—for instance, a sequence of events or a sequence of steps. There is a sequence of steps to follow when waxing a car. The first is to wash the car. The second is to apply the wax. The third is to buff to a brilliant luster. This sequence has a *first* step, a *second* step, and a *third* step.

In mathematics we use the term **sequence** to describe a function whose domain consists of the positive integers. For example, the function

$$f(n) = n^2,$$

where n is any positive integer, is a sequence. With sequences, instead of using the function-value notation $f(n)$, we commonly use the letter a along with subscript notation, so we write $f(n)$ as a_n. For example, $f(1)$ becomes a_1 and $f(2)$ becomes a_2. If $a_n = n^2$, we have $a_1 = 1^2 = 1$, $a_2 = 2^2 = 4$, and $a_3 = 3^2 = 9$.

When we list the function values of a sequence in the order

$$a_1, a_2, a_3, \ldots, a_n, \ldots,$$

this arrangement also defines the sequence, and the values are called **terms** of the sequence. The *first term* is a_1, the *second term* is a_2, and so on. The ***n*th term**, or **general term**, is a_n. When specific values for $a_1, a_2, \ldots$ are listed, a_n is usually indicated by a rule for the sequence. For example, if the general term is $a_n = n^2$, then the sequence is

$$1, 4, 9, \ldots, n^2, \ldots. \tag{1}$$

We also use the term *sequence* when the domain is a *finite* set of consecutive positive integers. For example, if $a_n = n^2$ and $n = 3, 4, 5$, and 6, then the sequence is

$$9, 16, 25, 36. \tag{2}$$

The sequence in (1) is called an *infinite sequence*, and the sequence in (2) is called a *finite sequence* because it has a first term *and* a last term.

EXAMPLE 1

Find the first four terms of the sequence with the given general term.

a. $a_n = \dfrac{2n}{3n+1}$.

In the expression $\dfrac{2n}{3n+1}$, we successively replace n by the integers 1, 2, 3, and 4.

$$n = 1, \qquad a_1 = \frac{2(1)}{3(1)+1} = \frac{2}{4} = \frac{1}{2}.$$

$$n = 2, \qquad a_2 = \frac{2(2)}{3(2)+1} = \frac{4}{7}.$$

$$n = 3, \qquad a_3 = \frac{2(3)}{3(3)+1} = \frac{6}{10} = \frac{3}{5}.$$

$$n = 4, \qquad a_4 = \frac{2(4)}{3(4)+1} = \frac{8}{13}.$$

Thus the first four terms are $\frac{1}{2}$, $\frac{4}{7}$, $\frac{3}{5}$, and $\frac{8}{13}$.

b. $a_n = (-1)^{n+1}2^n$.

$$n = 1, \qquad a_1 = (-1)^{1+1}2^1 = (1)(2) = 2.$$

$$n = 2, \qquad a_2 = (-1)^{2+1}2^2 = (-1)(4) = -4.$$

$$n = 3, \qquad a_3 = (-1)^{3+1}2^3 = (1)(8) = 8.$$

$$n = 4, \qquad a_4 = (-1)^{4+1}2^4 = (-1)(16) = -16.$$

The first four terms are 2, -4, 8, and -16. Note that these terms alternate in sign because of the factor $(-1)^{n+1}$.

EXAMPLE 2

Find a general term for the given sequence.

a. $2, 4, 6, 8, \ldots.$

Here $a_1 = 2$, $a_2 = 4$, $a_3 = 6$, and $a_4 = 8$. Looking for a pattern, we observe that $a_1 = 2(1)$, $a_2 = 2(2)$, $a_3 = 2(3)$, and $a_4 = 2(4)$. That is, the value of each term is twice the number of that term. Thus a general term is given by $a_n = 2n$.

b. $x, x^2, x^3, x^4, \ldots$.

By inspecting the terms, we see that a general term is $a_n = x^n$.

If we have a sequence, then an indicated sum of its terms is called a **series.** For example, given the sequence

$$2, 3, 4,$$

the corresponding series is

$$2 + 3 + 4.$$

Similarly, the series that corresponds to

$$x, x^2, x^3, \ldots, x^n$$

is

$$x + x^2 + x^3 + \cdots + x^n.$$

To represent such sums we can use **sigma** (or **summation**) **notation**, so named because the Greek letter Σ (sigma) is used. For example,

$$\sum_{k=1}^{3} (2k + 5)$$

denotes the sum of those numbers obtained from the expression $2k + 5$ by first replacing k by 1, then 2, and finally by 3. That is, k takes on consecutive integer values from 1 to 3. Thus

$$\begin{aligned}\sum_{k=1}^{3} (2k + 5) &= [2(1) + 5] + [2(2) + 5] + [2(3) + 5] \\ &= 7 + 9 + 11 = 27.\end{aligned}$$

The expression $\sum_{k=1}^{3} (2k + 5)$ may be read *the sum of* $2k + 5$ *from* $k = 1$ *to* $k = 3$. The letter k is called the **index of summation**; the numbers 1 and 3 are called the **limits of summation** (1 is the *lower limit* and 3 is the *upper limit*). The letter used for the index is a "dummy" symbol in the sense that it does not affect the sum of the terms. Any other letter may be used. For example,

$$\sum_{j=1}^{3} (2j + 5) = 7 + 9 + 11 = \sum_{k=1}^{3} (2k + 5).$$

With sigma notation, the series $2+3+4$ may be written

$$\sum_{n=2}^{4} n,$$

and the series $x+x^2+\cdots+x^n$ may be written

$$\sum_{k=1}^{n} x^k.$$

EXAMPLE 3

Evaluate each of the following.

a. $\displaystyle\sum_{k=4}^{7} \frac{k^2+3}{2}.$

Here the sum begins with $k=4$.

$$\sum_{k=4}^{7} \frac{k^2+3}{2} = \frac{4^2+3}{2} + \frac{5^2+3}{2} + \frac{6^2+3}{2} + \frac{7^2+3}{2}$$
$$= \frac{19}{2} + \frac{28}{2} + \frac{39}{2} + \frac{52}{2} = \frac{138}{2} = 69.$$

b. $\displaystyle\sum_{n=0}^{2} (-1)^n(2^n-1).$

Here the sum begins with $n=0$.

$$\sum_{n=0}^{2} (-1)^n(2^n-1) = (-1)^0(2^0-1) + (-1)^1(2^1-1) + (-1)^2(2^2-1)$$
$$= (1)(0) + (-1)(1) + (1)(3)$$
$$= 0 - 1 + 3 = 2.$$

c. $\displaystyle\sum_{n=1}^{4} 2.$

The 2 can be thought of as $2+0n$. Thus

$$\sum_{n=1}^{4} 2 = 2+2+2+2 = 8.$$

Exercise 14.1

In Problems **1–10**, *find the first four terms of the sequence that has the given general term.*

1. $a_n = 3n$.

2. $a_n = \frac{1}{2}n$.

3. $a_n = n^2 + n$.

4. $a_n = 2n^2 + 4$.

5. $a_n = \dfrac{n^2 - 1}{n^2 + 2}$.

6. $a_n = \dfrac{n}{2^n}$.

7. $a_n = x^{2n}$.

8. $a_n = (2n)^2$.

9. $a_n = (-1)^{n+1} n^2$.

10. $a_n = (-1)^n (2n)^2$.

In Problems **11–18**, *find a general term for the given sequence.*

11. $1, 2, 3, 4, \ldots$.

12. $0, -1, -2, -3, \ldots$.

13. $3, 6, 9, 12, \ldots$.

14. $1, \frac{3}{2}, 2, \frac{5}{2}, \ldots$.

15. $1, 3, 9, 27, \ldots$.

16. $x^2, x^4, x^6, x^8, \ldots$.

17. $\frac{1}{2}, -\frac{1}{3}, \frac{1}{4}, -\frac{1}{5}, \ldots$.

18. $-\frac{1}{2}, \frac{1}{3}, -\frac{1}{4}, \frac{1}{5}, \ldots$.

In Problems **19–28**, *evaluate the given sum.*

19. $\sum_{k=1}^{5} (k + 4)$.

20. $\sum_{k=4}^{7} (5 - 2k)$.

21. $\sum_{n=0}^{3} (3n^2 - 7)$.

22. $\sum_{n=2}^{4} \dfrac{n + 1}{n - 1}$.

23. $\sum_{n=1}^{3} 3$.

24. $\sum_{n=1}^{5} 1$.

25. $\sum_{k=3}^{4} \dfrac{(-1)^k (k + 1)}{2^k}$.

26. $\sum_{n=1}^{4} (n^2 + n)$.

27. $\sum_{n=1}^{3} 2\left(\dfrac{1}{2}\right)^n$.

28. $\sum_{n=0}^{3} (-1)^n \dfrac{1}{n + 1}$.

14.2 ARITHMETIC PROGRESSIONS

If you look for the pattern of the terms in the sequence

$$1, 3, 5, 7,$$

you may notice that by adding 2 to any term, you get the next term. This means that the difference between any term and the term preceding it is 2:

$$3 - 1 = 2, \qquad 5 - 3 = 2, \qquad 7 - 5 = 2.$$

We call the sequence 1, 3, 5, 7 an *arithmetic progression* with *common difference* 2. Here is a more general definition.

A sequence in which each term after the first can be obtained by adding the constant d to the preceding term is called an **arithmetic progression** with **common difference** d.

EXAMPLE 1

Suppose the first term of an arithmetic progression is 6 and the common difference d is 5. Then

$$a_1 = 6,$$
$$a_2 = 6 + 5 = 11,$$
$$a_3 = 11 + 5 = 16,$$
$$a_4 = 16 + 5 = 21, \quad \text{and so on.}$$

Thus the arithmetic progression is

$$6, 11, 16, 21, \ldots.$$

EXAMPLE 2

a. In the arithmetic progression

$$\frac{1}{3}, \frac{2}{3}, 1, \frac{4}{3}, \ldots$$

we can find d by choosing *any* two consecutive terms and subtracting the first one from the second one. For example, subtracting the first term from the second term gives

$$d = \frac{2}{3} - \frac{1}{3} = \frac{1}{3}.$$

b. The sequence

$$2, -1, -4, -7, \ldots$$

is an arithmetic progression, because if we add -3 to *any* term, we get the next term:

$$2 + (-3) = -1, \qquad -1 + (-3) = -4, \qquad \text{and so on.}$$

We can find a formula for the nth term of an arithmetic progression with first term a_1 and common difference d. In this progression we have

$$a_2 = a_1 + d,$$
$$a_3 = a_2 + d = (a_1 + d) + d = a_1 + 2d,$$
$$a_4 = a_3 + d = (a_1 + 2d) + d = a_1 + 3d.$$

Remaining terms can be obtained in a similar manner. Thus the arithmetic progression is

$$a_1, \quad a_1 + d, \quad a_1 + 2d, \quad a_1 + 3d, \ldots.$$

Each term is the sum of a_1 and a number that is a multiple of d. That number is d times one *less* than the number of the term. For example, the tenth term is

$$a_{10} = a_1 + (10 - 1)d = a_1 + 9d.$$

More generally, we have the following formula for the nth term of an arithmetic progression.

$$a_n = a_1 + (n - 1)d \qquad (1)$$

gives the nth term, a_n, of an arithmetic progression with first term a_1 and common difference d.

In Example 2, the nth term of the arithmetic progression in part a is $a_n = \frac{1}{3} + (n - 1)\frac{1}{3} = \frac{1}{3} + \frac{1}{3}(n - 1)$; in part b it is $a_n = 2 + (n - 1)(-3) = 2 - 3(n - 1)$.

EXAMPLE 3

Find the eighteenth term of the sequence

$$7, 13, 19, \ldots.$$

This sequence is an arithmetic progression with $a_1 = 7$ and $d = 6$. To find a_{18}, we use Eq. 1 with $n = 18$.

$$a_n = a_1 + (n - 1)d,$$
$$a_{18} = 7 + (18 - 1)6 = 7 + (17)6 = 109.$$

EXAMPLE 4

The third term of an arithmetic progression is 2 and the tenth term is −26. Find a_1 and d.

We have $a_3 = 2$ and $a_{10} = -26$. Using Eq. 1 with $n = 3$ and $a_3 = 2$ gives

$$2 = a_1 + (3 - 1)d,$$
$$2 = a_1 + 2d. \qquad (2)$$

Using Eq. 1 with $n = 10$ and $a_{10} = -26$ gives

$$-26 = a_1 + (10 - 1)d,$$
$$-26 = a_1 + 9d. \qquad (3)$$

Equations 2 and 3 form a system of equations.

$$\begin{cases} 2 = a_1 + 2d, & (4) \\ -26 = a_1 + 9d. & (5) \end{cases}$$

Multiplying Eq. 4 by -1 and adding it to Eq. 5, we have

$$-28 = 7d,$$
$$-4 = d.$$

Substituting -4 for d in Eq. 4 gives

$$2 = a_1 + 2(-4),$$
$$2 = a_1 - 8,$$
$$10 = a_1.$$

Thus $a_1 = 10$ and $d = -4$.

We can find a formula for the sum S_n of the first n terms of an arithmetic progression. Suppose that $a_1, a_2, \ldots$ is an arithmetic progression. Then $a_2 = a_1 + d$, $a_3 = a_1 + 2d$, and so on. Thus

$$S_n = a_1 + a_2 + a_3 + \cdots + a_n$$

or

$$S_n = a_1 + (a_1 + d) + (a_1 + 2d) + \cdots + [a_1 + (n - 1)d]. \tag{6}$$

In an arithmetic progression, we can find a term by subtracting d from the term that follows it. Thus the terms in the sum S_n can be written in reverse order to give

$$S_n = a_n + (a_n - d) + (a_n - 2d) + \cdots + [a_n - (n - 1)d]. \tag{7}$$

Adding corresponding sides of Eq. 6 and Eq. 7 gives

$$\begin{array}{rcccccc} S_n = & a_1 & + & (a_1 + d) & + \cdots + & [a_1 + (n - 1)d] \\ S_n = & a_n & + & (a_n - d) & + \cdots + & [a_n - (n - 1)d] \\ \hline 2S_n = & (a_1 + a_n) & + & (a_1 + a_n) & + \cdots + & (a_1 + a_n). \end{array}$$

On the right side of the last equation, the number $a_1 + a_n$ occurs n times. Thus

$$2S_n = n(a_1 + a_n).$$

By dividing both sides by 2, we have the following result.

The sum S_n of the first n terms of an arithmetic progression is given by

$$S_n = \frac{n}{2}(a_1 + a_n), \tag{8}$$

where a_1 is the first term and a_n is the nth term.

EXAMPLE 5

Find the sum of the first eight terms of the series

$$1 + 4 + 7 + \cdots.$$

The terms form an arithmetic progression with $a_1 = 1$ and $d = 3$. The eighth term is given by

$$a_n = a_1 + (n-1)d,$$
$$a_8 = 1 + (8-1)3 = 22.$$

We can now find the sum by using Eq. 8.

$$S_8 = \frac{8}{2}(a_1 + a_8) = 4(1 + 22) = 92.$$

EXAMPLE 6

Find the sum of the odd integers between 20 *and* 60.

The odd integers between 20 and 60 form an arithmetic progression with $d = 2$. Here $a_1 = 21$ and $a_n = 59$. To find n, we have

$$a_n = a_1 + (n-1)d,$$
$$59 = 21 + (n-1)2,$$
$$38 = 2n - 2,$$
$$40 = 2n,$$
$$20 = n.$$

That is, 59 is the twentieth term of the arithmetic progression. Using Eq. 8 gives

$$S_{20} = \frac{20}{2}(21 + 59) = 800.$$

We can find another formula for S_n besides Eq. 8. Replacing a_n in that equation by $a_1 + (n-1)d$, we have

$$\begin{aligned} S_n &= \frac{n}{2}(a_1 + a_n) \\ &= \frac{n}{2}[a_1 + a_1 + (n-1)d]. \end{aligned}$$

Thus

$$\boxed{S_n = \frac{n}{2}[2a_1 + (n-1)d].}$$

Let's redo Example 5 by using this formula. We have

$$S_8 = \frac{8}{2}[2(1) + (8-1)3] = 4[2 + 21] = 92.$$

Exercise 14.2

In Problems **1–10**, *for those sequences that are arithmetic progressions, find the indicated term.*

1. 2, 6, 10, . . . ; a_6.

2. 4, 9, 14, . . . ; a_5.

3. 13, 1, −11, . . . ; a_8.

4. 13, 0, −13, . . . ; a_7.

5. 4, 1, −1, . . . ; a_5.

6. 4, 1, $\frac{1}{4}$, . . . ; a_7.

7. $\frac{1}{3}$, $\frac{2}{3}$, 1, . . . ; a_{11}.

8. a, $-a$, $-3a$, . . . ; a_9.

9. .4, .8, 1.2, . . . ; a_{10}.

10. −3, −1, 1, . . . ; a_6.

In Problems **11–20**, *find the indicated sum.*

11. $2+4+6+\cdots$; S_{10}.

12. $1+5+9+\cdots$; S_9.

13. $15+10+5+\cdots$; S_{16}.

14. $-\frac{1}{3}+\frac{1}{3}+1+\cdots$; S_{12}.

15. $2+0-2-4-\cdots$; S_{12}.

16. $14+7+0+\cdots$; S_9.

17. $-10+6+22+\cdots$; S_{10}.

18. $\frac{3}{6}+\frac{2}{6}+\frac{1}{6}+\cdots$; S_9.

19. $3+6+9+\cdots+27$.

20. $7+9+11+\cdots+33$.

21. The fourth term of an arithmetic progression with $d = 14$ is 86. Find (a) the first term and (b) the eighth term.

22. The first term of an arithmetic progression is 16. The tenth term is 10. Find d.

23. The sixteenth term of an arithmetic progression is 28. The first term is −4. Find d.

24. In an arithmetic progression, $a_1 = 6$, $a_n = 26$, and $d = 4$. Find n.

25. The fourth term of an arithmetic progression is 11 and the eleventh term is 32. Find a_1 and d.

26. The third term of an arithmetic progression is −2 and the tenth term is 12. Find the first term.

27. Find the sum of the first 100 positive integers.

28. Find the sum of the odd integers between 30 and 80.

In Problems **29–32**, *find the values of the indicated quantities for the arithmetic progression that has the given properties.*

29. $a_{13} = 42$, $d = 2$; a_{16}, S_{16}.

30. $a_9 = 20$, $d = 1$; a_{16}, S_{16}.

31. $a_1 = -30$, $d = 3$, $S_n = 69$; n, a_n.

32. $a_1 = 12$, $a_n = 42$, $d = 2$; n, S_n.

33. The length of the first swing of a pendulum is 10 cm, and because of resistance, each succeeding swing is $\frac{1}{4}$ cm less. What is the length of the thirteenth pendulum swing?

34. How many swings of the pendulum in Problem 33 are completed before the pendulum comes to rest?

35. If a person saves 1¢ the first day, 2¢ the next day, 3¢ the next day, and so on, how much money will the person have at the end of 30 days?

36. In a vacuum, an object falls approximately 4.9 m the first second, 14.7 m the next, 24.5 m the next, and so on. How far does it fall in 10 s?

37. A display in a supermarket consists of boxes piled up in the form of a pyramid. There are 20 boxes on the bottom layer and each successive layer has two fewer boxes. How many boxes are in the display?

14.3 GEOMETRIC PROGRESSIONS

In the sequence

$$3, 6, 12, 24, 48,$$

each term, after the first, can be obtained by multiplying the preceding term by 2. This means that the ratio (or quotient) of every two consecutive terms is 2:

$$\frac{6}{3}=2, \quad \frac{12}{6}=2, \quad \text{and so on.}$$

We call the sequence 3, 6, 12, 24, 48 a *geometric progression* with *common ratio* 2. Here is a more general definition.

> A sequence in which each term, after the first, can be obtained by multiplying the preceding term by the constant r is called a **geometric progression** with **common ratio** r.

EXAMPLE 1

Suppose the first term of a geometric progression is 3 and the common ratio r is $-\frac{1}{2}$. Then

$$\begin{aligned} a_1 &= 3, \\ a_2 &= 3\left(-\frac{1}{2}\right) = -\frac{3}{2}, \\ a_3 &= \left(-\frac{3}{2}\right)\left(-\frac{1}{2}\right) = \frac{3}{4}, \\ a_4 &= \frac{3}{4}\left(-\frac{1}{2}\right) = -\frac{3}{8}, \quad \text{and so on.} \end{aligned}$$

Thus the geometric progression is

$$3, -\frac{3}{2}, \frac{3}{4}, -\frac{3}{8}, \ldots .$$

EXAMPLE 2

a. The sequence

$$4, .4, .04, .004$$

is a geometric progression with $r = .1$. For example, the ratio of the second term to the first term is $.4/4 = .1$.

b. The sequence

$$\frac{1}{2}, 1, 2, 4, \ldots$$

is a geometric progression with $r = 2$. Check this out.

We can find a formula for the nth term of a geometric progression. Suppose a geometric progression has first term a_1 and common ratio r. Then

$$\begin{aligned} a_2 &= a_1 r, \\ a_3 &= a_2 r = (a_1 r)r = a_1 r^2, \\ a_4 &= a_3 r = (a_1 r^2)r = a_1 r^3, \quad \text{and so on.} \end{aligned}$$

Thus the geometric progression is

$$a_1, a_1 r, a_1 r^2, a_1 r^3, \ldots .$$

In each term the exponent for r is one less than the number of the term. For example, the tenth term is $a_{10} = a_1 r^{10-1} = a_1 r^9$. More generally, we have the following formula.

$$a_n = a_1 r^{n-1} \tag{1}$$

gives the nth term, a_n, of a geometric progression with first term a_1 and common ratio r.

EXAMPLE 3

Find the fifth term of the geometric progression

$$2, -\frac{1}{2}, \frac{1}{8}, \ldots .$$

Here $a_1 = 2$ and $r = (-\frac{1}{2})/2 = -\frac{1}{4}$. Using Eq. 1 with $n = 5$ gives

$$\begin{aligned} a_n &= a_1 r^{n-1}, \\ a_5 &= 2\left(-\frac{1}{4}\right)^{5-1} = 2\left(-\frac{1}{4}\right)^4 = \frac{1}{128}. \end{aligned}$$

EXAMPLE 4

The first term of a geometric progression is 3 and the sixth term is $\frac{3}{32}$. Find the common ratio r.

Here $a_1 = 3$ and $a_6 = \frac{3}{32}$. Using Eq. 1 with $n = 6$, we have

$$a_1 r^{n-1} = a_n,$$

$$3r^5 = \frac{3}{32},$$

$$r^5 = \frac{1}{32},$$

$$r = \frac{1}{2} \qquad \text{[taking the fifth root of both sides].}$$

EXAMPLE 5

A person invests a principal of \$100 *and earns interest of 6% at the end of each year. If none of the interest is withdrawn, how much is the investment worth at the end of 5 years?*

At the end of the first year, the value of the investment is the sum of the original investment and the interest:

$$\begin{aligned} 100 + 100(.06) &= 100(1 + .06) \qquad \text{[factoring]} \\ &= 100(1.06). \end{aligned}$$

This amount is the principal on which interest is earned for the second year. At the end of that year, the value of the investment is the sum of principal and interest, or

$$\begin{aligned} 100(1.06) + [100(1.06)].06 &= 100(1.06)[1 + .06] \qquad \text{[factoring]} \\ &= 100(1.06)^2. \end{aligned}$$

Similarly, we can show that the value of the investment at the end of the third year is $100(1.06)^3$, and so on. These values,

$$100(1.06),\ 100(1.06)^2,\ 100(1.06)^3, \ldots,$$

form a geometric progression where the common ratio is 1.06. In general, the value of the investment at the end of n years is

$$100(1.06)^n.$$

Thus, at the end of 5 years, the value is $100(1.06)^5$ dollars. (A calculator value for this is \$133.82.) Because the interest is reinvested, the person is earning "interest on interest," called *compound interest.*

We can find a formula for the sum S_n of the first n terms of a geometric progression. Since

$$S_n = a_1 + a_1r + a_1r^2 + \cdots + a_1r^{n-1}, \tag{2}$$

multiplying both sides by r gives

$$rS_n = a_1r + a_1r^2 + a_1r^3 + \cdots + a_1r^n. \tag{3}$$

Subtracting Eq. 3 from Eq. 2 and factoring, we have

$$S_n - rS_n = a_1 - a_1r^n,$$

$$S_n(1 - r) = a_1(1 - r^n).$$

If $r \neq 1$, then dividing both sides by $1 - r$ gives the following result.

$$\mathbf{S_n = \frac{a_1(1 - r^n)}{1 - r}} \tag{4}$$

gives the sum S_n of the first n terms of a geometric progression with first term a_1 and common ratio r.

From Eq. 4 we can get another formula for S_n that involves a_n. We have

$$S_n = \frac{a_1(1 - r^n)}{1 - r} = \frac{a_1 - a_1r^n}{1 - r}$$

$$= \frac{a_1 - (a_1r^{n-1})r}{1 - r}.$$

From Eq. 1, $a_1r^{n-1} = a_n$, so we have

$$\mathbf{S_n = \frac{a_1 - ra_n}{1 - r}}. \tag{5}$$

If $r = 1$, then Eqs. 4 and 5 cannot be used. However, in that case we have

$$S_n = a_1 + a_1 + \cdots + a_1 = na_1.$$

EXAMPLE 6

a. *Find the sum of the first* 10 *terms of the geometric progression* 12, 6, 3, ...

Here $a_1 = 12$, $r = \frac{6}{12} = \frac{1}{2}$, and $n = 10$. From Eq. 4,

$$S_{10} = \frac{a_1(1 - r^{10})}{1 - r} = \frac{12[1 - (\frac{1}{2})^{10}]}{1 - \frac{1}{2}}$$

$$= \frac{12[1 - \frac{1}{1024}]}{\frac{1}{2}} = \frac{3069}{128}.$$

b. *Find the sum of the series* $1 + \frac{1}{3} + (\frac{1}{3})^2 + \cdots + (\frac{1}{3})^5$.

The terms of the series form a geometric progression with $a_1 = 1$ and $r = \frac{1}{3}$. This series is called a *geometric series*. Here $n = 6$ (not 5). From Eq. 4 we have

$$S_6 = \frac{a_1(1 - r^6)}{1 - r} = \frac{1[1 - (\frac{1}{3})^6]}{1 - \frac{1}{3}}$$

$$= \frac{\frac{728}{729}}{\frac{2}{3}} = \frac{364}{243}.$$

Since the last term in the sum is known to be $(\frac{1}{3})^5$, we could have used Eq. 5 to obtain the same result.

In advanced mathematics, a meaning is given to the "sum" of the infinite geometric series

$$a_1 + a_1r + a_1r^2 + \cdots + a_1r^{n-1} + \cdots.$$

The sum of the first n terms is

$$S_n = \frac{a_1(1 - r^n)}{1 - r}, \tag{6}$$

If $|r| < 1$, then as n gets larger and larger, r^n gets closer and closer to zero. Thus r^n has less and less effect on S_n; we can even think of r^n as 0 in Eq. 6. Replacing S_n by S and r^n by 0 in that equation, we have the following result.

> When $|r| < 1$, the sum S of an infinite geometric series is given by
>
> $$S = \frac{a_1}{1 - r}. \tag{7}$$

EXAMPLE 7

Find the sum $4 - 2 + 1 - \frac{1}{2} + \cdots$.

The series is geometric with $a_1 = 4$ and $r = -\frac{1}{2}$. Since $|r| < 1$, by Eq. 7 the sum S is given by

$$S = \frac{a_1}{1 - r} = \frac{4}{1 - (-\frac{1}{2})} = \frac{4}{\frac{3}{2}} = \frac{8}{3}.$$

Exercise 14.3

In Problems **1–10**, *for those sequences that are geometric progressions, find the indicated term.*

1. $3, 6, 12, \ldots; \quad a_5$.

2. $4, 1, \frac{1}{4}, \ldots; \quad a_7$.

3. $6, -3, \frac{3}{2}, \ldots; \quad a_6$.

4. $-1, 3, -9, \ldots; \quad a_6$.

5. $2, 4, 6, \ldots; \quad a_5$.

6. $\frac{1}{3}, \frac{2}{3}, 1, \ldots; \quad a_5$.

7. $-4, 2, -1, \ldots; \quad a_9$.

8. $.3, .03, .003, \ldots; \quad a_7$.

9. $\frac{3}{2}, \frac{9}{4}, \frac{27}{8}, \ldots; \quad a_5$.

10. $12, -4, \frac{4}{3}, \ldots; \quad a_6$.

In Problems **11–18**, *find the indicated sum.*

11. $3 + 9 + 27 + \cdots; \quad S_6$.

12. $1 - \frac{1}{4} + \frac{1}{16} - \cdots; \quad S_6$.

13. $6 - 12 + 24 - \cdots; \quad S_6$.

14. $.1 + .02 + .004 + \cdots; \quad S_5$.

15. $.3 + .03 + .003 + \cdots; \quad S_7$.

16. $24 + 12 + 6 + \cdots; \quad S_7$.

17. $1 + \frac{1}{2} + \frac{1}{4} + \cdots + \frac{1}{128}$.

18. $1 - \frac{1}{2} + \frac{1}{4} - \cdots + \frac{1}{64}$.

19. The first term of a geometric progression is 16. The fourth term is $\frac{1}{4}$. Find r.

20. The first term of a geometric progression is 18 and the fourth term is $\frac{2}{81}$. Find the second term.

21. A distant star now has a surface temperature of 10,000°C, and observations indicate that the temperature decreases by 10% every 1000 years. What will be the temperature 4000 years from now?

22. Find $\sum_{n=1}^{6} \left(\frac{1}{2}\right)^n$.

In Problems **23–28**, *find the indicated quantities for the geometric progression that has the given properties.*

23. $r = \frac{1}{2}, S_6 = 126; \quad a_1, a_6$.

24. $a_1 = 1, a_6 = 32; \quad r, S_6$.

25. $a_1 = \frac{3}{4}, a_n = -96, S_n = -\frac{255}{4}; \quad r$.

26. $a_1 = \frac{1}{2}, a_{10} = 256; \quad a_{12}, S_{12}$.

27. $r = \frac{1}{3}, S_5 = 121; \quad a_1, a_5$.

28. $a_6 = 1, a_8 = 9; \quad S_5$.

29. By means of a pump, air is being removed from a container in such a way that each second, one-tenth of the remaining air in the container is removed. After 5 s, what percentage of air is left? Give your answer to the nearest percentage.

30. A tank full of alcohol is emptied of one-fourth its contents. The tank is then filled with water. This is repeated three more times. What part of the tank is now alcohol?

31. Find the value of an investment of \$250 after n years if interest is compounded annually at the rate of 6.5%.

32. Bacteria growing in a culture double their number every day. If there were originally 10,000 bacteria, how many will there be after 8 days?

In Problems **33–36**, *find the sum.*

33. $3 + \frac{3}{2} + \frac{3}{4} + \cdots$.

34. $4 + 1 + \frac{1}{4} + \cdots$.

35. $100 - 10 + 1 - .1 + \cdots.$

36. $.02 + .002 + .0002 + \cdots.$

37. A ball is dropped from a height of 4 m. After the first bounce, the ball reaches a height of 2 m; after the second bounce a height of 1 m, and so on. What is the total distance traveled by the ball before coming to rest?

38. The tip of a pendulum moves through a distance of 6 cm, after which the distance is constantly decreased by 10% on each swing. Before coming to rest, through what total distance has the tip moved?

14.4 REVIEW

IMPORTANT TERMS AND SYMBOLS

sequence *(14.1)*
term of sequence *(14.1)*
series *(14.1)*
sigma notation *(14.1)*
Σ *(14.1)*
index *(14.1)*
lower limit *(14.1)*
upper limit *(14.1)*
arithmetic progression *(14.2)*
common difference *(14.2)*
geometric progression *(14.3)*
common ratio *(14.3)*

REVIEW PROBLEMS

In Problems **1** *and* **2**, *find the tenth term of the given arithmetic progression.*

1. 3, 8, 13, ...

2. 7, 4, 1, ...

In Problems **3** *and* **4**, *find the sixth term of the given geometric progression.*

3. $8, -2, \frac{1}{2}, \ldots.$

4. $3, 12, 48, \ldots.$

In Problems **5** *and* **6**, *evaluate.*

5. $\sum_{n=1}^{5} (n^2 - 2n).$

6. $\sum_{n=0}^{4} (-1)^n(2^n - 1).$

7. Find $1 + \frac{1}{2} + (\frac{1}{2})^2 + \cdots + (\frac{1}{2})^6.$

8. Find $1 + 3 + 5 + \cdots + 33.$

9. How many terms of the sequence $-16, -12, -8, \ldots$ must be added to give a sum of 44?

10. Find the sum of all odd integers between 30 and 60.

11. The twelfth term of an arithmetic progression is -12 and the twenty-third term is 20. Find the sixteenth term.

12. The first term of a geometric progression is 36 and the third term is $\frac{1}{4}$. Find the fifth term.

13. Find the sum $1 + \frac{1}{4} + \frac{1}{16} + \cdots$.

14. Find the sum $1 - \frac{1}{4} + \frac{1}{16} - \cdots$.

15. A ball is released from an initial height of 8 m. After each contact with the floor, the ball rebounds to a height that is three-fourths of the height from which it last fell. What height does the ball reach on the fifth bounce?

16. For the ball in Problem 15, what is the total distance that the ball travels before coming to rest?

17. A twelve-hour clock strikes once at 1 o'clock, twice at 2 o'clock, and so on. How many strikes does it make in 24 consecutive hours?

15 Conic Sections

15.1 THE DISTANCE FORMULA

We can find a formula that gives the distance between two points in a rectangular coordinate plane. Suppose the points are (x_1, y_1) and (x_2, y_2), as in Fig. 15-1. Let

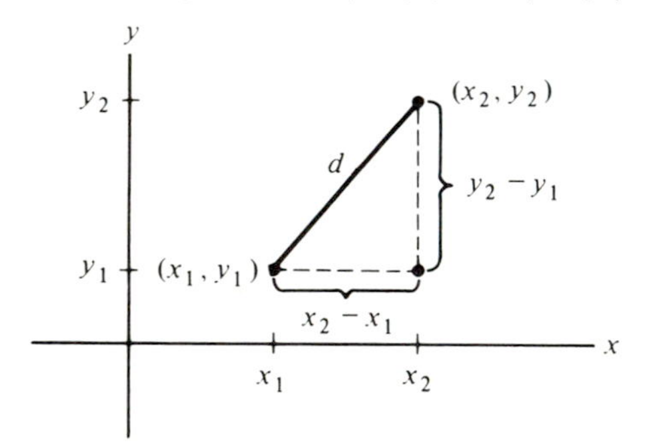

FIGURE 15-1

the length of the line segment joining them be d. By drawing line segments parallel to the axes, we have a right triangle with d as the length of the hypotenuse. The lengths of the other sides are $x_2 - x_1$ and $y_2 - y_1$. By the Pythagorean theorem,*

$$d^2 = (x_2 - x_1)^2 + (y_2 - y_1)^2.$$

---

* The Pythagorean theorem states that the square of the length of the hypotenuse of a right triangle is equal to the sum of the squares of the lengths of the other two sides.

Since d cannot be negative, we must have

$$d = \sqrt{(x_2 - x_1)^2 + (y_2 - y_1)^2},$$

which is called the **distance formula**.

EXAMPLE 1

Find the distance between the given points.

a. $(5, -2)$ and $(-3, 4)$.

Suppose $(x_1, y_1) = (5, -2)$ and $(x_2, y_2) = (-3, 4)$. Then

$$\begin{aligned} d &= \sqrt{(x_2 - x_1)^2 + (y_2 - y_1)^2} \\ &= \sqrt{(-3 - 5)^2 + [4 - (-2)]^2} \\ &= \sqrt{(-8)^2 + (6)^2} = \sqrt{64 + 36} \\ &= \sqrt{100} = 10. \end{aligned}$$

If we chose $(-3, 4)$ as (x_1, y_1) and $(5, -2)$ as (x_2, y_2), we would get the same result.

b. $(3, -6)$ and the origin.

Let $(x_1, y_1) = (0, 0)$ and $(x_2, y_2) = (3, -6)$. Then

$$\begin{aligned} d &= \sqrt{(x_2 - x_1)^2 + (y_2 - y_1)^2} \\ &= \sqrt{(3 - 0)^2 + (-6 - 0)^2} \\ &= \sqrt{9 + 36} = \sqrt{45} = 3\sqrt{5}. \end{aligned}$$

Exercise 15.1

Find the distance between the given points.

1. $(2, 3)$, $(5, 7)$.

2. $(4, 5)$, $(12, 11)$.

3. $(0, 5)$, $(2, -2)$.

4. $(1, 3)$, $(1, 4)$.

5. $(-1, -2)$, $(-3, -4)$.

6. $(-4, 4)$, origin.

7. $(2, -3)$, $(-5, -3)$.

8. $(-\frac{3}{2}, \frac{1}{2})$, $(\frac{1}{2}, \frac{3}{2})$.

9. $(1, -\frac{1}{2})$, origin.

10. $(4, 0)$, $(-1, \sqrt{11})$.

11. $(-5, 0)$, $(0, -12)$.

12. $(0, -2)$, $(-5, -2)$.

15.2 THE CIRCLE

In the rest of this chapter we'll consider four curves, each of which may be formed by the intersection of a plane with a right circular cone. These curves are called **conic sections**, or **conics**, and specifically they are the *circle*, the *parabola*, the *ellipse*, and the *hyperbola*. They are shown in Fig. 15-2, where the type of curve obtained depends on how the plane cuts the cone. In this section we'll look at the circle.

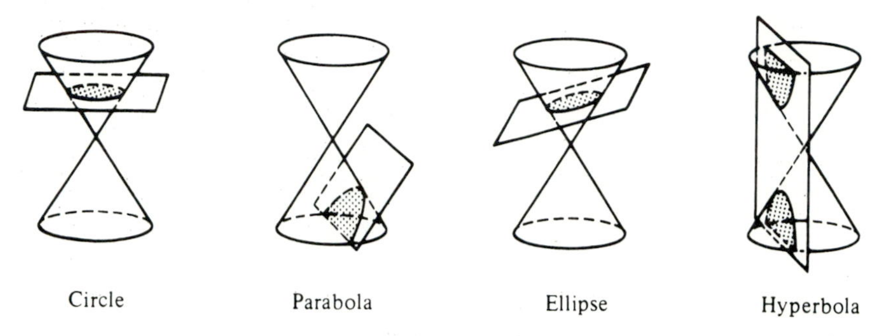

FIGURE 15-2

A **circle** consists of all points in a plane that are at a given distance from a fixed point in the plane. The fixed point is the **center**, and the given distance is the **radius** of the circle.

Suppose that a circle has a radius of r and has its center at (h, k), as in Fig. 15-3. Let (x, y) be any point on the circle. The distance of (x, y) from (h, k) must

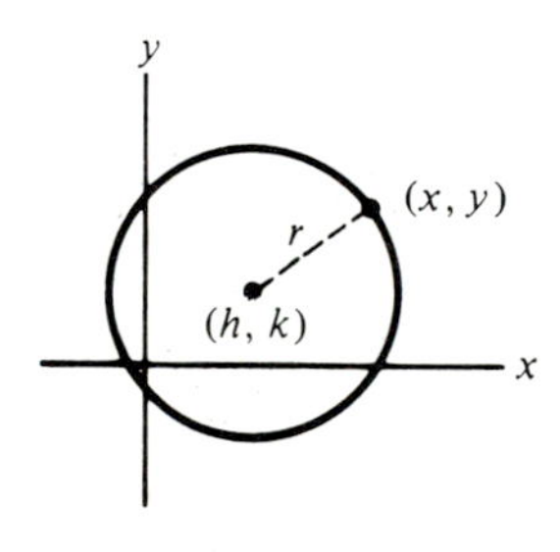

FIGURE 15-3

be r. Applying the distance formula to these points gives

$$\sqrt{(x-h)^2 + (y-k)^2} = r.$$

Squaring both sides, we have

$$(x-h)^2 + (y-k)^2 = r^2.$$

Any point on the circle must satisfy this equation. Also, it can be shown that all points satisfying the equation lie on the circle. In general,

$$(x - h)^2 + (y - k)^2 = r^2 \tag{1}$$

is the **standard form** of an equation of the circle with center (h, k) and radius r.

EXAMPLE 1

Find the standard form of an equation of the circle having the given center and radius.

a. Center $(-3, 2)$ and radius 4.

We use Eq. 1 with $h = -3$, $k = 2$, and $r = 4$.

$$(x - h)^2 + (y - k)^2 = r^2,$$
$$[x - (-3)]^2 + (y - 2)^2 = 4^2,$$
$$(x + 3)^2 + (y - 2)^2 = 16.$$

b. Center $(0, 0)$ and radius $\sqrt{3}$.

Here $h = 0$, $k = 0$, and $r = \sqrt{3}$.

$$(x - h)^2 + (y - k)^2 = r^2,$$
$$(x - 0)^2 + (y - 0)^2 = (\sqrt{3})^2,$$
$$x^2 + y^2 = 3.$$

In general, as Example 1b illustrates,

$$x^2 + y^2 = r^2 \tag{2}$$

is the standard form of an equation of the circle with center at the origin and radius r.

EXAMPLE 2

Describe and draw the graph of the given equation.

a. $(x - 1)^2 + (y + 4)^2 = 9$.

This has the form of Eq. 1. By writing this equation as

$$(x - 1)^2 + [y - (-4)]^2 = (3)^2,$$

we see that $h = 1$, $k = -4$, and $r = 3$. Thus the graph is a circle with center $(1, -4)$ and radius 3. See Fig. 15-4(a).

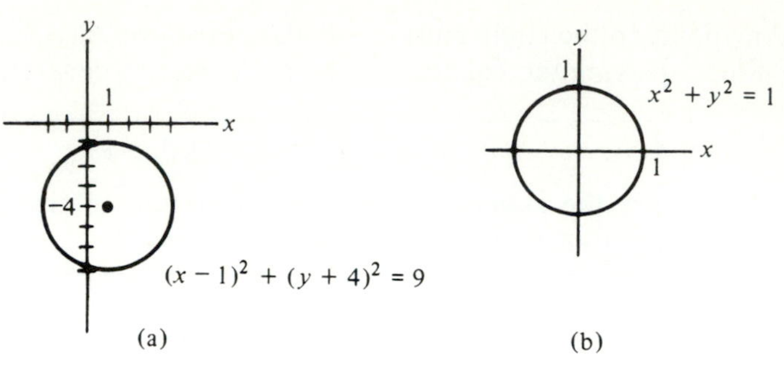

FIGURE 15-4

b. $x^2 + y^2 = 1$.

This has the form of Eq. 2 with $r^2 = 1$, so $r = 1$. The graph is a circle with center at the origin and radius 1. See Fig. 15-4(b).

If we expand the terms in the standard form of a circle, we obtain

$$(x - h)^2 + (y - k)^2 = r^2,$$

$$x^2 - 2hx + h^2 + y^2 - 2ky + k^2 = r^2,$$

$$x^2 + y^2 + (-2h)x + (-2k)y + (h^2 + k^2 - r^2) = 0 \qquad \text{[rearranging]}.$$

Since h, k, and r are constants, we can replace $-2h$, $-2k$, and $h^2 + k^2 - r^2$ by the constants D, E, and F, respectively. This gives the following special form.

$$\mathbf{x^2 + y^2 + Dx + Ey + F = 0,} \qquad (3)$$

where D, E, and F are constants, is called the **general form** of an equation of a circle.

EXAMPLE 3

Find the general form of an equation of the circle with center $(1, 4)$ *and radius* 2.

The standard form is

$$(x - 1)^2 + (y - 4)^2 = 2^2.$$

Expanding gives

$$x^2 - 2x + 1 + y^2 - 8y + 16 = 4,$$

$$x^2 + y^2 - 2x - 8y + 13 = 0,$$

which is the general form.

When a circle is given in general form, we can find the center and radius by going to standard form. This involves completing the square.* For example, the equation

$$x^2 + y^2 - 6x + 10y + 5 = 0 \tag{4}$$

is in general form. We now regroup:

$$(x^2 - 6x \qquad) + (y^2 + 10y \qquad) = -5.$$

To complete the square in x, we add $[\frac{1}{2}(-6)]^2$ or 9 to *both* sides; to complete the square in y, we add $[\frac{1}{2}(10)]^2$ or 25 to *both* sides. This gives

$$(x^2 - 6x + 9) + (y^2 + 10y + 25) = -5 + 9 + 25,$$

or the standard form

$$(x - 3)^2 + (y + 5)^2 = 29.$$

Thus the graph of Eq. 4 is a circle with center $(3, -5)$ and radius $\sqrt{29}$.

EXAMPLE 4

Describe the graphs of the following equations.

a. $x^2 + y^2 - y = 0$.

This has the form of Eq. 3, $x^2 + y^2 + Dx + Ey + F = 0$.

$$x^2 + y^2 - y = 0,$$

$$x^2 + (y^2 - y) = 0,$$

$$x^2 + \left(y^2 - y + \frac{1}{4}\right) = \frac{1}{4} \qquad \text{[completing the square]},$$

$$(x - 0)^2 + \left(y - \frac{1}{2}\right)^2 = \left(\frac{1}{2}\right)^2 \qquad \text{[standard form]}.$$

The graph is a circle with center $(0, \frac{1}{2})$ and radius $\frac{1}{2}$.

b. $2x^2 + 2y^2 + 8x - 3y + 5 = 0$.

This will have the form of Eq. 3 if we get the coefficients of the x^2- and y^2-terms to be 1. Thus we divide both sides by 2.

$$x^2 + y^2 + 4x - \frac{3}{2}y + \frac{5}{2} = 0,$$

$$(x^2 + 4x) + \left(y^2 - \frac{3}{2}y\right) = -\frac{5}{2} \qquad \text{[regrouping]},$$

$$(x^2 + 4x + 4) + \left(y^2 - \frac{3}{2}y + \frac{9}{16}\right) = -\frac{5}{2} + 4 + \frac{9}{16} \qquad \text{[completing the squares]},$$

---

* Completing the square was discussed in Section 6.2.

$$(x+2)^2 + \left(y - \frac{3}{4}\right)^2 = \frac{33}{16}.$$

The graph is a circle with center $\left(-2, \frac{3}{4}\right)$ and radius $\sqrt{\frac{33}{16}}$ or $\frac{\sqrt{33}}{4}$.

Every circle has an equation with general form $x^2 + y^2 + Dx + Ey + F = 0$. But an equation of this form does not always have a circle for its graph. The graph could be a point, or there may not be any graph at all. You will know what it is after completing the square. Example 5 will illustrate.

EXAMPLE 5

Describe the graphs of the following equations.

a. $x^2 + y^2 - 2x - 6y + 10 = 0$.

This equation has the form $x^2 + y^2 + Dx + Ey + F = 0$. Completing the squares in x and y gives

$$(x^2 - 2x + 1) + (y^2 - 6y + 9) = -10 + 1 + 9,$$
$$(x-1)^2 + (y-3)^2 = 0.$$

This equation implies a center at $(1, 3)$, but a radius of 0. Thus the graph is the single point $(1, 3)$.

b. $x^2 + y^2 - 2x - 6y + 11 = 0$.

Completing the squares in x and y gives

$$(x^2 - 2x + 1) + (y^2 - 6y + 9) = -11 + 1 + 9,$$
$$(x-1)^2 + (y-3)^2 = -1.$$

Since the left side of this equation is a sum of squares, it can never equal -1, a negative number. Thus no point satisfies the equation, so the equation has no graph.

In general, any equation of the form $x^2 + y^2 + Dx + Ey + F = 0$ will give rise to a circle, a point, or no graph.

Exercise 15.2

*In Problems **1–6**, find the standard and general forms of an equation of the circle with the given center C and radius r.*

1. $C = (2, 3)$, $r = 6$.

2. $C = (4, -5)$, $r = 2$.

3. $C = (-1, 6)$, $r = 4$.

4. $C = (-2, -3)$, $r = 1$.

5. $C = (0, 0)$, $r = \frac{1}{2}$.

6. $C = (3, 0)$, $r = \sqrt{3}$.

In Problems **7–12**, *give the center C and radius r of the circle with the given equation. Also, draw the circle.*

7. $x^2 + y^2 = 9$.

8. $(x - 1)^2 + (y - 2)^2 = 1$.

9. $(x - 3)^2 + (y + 4)^2 = 4$.

10. $(x + 6)^2 + (y + 1)^2 = 3^2$.

11. $(x + 2)^2 + y^2 = 1$.

12. $x^2 + (y - 3)^2 = 16$.

In Problems **13–24**, *describe the graph of the equation. For those that are circles, give the center C and radius r.*

13. $x^2 + y^2 - 2x - 4y - 4 = 0$.

14. $x^2 + y^2 + 4x - 6y + 9 = 0$.

15. $x^2 + y^2 + 6y + 5 = 0$.

16. $x^2 + y^2 - 12x + 27 = 0$.

17. $x^2 + y^2 + 2x - 2y + 3 = 0$.

18. $x^2 + y^2 + 6x - 2y - 15 = 0$.

19. $x^2 + y^2 - 14x + 4y + 37 = 0$.

20. $x^2 + y^2 + 4y - 8x + 21 = 0$.

21. $x^2 + y^2 - 3x + y + \frac{5}{2} = 0$.

22. $9x^2 + 9y^2 - 6x + 18y + 9 = 0$.

23. $2x^2 + 2y^2 - 4x + 7y + 2 = 0$.

24. $16x^2 + 16y^2 + 24x - 48y - 3 = 0$.

15.3 THE PARABOLA AND ELLIPSE

Another conic section is the *parabola*. From Sec. 12.3 we know that the graph of a quadratic function

$$y = ax^2 + bx + c \tag{1}$$

is a parabola that opens upward if $a > 0$, and downward if $a < 0$. See Fig. 15-5.

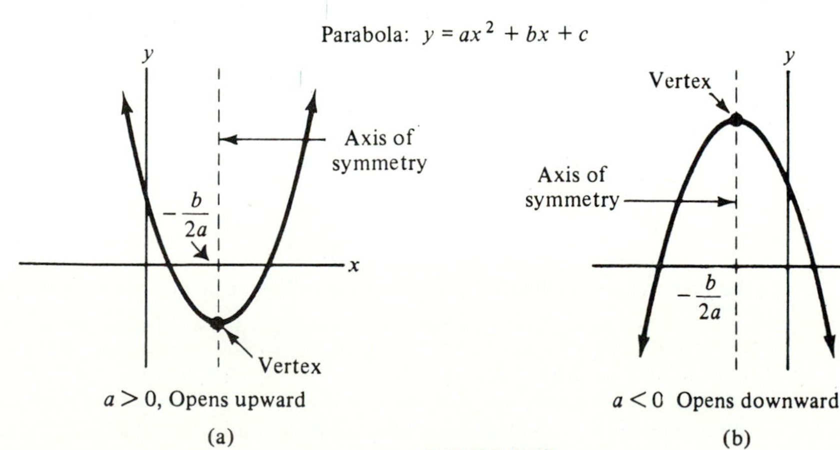

FIGURE 15-5

The vertex occurs at the point with x-coordinate $-\dfrac{b}{2a}$. The broken line indicates the axis of symmetry. In each case the axis of symmetry is parallel to the y-axis.

If we interchange the roles of x and y in Eq. 1, then we obtain the equation

$$x = ay^2 + by + c,$$

which gives x as a quadratic function of y. The graph of this equation is also a parabola, but its axis of symmetry is parallel to the x-axis. If $a > 0$, the parabola opens to the *right* [see Fig. 15-6(a)], and if $a < 0$, it opens to the *left* [Fig. 15-6(b)]. The vertex occurs at the point with y-coordinate $-\dfrac{b}{2a}$.

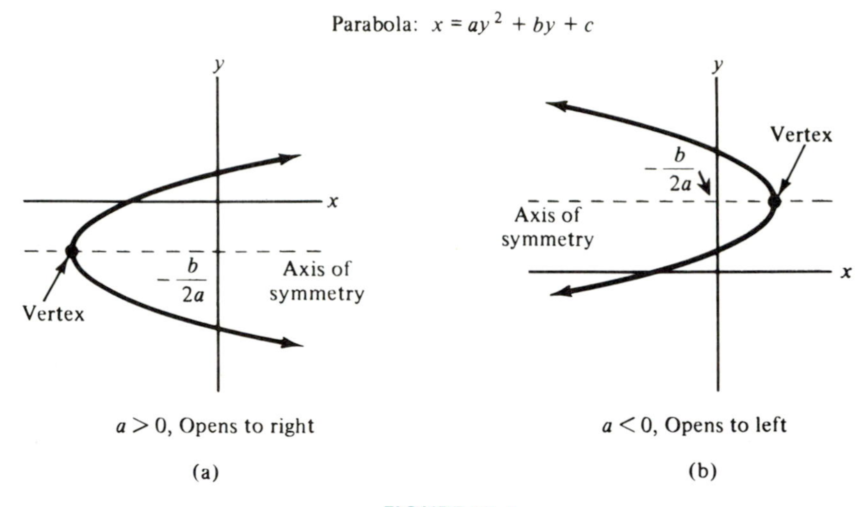

FIGURE 15-6

EXAMPLE 1

Identify and sketch the graph of the equation $y = -x^2 - 4x - 1$.

This equation has the form $y = ax^2 + bx + c$, where $a = -1 < 0$. Thus its graph is a parabola that opens downward. The vertex occurs at the point with x-coordinate

$$-\frac{b}{2a} = -\frac{-4}{2(-1)} = -2.$$

We find the y-coordinate by substituting -2 for x in the given equation:

$$y = -(-2)^2 - 4(-2) - 1 = -4 + 8 - 1 = 3.$$

Thus the vertex is $(-2, 3)$. We locate the vertex and then plot some points on one side of the axis of symmetry. By using this axis, we find corresponding points on the other side. Finally, we pass a parabola through all points (see Fig. 15-7).

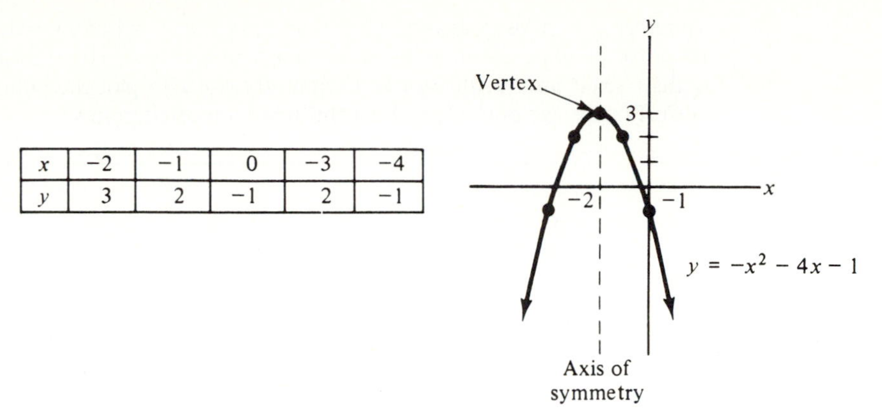

x	-2	-1	0	-3	-4
y	3	2	-1	2	-1

FIGURE 15-7

EXAMPLE 2

Identify and sketch the graph of $x = y^2 - 6y + 2$.

This equation has the form $x = ay^2 + by + c$, where $a = 1 > 0$. Thus its graph is a parabola with axis parallel to the x-axis and opening to the right. The vertex occurs at the point with y-coordinate

$$-\frac{b}{2a} = -\frac{-6}{2(1)} = 3.$$

By substituting 3 for y in the given equation, we find the x-coordinate of the vertex:

$$x = 3^2 - 6(3) + 2 = 9 - 18 + 2 = -7.$$

Thus the vertex is $(-7, 3)$. We locate the vertex and pass the axis of symmetry through it. (See Fig. 15-8). To plot some points of the parabola below the axis of

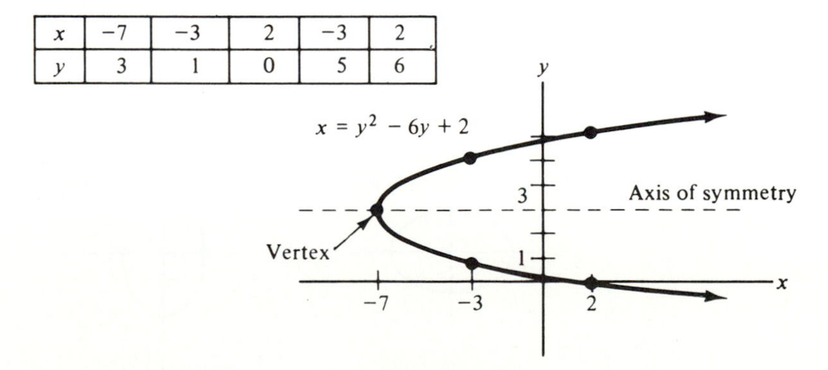

x	-7	-3	2	-3	2
y	3	1	0	5	6

FIGURE 15-8

symmetry, we choose values of y less than 3, such as $y = 1$ and $y = 0$, and then find the corresponding x-values by substitution into $x = y^2 - 6y + 2$. This gives the points $(-3, 1)$ and $(2, 0)$. By using symmetry, we also plot the points $(-3, 5)$ and $(2, 6)$. Finally, we draw a parabola through all plotted points.

The graph of $\frac{x^2}{3^2} + \frac{y^2}{2^2} = 1$ is shown in Fig. 15-9. This oval-shaped curve is

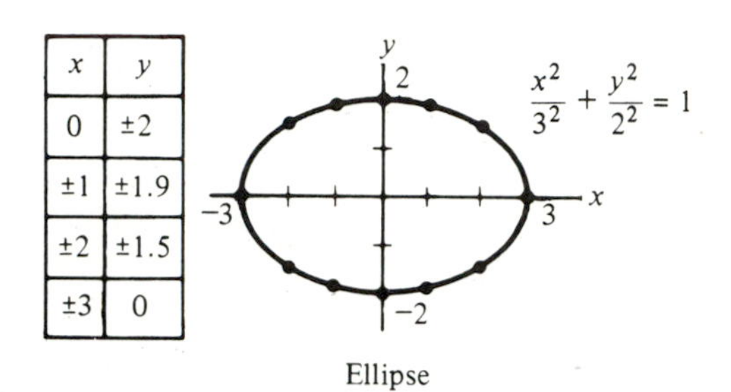

x	y
0	±2
±1	±1.9
±2	±1.5
±3	0

FIGURE 15-9

called an **ellipse**. From the graph we see that this ellipse has x-intercepts $(-3, 0)$ and $(3, 0)$ and y-intercepts $(0, -2)$ and $(0, 2)$.

More generally, the graph of any equation of the form

$$\frac{x^2}{a^2} + \frac{y^2}{b^2} = 1,$$

where a and b are positive constants and $a \neq b$, is an *ellipse* with x-intercepts $(-a, 0)$ and $(a, 0)$, and y-intercepts $(0, -b)$ and $(0, b)$. For convenience, we say that the x-intercepts are $-a$ and a, and the y-intercepts are $-b$ and b. In Fig. 15-10(a) we have $a > b$, and the ellipse is said to be *horizontal*. In Fig. 15-10(b) we have $b > a$, and the graph is a *vertical* ellipse. By plotting only the intercepts of an ellipse, you should be able to quickly sketch its graph.

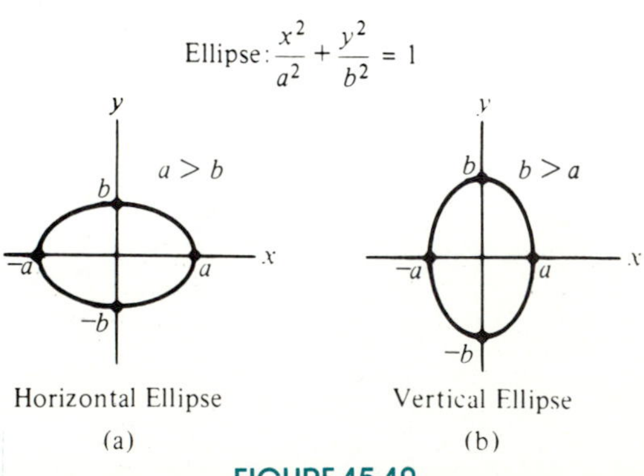

FIGURE 15-10

EXAMPLE 3

Identify and sketch the graph of each equation.

a. $\dfrac{x^2}{25}+\dfrac{y^2}{9}=1.$

This equation has the form $\dfrac{x^2}{a^2}+\dfrac{y^2}{b^2}=1$, where $a=5$, $b=3$, and $a>b$. Thus the graph is a *horizontal ellipse* with x-intercepts -5 and 5 and y-intercepts -3 and 3. See Fig. 15-11(a).

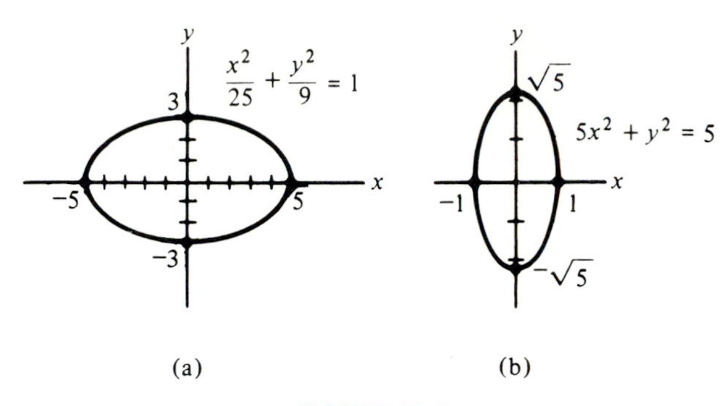

FIGURE 15-11

b. $5x^2+y^2=5.$

We can write this equation in the form of an ellipse if we first divide both sides by 5 so that the right side is 1.

$$5x^2+y^2=5,$$

$$\frac{5x^2}{5}+\frac{y^2}{5}=\frac{5}{5},$$

$$x^2+\frac{y^2}{5}=1,$$

which can be written as

$$\frac{x^2}{1^2}+\frac{y^2}{(\sqrt{5})^2}=1.$$

This equation has the form $\dfrac{x^2}{a^2}+\dfrac{y^2}{b^2}=1$, where $a=1$, $b=\sqrt{5}$, and $b>a$. Thus the graph is a *vertical ellipse* with x-intercepts -1 and 1 and y-intercepts $-\sqrt{5}$ and $\sqrt{5}$. See Fig. 15-11(b).

Exercise 15.3

Identify and sketch the graph of each equation.

1. $y = x^2 - 6x + 4.$
2. $x = -2y^2 + 4y + 3.$
3. $\frac{x^2}{9} + \frac{y^2}{25} = 1.$
4. $\frac{x^2}{16} + \frac{y^2}{9} = 1.$
5. $x = 4y^2 + 4y - 1.$
6. $\frac{x^2}{\sqrt{16}} + \frac{y^2}{\sqrt{36}} = 1.$
7. $y = (2 - x)x.$
8. $y = -3x^2 - 12x - 12.$
9. $\frac{x^2}{36} + \frac{y^2}{10} = 1.$
10. $16x = y^2.$
11. $x^2 + 4y^2 = 16.$
12. $y^2 = 36 - 3x^2.$
13. $y^2 + 6x = 0.$
14. $y - x^2 = 1.$
15. $8x^2 + y^2 = 8.$
16. $12y - 2x = 4y^2 + 5.$
17. $x + y = y^2.$
18. $x(x + 1) = x - 9y^2 + 9.$

15.4 THE HYPERBOLA

The graph of $\frac{x^2}{4^2} - \frac{y^2}{2^2} = 1$ is shown in Fig. 15-12; it is called a **hyperbola**. The x-intercepts are -4 and 4, and there are no y-intercepts.

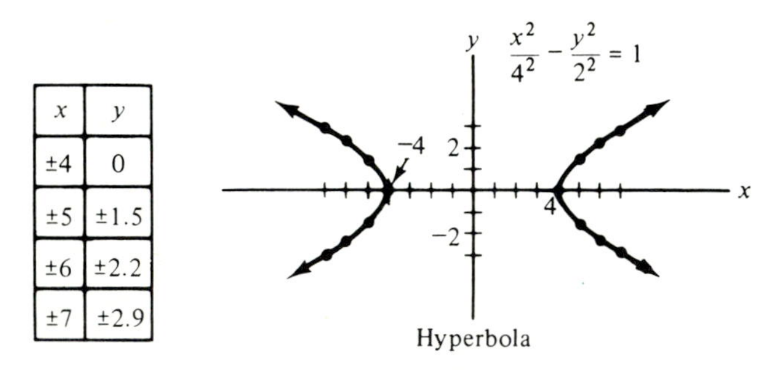

x	y
±4	0
±5	±1.5
±6	±2.2
±7	±2.9

FIGURE 15-12

More generally, the graph of any equation of the form

$$\frac{x^2}{a^2} - \frac{y^2}{b^2} = 1,$$

where a and b are positive constants is a (*horizontal*) *hyperbola*, as shown in Fig. 15-13(a). The x-intercepts are $-a$ and a, and there are no y-intercepts. In Fig.

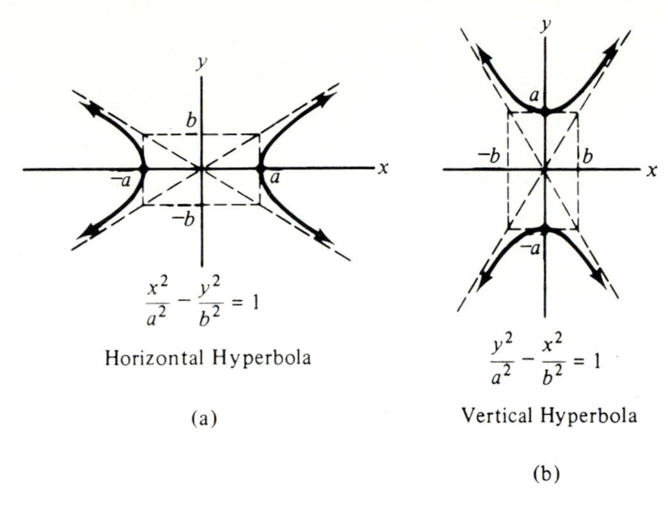

FIGURE 15-13

15-13(a), a broken rectangle of dimensions $2a$ by $2b$, centered at the origin, has been drawn. The diagonals of this rectangle have been extended (broken lines). Notice that as the points on the hyperbola get farther from the origin, they get very close to the extended diagonal lines. Such lines are called **asymptotes** and can be used as a guide for sketching the hyperbola. The rectangle and asymptotes are *not* part of the graph.

The graph of an equation of the form

$$\frac{y^2}{a^2} - \frac{x^2}{b^2} = 1,$$

where a and b are positive constants, is shown in Fig. 15-13(b). It is also a *hyperbola*. However, this hyperbola has y-intercepts of $-a$ and a, but no x-intercepts. It is called a *vertical* hyperbola. As before, the extended diagonals of the rectangle determined by a and b are asymptotes.

EXAMPLE 1

Identify and sketch the graph of each equation.

a. $\frac{x^2}{4} - \frac{y^2}{4} = 1.$

This equation has the form $\frac{x^2}{a^2} - \frac{y^2}{b^2} = 1$, where $a = 2$ and $b = 2$. Thus the graph is a *horizontal hyperbola* with x-intercepts -2 and 2. By first plotting the intercepts and drawing the asymptotes, we then sketch the hyperbola as in Fig. 15-14(a).

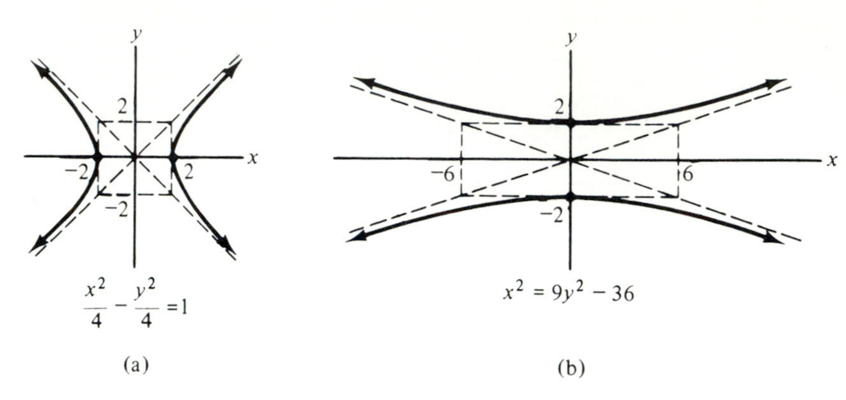

FIGURE 15-14

b. $x^2 = 9y^2 - 36$.

We'll try to write the equation in a familiar form. Since the squares of both x and y are involved, let's first get those terms on one side and the constant term on the other side.

$$x^2 = 9y^2 - 36,$$

$$x^2 - 9y^2 = -36 \qquad \text{[subtracting } 9y^2 \text{ from both sides]}.$$

Now we divide both sides by -36 in order to get the right side to be 1.

$$\frac{x^2}{-36} - \frac{9y^2}{-36} = \frac{-36}{-36},$$

$$-\frac{x^2}{36} + \frac{y^2}{4} = 1,$$

$$\frac{y^2}{4} - \frac{x^2}{36} = 1 \qquad \text{[rearranging]},$$

$$\frac{y^2}{2^2} - \frac{x^2}{6^2} = 1.$$

The last equation has the form $\frac{y^2}{a^2} - \frac{x^2}{b^2} = 1$, where $a = 2$ and $b = 6$. Thus its graph is a *vertical hyperbola* with y-intercepts -2 and 2. After plotting the intercepts and drawing the asymptotes, we sketch the hyperbola as in Fig. 15-14(b).

Here is a summary of forms of equations of circles, parabolas, ellipses, and hyperbolas that have been considered in this chapter.

Circle

center at origin: $x^2 + y^2 = r^2$

center at (h, k): $(x - h)^2 + (y - k)^2 = r^2$

Parabola

$y = ax^2 + bx + c$	$x = ay^2 + by + c$
$a > 0$, opens upward	$a > 0$, opens to the right
$a < 0$, opens downward	$a < 0$, opens to the left

Ellipse

$$\frac{x^2}{a^2} + \frac{y^2}{b^2} = 1 \qquad \begin{cases} \text{horizontal if } a > b \\ \text{vertical if } b > a \end{cases}$$

Hyperbola

$$\text{horizontal: } \frac{x^2}{a^2} - \frac{y^2}{b^2} = 1 \qquad \text{vertical: } \frac{y^2}{a^2} - \frac{x^2}{b^2} = 1$$

EXAMPLE 2

Identify and sketch the graph of each equation.

a. $\frac{x^2}{16} + \frac{y^2}{16} = 1.$

You may be tempted to say that this equation fits a form of an ellipse. But in each form of ellipse, the denominators are different. However, if we multiply both sides of the equation by 16, we get

$$16\left(\frac{x^2}{16} + \frac{y^2}{16}\right) = 16 \cdot 1,$$

$$x^2 + y^2 = 16.$$

This is an equation of a *circle* with center at the origin and radius 4. See Fig. 15-15(a).

b. $x^2 + 4y^2 = 1.$

We can write $4y^2$ as $\frac{y^2}{\frac{1}{4}}$. Thus

$$x^2 + \frac{y^2}{\frac{1}{4}} = 1,$$

$$\frac{x^2}{1^2} + \frac{y^2}{(\frac{1}{2})^2} = 1.$$

This equation has the form $\frac{x^2}{a^2} + \frac{y^2}{b^2} = 1$, where $a = 1$, $b = \frac{1}{2}$, and $a > b$. Thus its graph is a *horizontal ellipse* with x-intercepts -1 and 1 and y-intercepts $-\frac{1}{2}$ and $\frac{1}{2}$. See Fig. 15-15(b).

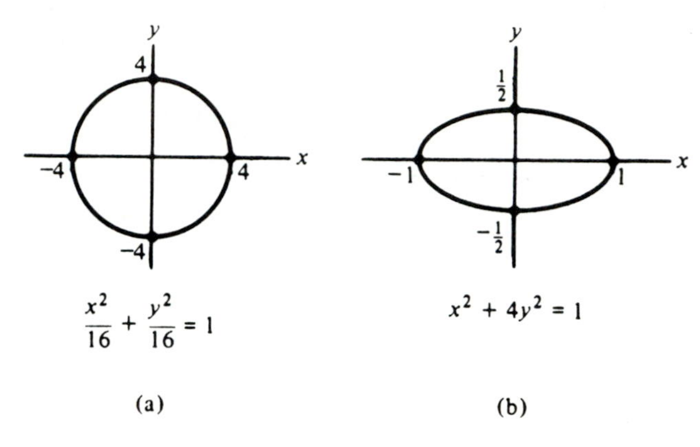

FIGURE 15-15

c. $3x + 2y^2 = 0$.

An equation whose literal terms consist of an x-term and a y^2-term has a parabola as its graph. Solving for x gives

$$3x = -2y^2,$$

$$x = -\frac{2}{3}y^2.$$

This equation has the form $x = ay^2 + by + c$, where $a = -\frac{2}{3} < 0$. Thus its graph is a parabola opening to the left. The vertex is at the point (x, y) where

$$y = -\frac{b}{2a} = -\frac{0}{2(-\frac{2}{3})} = 0$$

and

$$x = -\frac{2}{3}y^2 = -\frac{2}{3}(0)^2 = 0.$$

See Fig. 15-16(a).

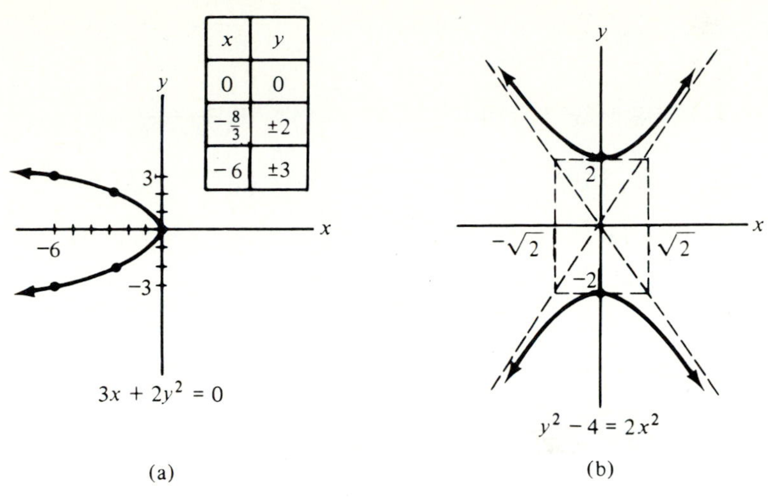

FIGURE 15-16

d. $y^2 - 4 = 2x^2$.

Writing the equation so that the x^2- and y^2-terms are on the left side and the constant is on the right side, we have

$$y^2 - 2x^2 = 4.$$

Dividing both sides by 4 gives

$$\frac{y^2}{4} - \frac{2x^2}{4} = \frac{4}{4},$$

$$\frac{y^2}{4} - \frac{x^2}{2} = 1,$$

$$\frac{y^2}{2^2} - \frac{x^2}{(\sqrt{2})^2} = 1.$$

This equation has the form $\frac{y^2}{a^2} - \frac{x^2}{b^2} = 1$, where $a = 2$, and $b = \sqrt{2}$. Thus its graph is a *vertical hyperbola* with y-intercepts -2 and 2. See Fig. 15-16(b).

Exercise 15.4

Identify and sketch the graph of each equation.

1. $\frac{x^2}{9} - \frac{y^2}{25} = 1.$

2. $\frac{y^2}{16} - \frac{x^2}{9} = 1.$

3. $y^2 - x^2 = 1.$

4. $2y + 3x^2 = 0.$

5. $4x^2 + 4y^2 = 1.$

6. $\frac{x^2}{36} - y = 0.$

7. $x + 8y + 2y^2 + 7 = 0.$

8. $\frac{x^2}{36} + \frac{y^2}{36} = 1.$

9. $9x^2 = 1 - y^2.$

10. $\frac{x^2}{4} - 1 = \frac{y^2}{4}.$

11. $25x^2 - 2y^2 = -100.$

12. $25x^2 - y^2 = 25.$

13. $\frac{x^2}{4} = -\frac{y}{6}.$

14. $\frac{x^2}{3} + \frac{y^2}{4} = 3.$

15. $x^2 = 5 - 20y^2.$

16. $9y^2 - 16x^2 - 144 = 0.$

15.5 REVIEW

IMPORTANT TERMS

distance formula *(15.1)*
conic sections *(15.2)*
circle *(15.2)*
center of circle *(15.2)*
radius *(15.2)*
standard form of circle *(15.2)*
general form of circle *(15.2)*
parabola *(15.3)*
ellipse *(15.3)*
hyperbola *(15.4)*
asymptote *(15.4)*

REVIEW PROBLEMS

In Problems **1–4**, *find the distance between the given points.*

1. $(1, 4), (-3, 2).$

2. $(-1, -1), (-6, -1).$

3. $(-8, 2), (0, -4).$

4. $(-3, -4), (-1, 1).$

In Problems **5–8**, *find the standard and general forms of an equation of a circle having the given center C and radius r.*

5. $C = (0, 0),\ r = 5.$

6. $C = (0, -2),\ r = \sqrt{2}.$

7. $C = (1, -1),\ r = \frac{1}{2}.$

8. $C = (\frac{1}{2}, \frac{1}{4}),\ r = 1.$

In Problems **9–16**, *describe the graph of the given equation. For those that are circles, give the center C and radius r.*

9. $x^2 + y^2 = 16.$

10. $(x - 4)^2 + (y - 3)^2 = 1.$

11. $(x + 2)^2 + (y - 1)^2 = 7.$

12. $x^2 + (y - 1)^2 - 12 = 0.$

13. $x^2 + y^2 + 5x - 6y + 2 = 0.$

14. $x^2 + y^2 - x - 1 = 0.$

15. $9x^2 + 9y^2 - 18x - 6y + 10 = 0.$

16. $3x^2 + 3y^2 - 6x + 12y + 16 = 0.$

In Problems **17–26**, *identify and sketch the graph of each equation.*

17. $9x^2 - 100y^2 = 900.$

18. $3x - 4y^2 = 0.$

19. $36x^2 + y^2 = 36.$

20. $36y^2 - x^2 = 4.$

21. $5x^2 + 2y = 0.$

22. $x^2 + \frac{1}{2}y^2 = 1.$

23. $6x^2 - 1 = -6y^2.$

24. $\frac{1}{49}x^2 - \frac{1}{49}y^2 = 1.$

25. $x + 2y = y^2 - 3.$

26. $3x^2 + 3y^2 = 12.$

Appendix A

Tables of Powers—Roots—Reciprocals

n	n^2	$\sqrt{n}$	$\sqrt{10n}$	n^3	$\sqrt[3]{n}$	$\sqrt[3]{10n}$	$\sqrt[3]{100n}$	$1/n$
1.0	1.0000	1.0000	3.1623	1.0000	1.0000	2.1544	4.6416	1.0000
1.1	1.2100	1.0488	3.3166	1.3310	1.0323	2.2240	4.7914	0.9091
1.2	1.4400	1.0954	3.4641	1.7280	1.0627	2.2894	4.9324	0.8333
1.3	1.6900	1.1402	3.6056	2.1970	1.0914	2.3513	5.0658	0.7692
1.4	1.9600	1.1832	3.7417	2.7440	1.1187	2.4101	5.1925	0.7143
1.5	2.2500	1.2247	3.8730	3.3750	1.1447	2.4662	5.3133	0.6667
1.6	2.5600	1.2649	4.0000	4.0960	1.1696	2.5198	5.4288	0.6250
1.7	2.8900	1.3038	4.1231	4.9130	1.1935	2.5713	5.5397	0.5882
1.8	3.2400	1.3416	4.2426	5.8320	1.2164	2.6207	5.6462	0.5556
1.9	3.6100	1.3784	4.3589	6.8590	1.2386	2.6684	5.7489	0.5263
2.0	4.0000	1.4142	4.4721	8.0000	1.2599	2.7144	5.8480	0.5000
2.1	4.4100	1.4491	4.5826	9.2610	1.2806	2.7589	5.9439	0.4762
2.2	4.8400	1.4832	4.6904	10.6480	1.3006	2.8020	6.0368	0.4545
2.3	5.2900	1.5166	4.7958	12.1670	1.3200	2.8439	6.1269	0.4348
2.4	5.7600	1.5492	4.8990	13.8240	1.3389	2.8845	6.2145	0.4167
2.5	6.2500	1.5811	5.0000	15.6250	1.3572	2.9240	6.2996	0.4000
2.6	6.7600	1.6125	5.0990	17.5760	1.3751	2.9625	6.3825	0.3846
2.7	7.2900	1.6432	5.1962	19.6830	1.3925	3.0000	6.4633	0.3704
2.8	7.8400	1.6733	5.2915	21.9520	1.4095	3.0366	6.5421	0.3571
2.9	8.4100	1.7029	5.3852	24.3890	1.4260	3.0723	6.6191	0.3448

n	n^2	$\sqrt{n}$	$\sqrt{10n}$	n^3	$\sqrt[3]{n}$	$\sqrt[3]{10n}$	$\sqrt[3]{100n}$	$1/n$
3.0	9.0000	1.7321	5.4772	27.0000	1.4422	3.1072	6.6943	0.3333
3.1	9.6100	1.7607	5.5678	29.7910	1.4581	3.1414	6.7679	0.3226
3.2	10.2400	1.7889	5.6569	32.7680	1.4736	3.1748	6.8399	0.3125
3.3	10.8900	1.8166	5.7446	35.9370	1.4888	3.2075	6.9104	0.3030
3.4	11.5600	1.8439	5.8310	39.3040	1.5037	3.2396	6.9795	0.2941
3.5	12.2500	1.8708	5.9161	42.8750	1.5183	3.2711	7.0473	0.2857
3.6	12.9600	1.8974	6.0000	46.6560	1.5326	3.3019	7.1138	0.2778
3.7	13.6900	1.9235	6.0828	50.6530	1.5467	3.3322	7.1791	0.2703
3.8	14.4400	1.9494	6.1644	54.8720	1.5605	3.3620	7.2432	0.2632
3.9	15.2100	1.9748	6.2450	59.3190	1.5741	3.3912	7.3061	0.2564
4.0	16.0000	2.0000	6.3246	64.0000	1.5874	3.4200	7.3681	0.2500
4.1	16.8100	2.0248	6.4031	68.9210	1.6005	3.4482	7.4290	0.2439
4.2	17.6400	2.0494	6.4807	74.0880	1.6134	3.4760	7.4889	0.2381
4.3	18.4900	2.0736	6.5574	79.5070	1.6261	3.5034	7.5478	0.2326
4.4	19.3600	2.0976	6.6333	85.1840	1.6386	3.5303	7.6059	0.2273
4.5	20.2500	2.1213	6.7082	91.1250	1.6510	3.5569	7.6631	0.2222
4.6	21.1600	2.1448	6.7823	97.3360	1.6631	3.5830	7.7194	0.2174
4.7	22.0900	2.1679	6.8557	103.823	1.6751	3.6088	7.7750	0.2128
4.8	23.0400	2.1909	6.9282	110.592	1.6869	3.6342	7.8297	0.2083
4.9	24.0100	2.2136	7.0000	117.649	1.6985	3.6593	7.8837	0.2041
5.0	25.0000	2.2361	7.0711	125.000	1.7100	3.6840	7.9370	0.2000
5.1	26.0100	2.2583	7.1414	132.651	1.7213	3.7084	7.9896	0.1961
5.2	27.0400	2.2804	7.2111	140.608	1.7325	3.7325	8.0415	0.1923
5.3	28.0900	2.3022	7.2801	148.877	1.7435	3.7563	8.0927	0.1887
5.4	29.1600	2.3238	7.3485	157.464	1.7544	3.7798	8.1433	0.1852
5.5	30.2500	2.3452	7.4162	166.375	1.7652	3.8030	8.1932	0.1818
5.6	31.3600	2.3664	7.4833	175.616	1.7758	3.8259	8.2426	0.1786
5.7	32.4900	2.3875	7.5498	185.193	1.7863	3.8485	8.2913	0.1754
5.8	33.6400	2.4083	7.6158	195.112	1.7967	3.8709	8.3396	0.1724
5.9	34.8100	2.4290	7.6811	205.379	1.8070	3.8930	8.3872	0.1695
6.0	36.0000	2.4495	7.7460	216.000	1.8171	3.9149	8.4343	0.1667
6.1	37.2100	2.4698	7.8102	226.981	1.8272	3.9365	8.4809	0.1639
6.2	38.4400	2.4900	7.8740	238.328	1.8371	3.9579	8.5270	0.1613
6.3	39.6900	2.5100	7.9372	250.047	1.8469	3.9791	8.5726	0.1587
6.4	40.9600	2.5298	8.0000	262.144	1.8566	4.0000	8.6177	0.1563
6.5	42.2500	2.5495	8.0623	274.625	1.8663	4.0207	8.6624	0.1538
6.6	43.5600	2.5690	8.1240	287.496	1.8758	4.0412	8.7066	0.1515
6.7	44.8900	2.5884	8.1854	300.763	1.8852	4.0615	8.7503	0.1493
6.8	46.2400	2.6077	8.2462	314.432	1.8945	4.0817	8.7937	0.1471
6.9	47.6100	2.6268	8.3066	328.509	1.9038	4.1016	8.8366	0.1449
7.0	49.0000	2.6458	8.3666	343.000	1.9129	4.1213	8.8790	0.1429
7.1	50.4100	2.6646	8.4261	357.911	1.9220	4.1408	8.9211	0.1408
7.2	51.8400	2.6833	8.4853	373.248	1.9310	4.1602	8.9628	0.1389

n	n^2	$\sqrt{n}$	$\sqrt{10n}$	n^3	$\sqrt[3]{n}$	$\sqrt[3]{10n}$	$\sqrt[3]{100n}$	$1/n$
7.3	53.2900	2.7019	8.5440	389.017	1.9399	4.1793	9.0041	0.1370
7.4	54.7600	2.7203	8.6023	405.224	1.9487	4.1983	9.0450	0.1351
7.5	56.2500	2.7386	8.6603	421.875	1.9574	4.2172	9.0856	0.1333
7.6	57.7600	2.7568	8.7178	438.976	1.9661	4.2358	9.1258	0.1316
7.7	59.2900	2.7749	8.7750	456.533	1.9747	4.2543	9.1657	0.1299
7.8	60.8400	2.7928	8.8318	474.552	1.9832	4.2727	9.2052	0.1282
7.9	62.4100	2.8107	8.8882	493.039	1.9916	4.2908	9.2443	0.1266
8.0	64.0000	2.8284	8.9443	512.000	2.0000	4.3089	9.2832	0.1250
8.1	65.6100	2.8460	9.0000	531.441	2.0083	4.3267	9.3217	0.1235
8.2	67.2400	2.8636	9.0554	551.368	2.0165	4.3445	9.3599	0.1220
8.3	68.8900	2.8810	9.1104	571.787	2.0247	4.3621	9.3978	0.1205
8.4	70.5600	2.8983	9.1652	592.704	2.0328	4.3795	9.4354	0.1190
8.5	72.2500	2.9155	9.2195	614.125	2.0408	4.3968	9.4727	0.1176
8.6	73.9600	2.9326	9.2736	636.056	2.0488	4.4140	9.5097	0.1163
8.7	75.6900	2.9496	9.3274	658.503	2.0567	4.4310	9.5464	0.1149
8.8	77.4400	2.9665	9.3808	681.472	2.0646	4.4480	9.5828	0.1136
8.9	79.2100	2.9833	9.4340	704.969	2.0723	4.4647	9.6190	0.1124
9.0	81.000	3.0000	9.4868	729.000	2.0801	4.4814	9.6549	0.1111
9.1	82.8100	3.0166	9.5394	753.571	2.0878	4.4979	9.6905	0.1099
9.2	84.6400	3.0332	9.5917	778.688	2.0954	4.5144	9.7259	0.1087
9.3	86.4900	3.0496	9.6436	804.357	2.1029	4.5307	9.7610	0.1075
9.4	88.3600	3.0659	9.6954	830.584	2.1105	4.5468	9.7959	0.1064
9.5	90.2500	3.0822	9.7468	857.375	2.1179	4.5629	9.8305	0.1053
9.6	92.1600	3.0984	9.7980	884.736	2.1253	4.5789	9.8648	0.1042
9.7	94.0900	3.1145	9.8489	912.673	2.1327	4.5947	9.8990	0.1031
9.8	96.0400	3.1305	9.8995	941.192	2.1400	4.6104	9.9329	0.1020
9.9	98.0100	3.1464	9.9499	970.299	2.1472	4.6261	9.9666	0.1010
10.0	100.000	3.1623	10.000	1000.00	2.1544	4.6416	10.0000	0.1000

Appendix B

Table of e^x and e^{-x}

x	e^x	e^{-x}	x	e^x	e^{-x}
0.00	1.0000	1.0000	0.80	2.2255	0.4493
0.05	1.0513	0.9512	0.85	2.3396	0.4274
0.10	1.1052	0.9048	0.90	2.4596	0.4066
0.15	1.1618	0.8607	0.95	2.5857	0.3867
0.20	1.2214	0.8187	1.0	2.7183	0.3679
0.25	1.2840	0.7788	1.1	3.0042	0.3329
0.30	1.3499	0.7408	1.2	3.3201	0.3012
0.35	1.4191	0.7047	1.3	3.6693	0.2725
0.40	1.4918	0.6703	1.4	4.0552	0.2466
0.45	1.5683	0.6376	1.5	4.4817	0.2231
0.50	1.6487	0.6065	1.6	4.9530	0.2019
0.55	1.7333	0.5769	1.7	5.4739	0.1827
0.60	1.8221	0.5488	1.8	6.0496	0.1653
0.65	1.9155	0.5220	1.9	6.6859	0.1496
0.70	2.0138	0.4966	2.0	7.3891	0.1353
0.75	2.1170	0.4724	2.1	8.1662	0.1225

x	e^x	e^{-x}	x	e^x	e^{-x}
2.2	9.0250	0.1108	3.9	49.402	0.0202
2.3	9.9742	0.1003	4.0	54.598	0.0183
2.4	11.023	0.0907	4.1	60.340	0.0166
2.5	12.182	0.0821	4.2	66.686	0.0150
2.6	13.464	0.0743	4.3	73.700	0.0136
2.7	14.880	0.0672	4.4	81.451	0.0123
2.8	16.445	0.0608	4.5	90.017	0.0111
2.9	18.174	0.0550	4.6	99.484	0.0101
3.0	20.086	0.0498	4.7	109.55	0.0091
3.1	22.198	0.0450	4.8	121.51	0.0082
3.2	24.533	0.0408	4.9	134.29	0.0074
3.3	27.113	0.0369	5	148.41	0.0067
3.4	29.964	0.0334	6	403.43	0.0025
3.5	33.115	0.0302	7	1096.6	0.0009
3.6	36.598	0.0273	8	2981.0	0.0003
3.7	40.447	0.0247	9	8103.1	0.0001
3.8	44.701	0.0224	10	22026	0.00005

Appendix C

Table of Natural Logarithms

In the body of the table the first two digits (and decimal point) of most entries are carried over from a preceding entry in the first column. For example, $\ln 3.32 = 1.19996$. However, an asterisk (*) indicates that the first two digits are those of a following entry in the first column. For example, $\ln 3.33 = 1.20297$.

To extend this table for a number less than 1.0 or greater than 10.09, write the number in the form $x = y \cdot 10^n$ where $1.0 \le y < 10$ and use the fact that $\ln x = \ln y + n \ln 10$. Some values of $n \ln 10$ are

$$
\begin{aligned}
1 \ln 10 &= 2.30259, & 6 \ln 10 &= 13.81551,\\
2 \ln 10 &= 4.60517, & 7 \ln 10 &= 16.11810,\\
3 \ln 10 &= 6.90776, & 8 \ln 10 &= 18.42068,\\
4 \ln 10 &= 9.21034, & 9 \ln 10 &= 20.72327,\\
5 \ln 10 &= 11.51293, & 10 \ln 10 &= 23.02585.
\end{aligned}
$$

For example,

$$
\begin{aligned}
\ln 332 &= \ln[(3.32)(10^2)] = \ln 3.32 + 2 \ln 10\\
&= 1.19996 + 4.60517 = 5.80513
\end{aligned}
$$

and

$$
\begin{aligned}
\ln .0332 &= \ln[(3.32)(10^{-2})] = \ln 3.32 - 2 \ln 10\\
&= 1.19996 - 4.60517 = -3.40521.
\end{aligned}
$$

Properties of logarithms may be used to find the logarithm of a number such as $\frac{3}{8}$.

$$
\begin{aligned}
\ln \tfrac{3}{8} &= \ln 3 - \ln 8 = 1.09861 - 2.07944\\
&= -.98083.
\end{aligned}
$$

N	0	1	2	3	4	5	6	7	8	9
1.0	0.0 0000	0995	1980	2956	3922	4879	5827	6766	7696	8618
1.1	9531	*0436	*1333	*2222	*3103	*3976	*4842	*5700	*6551	*7395
1.2	0.1 8232	9062	9885	*0701	*1511	*2314	*3111	*3902	*4686	*5464
1.3	0.2 6236	7003	7763	8518	9267	*0010	*0748	*1481	*2208	*2930
1.4	0.3 3647	4359	5066	5767	6464	7156	7844	8526	9204	9878
1.5	0.4 0547	1211	1871	2527	3178	3825	4469	5108	5742	6373
1.6	7000	7623	8243	8858	9470	*0078	*0672	*1282	*1879	*2473
1.7	0.5 3063	3649	4232	4812	5389	5962	6531	7098	7661	8222
1.8	8779	9333	9884	*0432	*0977	*1519	*2058	*2594	*3127	*3658
1.9	0.6 4185	4710	5233	5752	6269	6783	7294	7803	8310	8813
2.0	9315	9813	*0310	*0804	*1295	*1784	*2271	*2755	*3237	*3716
2.1	0.7 4194	4669	5142	5612	6081	6547	7011	7473	7932	8390
2.2	8846	9299	9751	*0200	*0648	*1093	*1536	*1978	*2418	*2855
2.3	0.8 3291	3725	4157	4587	5015	5442	5866	6289	6710	7129
2.4	7547	7963	8377	8789	9200	9609	*0016	*0422	*0826	*1228
2.5	0.9 1629	2028	2426	2822	3216	3609	4001	4391	4779	5166
2.6	5551	5935	6317	6698	7078	7456	7833	8208	8582	8954
2.7	9325	9695	*0063	*0430	*0796	*1160	*1523	*1885	*2245	*2604
2.8	1.0 2962	3318	3674	4028	4380	4732	5082	5431	5779	6126
2.9	6471	6815	7158	7500	7841	8181	8519	8856	9192	9527
3.0	9861	*0194	*0526	*0856	*1186	*1514	*1841	*2168	*2493	*2817
3.1	1.1 3140	3462	3783	4103	4422	4740	5057	5373	5688	6002
3.2	6315	6627	6938	7248	7557	7865	8173	8479	8784	9089
3.3	9392	9695	9996	*0297	*0597	*0896	*1194	*1491	*1788	*2083
3.4	1.2 2378	2671	2964	3256	3547	3837	4127	4415	4703	4990
3.5	5276	5562	5846	6130	6413	6695	6976	7257	7536	7815
3.6	8093	8371	8647	8923	9198	9473	9746	*0019	*0291	*0563
3.7	1.3 0833	1103	1372	1641	1909	2176	2442	2708	2972	3237
3.8	3500	3763	4025	4286	4547	4807	5067	5325	5584	5841
3.9	6098	6354	6609	6864	7118	7372	7624	7877	8128	8379
4.0	8629	8879	9128	9377	9624	9872	*0118	*0364	*0610	*0854
4.1	1.4 1099	1342	1585	1828	2070	2311	2552	2792	3031	3270
4.2	3508	3746	3984	4220	4456	4692	4927	5161	5395	5629
4.3	5862	6094	6326	6557	6787	7018	7247	7476	7705	7933
4.4	8160	8387	8614	8840	9065	9290	9515	9739	9962	*0185
4.5	1.5 0408	0630	0851	1072	1293	1513	1732	1951	2170	2388
4.6	2606	2823	3039	3256	3471	3687	3902	4116	4330	4543
4.7	4756	4969	5181	5393	5604	5814	6025	6235	6444	6653
4.8	6862	7070	7277	7485	7691	7898	8104	8309	8515	8719
4.9	8924	9127	9331	9534	9737	9939	*0141	*0342	*0543	*0744
5.0	1.6 0944	1144	1343	1542	1741	1939	2137	2334	2531	2728
5.1	2924	3120	3315	3511	3705	3900	4094	4287	4481	4673
5.2	4866	5058	5250	5441	5632	5823	6013	6203	6393	6582
5.3	6771	6959	7147	7335	7523	7710	7896	8083	8269	8455
5.4	8640	8825	9010	9194	9378	9562	9745	9928	*0111	*0293
N	0	1	2	3	4	5	6	7	8	9

N	0	1	2	3	4	5	6	7	8	9
5.5	1.7 0475	0656	0838	1019	1199	1380	1560	1740	1919	2098
5.6	2277	2455	2633	2811	2988	3166	3342	3519	3695	3871
5.7	4047	4222	4397	4572	4746	4920	5094	5267	5440	5613
5.8	5786	5958	6130	6302	6473	6644	6815	6985	7156	7326
5.9	7495	7665	7834	8002	8171	8339	8507	8675	8842	9009
6.0	1.7 9176	9342	9509	9675	9840	*0006	*0171	*0336	*0500	*0665
6.1	1.8 0829	0993	1156	1319	1482	1645	1808	1970	2132	2294
6.2	2455	2616	2777	2938	3098	3258	3418	3578	3737	3896
6.3	4055	4214	4372	4530	4688	4845	5003	5160	5317	5473
6.4	5630	5786	5942	6097	6253	6408	6563	6718	6872	7026
6.5	7180	7334	7487	7641	7794	7947	8099	8251	8403	8555
6.6	8707	8858	9010	9160	9311	9462	9612	9762	9912	*0061
6.7	1.9 0211	0360	0509	0658	0806	0954	1102	1250	1398	1545
6.8	1692	1839	1986	2132	2279	2425	2571	2716	2862	3007
6.9	3152	3297	3442	3586	3730	3874	4018	4162	4305	4448
7.0	4591	4734	4876	5019	5161	5303	5445	5586	5727	5869
7.1	6009	6150	6291	6431	6571	6711	6851	6991	7130	7269
7.2	7408	7547	7685	7824	7962	8100	8238	8376	8513	8650
7.3	8787	8924	9061	9198	9334	9470	9606	9742	9877	*0013
7.4	2.0 0148	0283	0418	0553	0687	0821	0956	1089	1223	1357
7.5	1490	1624	1757	1890	2022	2155	2287	2419	2551	2683
7.6	2815	2946	3078	3209	3340	3471	3601	3732	3862	3992
7.7	4122	4252	4381	4511	4640	4769	4898	5027	5156	5284
7.8	5412	5540	5668	5796	5924	6051	6179	6306	6433	6560
7.9	6686	6813	6939	7065	7191	7317	7443	7568	7694	7819
8.0	7944	8069	8194	8318	8443	8567	8691	8815	8939	9063
8.1	9186	9310	9433	9556	9679	9802	9924	*0047	*0169	*0291
8.2	2.1 0413	0535	0657	0779	0900	1021	1142	1263	1384	1505
8.3	1626	1746	1866	1986	2106	2226	2346	2465	2585	2704
8.4	2823	2942	3061	3180	3298	3417	3535	3653	3771	3889
8.5	4007	4124	4242	4359	4476	4593	4710	4827	4943	5060
8.6	5176	5292	5409	5524	5640	5756	5871	5987	6102	6217
8.7	6332	6447	6562	6677	6791	6905	7020	7134	7248	7361
8.8	7475	7589	7702	7816	7929	8042	8155	8267	8380	8493
8.9	8605	8717	8830	8942	9054	9165	9277	9389	9500	9611
9.0	9722	9834	9944	*0055	*0166	*0276	*0387	*0497	*0607	*0717
9.1	2.2 0827	0937	1047	1157	1266	1375	1485	1594	1703	1812
9.2	1920	2029	2138	2246	2354	2462	2570	2678	2786	2894
9.3	3001	3109	3216	3324	3431	3538	3645	3751	3858	3965
9.4	4071	4177	4284	4390	4496	4601	4707	4813	4918	5024
9.5	5129	5234	5339	5444	5549	5654	5759	5863	5968	6072
9.6	6176	6280	6384	6488	6592	6696	6799	6903	7006	7109
9.7	7213	7316	7419	7521	7624	7727	7829	7932	8034	8136
9.8	8238	8340	8442	8544	8646	8747	8849	8950	9051	9152
9.9	9253	9354	9455	9556	9657	9757	9858	9958	*0058	*0158
10.0	2.3 0259	0358	0458	0558	0658	0757	0857	0956	1055	1154
N	0	1	2	3	4	5	6	7	8	9

Answers

Exercise 1.1

1. Set. **3.** Positive. **5.** Zero. **7.** Irrational. **9.** 4. **11.** -3. **13.** 6. **15.** -12. **17.** True. **19.** True. **21.** False.

Exercise 1.2

1. Greater. **3.** $4 > 3$. **5.** Absolute value. **7.** Zero, negative. **9.** $>$. **11.** $<$. **13.** $<$. **15.** 270. **17.** 0. **19.** 2. **21.** -5. **23.** $0, -3, 4, 5, -6$.

Exercise 1.3

1. Distributive law. **3.** Associative law. **5.** Commutative law. **7.** Distributive law. **9.** $18x$. **11.** $x + 10$. **13.** $8 + 4x$. **15.** $24xy$. **17.** $24xyz$. **19.** $5ax + 15a$. **21.** $6x - 4$. **23.** $2xy - 12x$. **25.** $10x - 5xy - 15xz$.

Exercise 1.4

1. -5. **3.** -5. **5.** -2. **7.** -10. **9.** -1. **11.** 1. **13.** 12. **15.** -21. **17.** 8. **19.** 2. **21.** -5. **23.** 0. **25.** $5xy$. **27.** 2. **29.** $-7x$. **31.** $-15x$. **33.** $36xyz$. **35.** $-6x + 9xy$.

37. $-7y+14z-7w$. **39.** $-\frac{x}{yz}$. **41.** $\frac{x}{yz}$. **43.** 1. **45.** 34. **47.** -63. **49.** 12. **51.** -1. **53.** -2. **55.** 0. **57.** $15-(-3)$, 18°C. **59.** 3. **61.** $\frac{5}{4}$.

Exercise 1.5

1. 1. **3.** $\frac{18}{77}$. **5.** $\frac{21}{8}$. **7.** $-\frac{3xy}{2}$. **9.** $\frac{3}{2p}$. **11.** $-\frac{4}{3}$. **13.** $\frac{5}{3}$. **15.** $-\frac{9}{2}$. **17.** $\frac{y}{14}$. **19.** $-6x$. **21.** 1. **23.** $\frac{21}{10}$. **25.** $-\frac{3}{4}$. **27.** 32. **29.** $\frac{72}{x}$. **31.** $\frac{2x}{5y}$. **33.** $\frac{18xz}{24yz}$. **35.** $\frac{3}{2x}$. **37.** 2 diopters. **39.** $56\frac{2}{3}$°C.

REVIEW PROBLEMS—CHAPTER 1

1. Commutative law. **3.** Associative law. **5.** Commutative law. **7.** Distributive law. **9.** True. **11.** True. **13.** False. **15.** False. **17.** $-\frac{1}{2}$. **19.** $3xy+15x$. **21.** $-10y$. **23.** -1. **25.** -12. **27.** 1. **29.** $-20x+12$. **31.** $x-1$. **33.** -8. **35.** $3xz$. **37.** 15. **39.** 5. **41.** $56xyz$. **43.** $\frac{2}{63}$. **45.** 12. **47.** 1. **49.** $-12x$. **51.** $8y-xy$. **53.** $-8zx+6zy+16z$. **55.** $-\frac{3x}{7}$. **57.** $-3xy$. **59.** $\frac{5}{9x}$. **61.** $-\frac{1}{27}$. **63.** (a) -6, (b) $-\frac{13}{6}$. **65.** (a) 11, (b) 5, (c) 11, (d) -5.

Exercise 2.1

1. 64. **3.** -8. **5.** -81. **7.** -72. **9.** $\frac{1}{3}$. **11.** -64. **13.** x^{13}. **15.** y^9. **17.** x^7. **19.** $(x-2)^8$. **21.** $\frac{x^7}{y^9}$. **23.** $14x^8$. **25.** $-12x^4$. **27.** x^{16}. **29.** x^9. **31.** t^{2n}. **33.** x^{29}. **35.** x^4. **37.** $\frac{1}{x}$. **39.** $-y^6$. **41.** $\frac{1}{x^6}$. **43.** x^{13}. **45.** x. **47.** $\frac{1}{x^5}$. **49.** a^6b^6. **51.** $16x^4$. **53.** $45y^6$. **55.** $\frac{a^3}{b^3}$. **57.** x^4y^8. **59.** $\frac{8y^3}{z^3}$. **61.** $\frac{4}{9}a^4b^6c^{12}$. **63.** $\frac{x^6}{y^{15}}$. **65.** $\frac{x^8y^{12}}{16z^{16}}$. **67.** $-x^{13}$. **69.** $16x^8y^4$. **71.** $-\frac{x^5y^5}{t^4}$. **73.** $27x^{10}$. **75.** 64. **77.** -26. **79.** Original volume is multiplied by a factor of 8. **81.** (a) 4, (b) 8.

Exercise 2.2

1. 1. **3.** $\frac{1}{8}$. **5.** 27. **7.** 3. **9.** -4. **11.** -27. **13.** $\frac{5}{4}$. **15.** 1. **17.** $\frac{1}{x^6}$. **19.** x^3. **21.** $\frac{3}{y^4}$. **23.** $16x$. **25.** $\frac{1}{x^5y^7}$. **27.** $\frac{2a^2c^5}{b^4}$. **29.** $\frac{x^9z^4}{w^2y^{12}}$. **31.** x. **33.** $\frac{1}{x^5}$. **35.** $\frac{6}{x^3}$. **37.** $\frac{y^{20}}{x^4}$. **39.** $\frac{2y^4}{x^2}$. **41.** $\frac{1}{9t^2}$. **43.** t^4. **45.** $\frac{y^5}{x^8}$. **47.** x^4. **49.** $\frac{y^4}{x^4}$. **51.** $\frac{5y^2}{8x^2}$. **53.** $-xz$. **55.** $\frac{4x^2y^{12}}{z^4}$. **57.** 2.6×10^2. **59.** 6.3×10^{-6}. **61.** 4.0×10^6. **63.** 2×10^{-1}.

Exercise 2.3

1. 6. **3.** 2. **5.** 7. **7.** −3. **9.** −4. **11.** 2. **13.** 0. **15.** −5. **17.** 2. **19.** .2. **21.** $\frac{1}{4}$.
23. −5. **25.** 11. **27.** $\frac{1}{3}$. **29.** 5. **31.** 8. **33.** $5\sqrt{2}$. **35.** $2\sqrt{3}$. **37.** $2\sqrt{2}$. **39.** 120.
41. $2\sqrt[4]{3}$. **43.** $2\sqrt[5]{2}$. **45.** $-5\sqrt[3]{4}$. **47.** $\frac{\sqrt{14}}{3}$. **49.** $\frac{\sqrt{15}}{15}$. **51.** $\frac{\sqrt[3]{10}}{3}$. **53.** 5. **55.** $-\frac{1}{2}$.
57. 4. **59.** 4. **61.** 13. **63.** 25. **65.** 3. **67.** $\frac{3}{2}$. **69.** 0.

REVIEW PROBLEMS—CHAPTER 2

In Problems **1–23**, T = true and F = false.

1. F. **3.** F. **5.** F. **7.** F. **9.** T. **11.** F. **13.** F. **15.** T. **17.** F. **19.** T. **21.** T. **23.** F.
25. 1. **27.** $\frac{1}{5}$. **29.** 9. **31.** x^{13}. **33.** $\frac{x^{10}}{y^{50}}$. **35.** $-32x^5y^{20}$. **37.** x^{14}. **39.** $-x^5$. **41.** $\frac{1}{25}$. **43.** 3.
45. 0. **47.** 13. **49.** $\frac{1}{9}$. **51.** .15. **53.** $3\sqrt{5}$. **55.** $3\sqrt[4]{10}$. **57.** $\frac{\sqrt{35}}{5}$. **59.** 2. **61.** 6. **63.** $\frac{2}{x^4}$.
65. $\frac{1}{16t^4}$. **67.** $\frac{y^6}{x^6}$. **69.** $\frac{z^2}{x^3y^5}$. **71.** $\frac{1}{x^2y^2z^2}$. **73.** (a) 9, (b) $\frac{1}{4}$.

Exercise 3.1

1. $15x^2$. **3.** $-12y$. **5.** $7x - 5$. **7.** $40x + 12y$. **9.** $-14x + 12y$. **11.** $8y - 8$. **13.** $4x - 3y$.
15. $7a - 8b + c$. **17.** $5x^2 + 15$. **19.** $2xy + 5z - 7$. **21.** $2x^2 - 9x - 13$. **23.** $-10 + 4x$.
25. $5x + 50$. **27.** $-a + 6$. **29.** $29x^2 - 22$. **31.** $8a - 11b - 13c$. **33.** $18x - 24$.
35. $9 - 100x + 60y$. **37.** $-2x - 4$. **39.** $M = 12{,}600 - 230x$. **41.** -2000θ.

Exercise 3.2

1. $6x^2y$. **3.** $3a^3b^2$. **5.** $-8x^4y^4$. **7.** $3xy$. **9.** $\frac{6y^4}{x^3}$. **11.** $a^3b^3c^2$. **13.** x^3yz^4. **15.** $24x^4y^5$.
17. $90x^{10}$. **19.** $\frac{a^8b^{11}}{c^6}$. **21.** $ab^2c^4d^2$. **23.** $x^3 - 4x^2 + 7x$. **25.** $-3a^2b + a^3b^2 - a^3b$. **27.** $x + 4x^2$.
29. $-5x^3y + 5xy^3 - 5x^2y^2$. **31.** $4x^5y^2 + 8x^4y^4 - 12x^6y^2$. **33.** $x^2 + 7x + 10$. **35.** $3y^2 - 4y - 4$.
37. $9x^2 - 6x + 1$. **39.** $16x^2 + 8xy + y^2$. **41.** $6x^4y + 6x^3y^2$. **43.** $t^3 - 8$.
45. $x^4 - 5x^3 - x^2 + 10x - 2$. **47.** $x^2 + 2xy + y^2 - 1$. **49.** $16x^3 - 4x$. **51.** $2xy + y^2$. **53.** $6x$.
55. $x^2 + x - 6$. **57.** $-15x^2y^2$. **59.** $-2x^2 + 3x - 1$. **61.** 0. **63.** $Pv + Pb + av + ab$.

Exercise 3.3

1. $\frac{b}{2}$. **3.** $-\frac{2b^2}{a}$. **5.** $-c$. **7.** $\frac{x}{2yz^2}$. **9.** $-\frac{ac}{b^2}$. **11.** $9x^2y$. **13.** $\frac{x^2}{9y^2}$. **15.** $-\frac{2y}{7}$.
17. $2x^2 - 3x + 4$. **19.** $\frac{5}{x} - x + 2$. **21.** $2x^2 - 3 + \frac{2}{5x}$. **23.** $-2 - \frac{y}{x}$. **25.** $1 + \frac{x}{y}$. **27.** $\frac{3}{y} - x + \frac{1}{x^2y^2}$.

29. $-2xy + \frac{2y^2}{3x} - \frac{7}{3} + \frac{4}{3x}$. **31.** $-10xy + \frac{5y}{2} - \frac{1}{y}$. **33.** $\frac{2}{xy} - 4y^2 - 3xy$. **35.** $-2x^6 - \frac{6x}{y^2} + \frac{1}{y^4}$.
37. $x - 2 + \frac{1}{x}$. **39.** $\frac{xz^2w}{y^3}$. **41.** $\frac{8x^4y^{15}z^2}{3}$. **43.** $\frac{4x^8y^{10}}{z^4}$.

Exercise 3.4

1. (a) 1, (b) −4. **3.** (a) 2, (b) 7. **5.** No. **7.** (a) 4, (b) $-\frac{2}{3}$. **9.** (a) 5, (b) −2.
11. $2x + 7 + \frac{10}{x-2}$. **13.** $1 + \frac{1}{x+2}$. **15.** $x - 2 - \frac{3}{4x+1}$. **17.** $2x - 3 + \frac{4}{2x+3}$. **19.** $x^2 + 3x - 2$.
21. $-x^3 + x^2 - 2x + 4 - \frac{19}{3x+5}$. **23.** $2x^2 + 1 - \frac{2}{4x-1}$. **25.** $x^3 + x^2 - x - 1$.
27. $-x^3 + 3x^2 - 9x + 27$. **29.** $x + 2 + \frac{2x+1}{x^2 - x + 1}$.
31. Because (quotient)(divisor) + remainder = dividend, if the remainder is 0, then the dividend is the product of the quotient and divisor.

REVIEW PROBLEMS—CHAPTER 3

1. $11x - 2y - 3$. **3.** $-2a + 19b + 4$. **5.** $-8xy + 26$. **7.** $9x - 18$. **9.** $3x^2 + 18x - 36$.
11. $2x^3y^4z^7$. **13.** $72a^5b^4$. **15.** $4x^8y^7$. **17.** $\frac{2x^5}{y^3}$. **19.** $2xy$. **21.** $x^3 - 2x^2 + 4x$.
23. $-2a^4b^2 + 2a^3b^2 - 3a^2b$. **25.** $x^2 - x - 12$. **27.** $x^2 - 4$. **29.** $x^2 - 5x + 6$. **31.** $x^2 + 4xy + 4y^2$.
33. $x^3 - 4x^2 + 3x - 12$. **35.** $6x^4 - 11x^3 + 3x^2 + 15x - 5$. **37.** $2x^2 - xy + 5x - 3y^2 - 5y + 2$.
39. $x^2 - 12x + 18$. **41.** $\frac{ay^2}{x}$. **43.** $\frac{y^{10}}{z^2}$. **45.** $-\frac{3}{8x}$. **47.** $x - 5 + \frac{7}{x}$. **49.** $\frac{x}{y} - 5y + \frac{7}{y}$.
51. $-x - 2xy^3w^2 + 2x^2y^2w$. **53.** $3x^2 + 3x - 1 - \frac{2}{2x-1}$. **55.** $x^3 - 2x^2 + 6x - 10$.
57. $-x^3 + 2x^2 - x + 2 - \frac{2}{2x+1}$.

Exercise 4.1

1. $x = -3$. **3.** $y = -30$. **5.** $x = -2$. **7.** $x = -\frac{15}{16}$. **9.** $S = 3$. **11.** $x = -\frac{1}{4}$. **13.** $y = 3$.
15. $x = 0$. **17.** $x = -3$. **19.** $x = 2$. **21.** $y = 11$. **23.** $z = 2$. **25.** $x = -2$. **27.** $x = -\frac{5}{2}$.
29. $r = -\frac{32}{11}$. **31.** $t = 4$. **33.** $x = -30$. **35.** $x = -\frac{24}{7}$. **37.** $x = -31$. **39.** $x = -\frac{23}{20}$. **41.** $z = \frac{1}{13}$.
43. $x = \frac{8}{9}$. **45.** 273 cm³. **47.** 120 m. **49.** $r = \frac{C}{2\pi}$. **51.** $r = \frac{I}{Pt}$. **53.** $t = \frac{A-P}{Pr}$. **55.** Yes.
57. 5 ft. **59.** 40 ft/s. **61.** 3. **63.** $\frac{V + Ex_a}{E}$. **65.** 92.3.

Exercise 4.2

1. $x + y = 8$. **3.** $13 = x + 4$. **5.** $z = (x - y) - 2$. **7.** $s = vt$. **9.** $3x = y$. **11.** $10x = 3x + 6$.
13. $M = \frac{1}{2}J$. **15.** $l = 100w$. **17.** $2(x - 12)$. **19.** $l = 6w - 4$. **21.** $E = mc^2$. **23.** $3 - 15x = 3 + 4x$.

25. $F = \frac{9}{5}C + 32$. **27.** $.085x$. **29.** $100 - x$. **31.** $.25x$ ml. **33.** $60q$. **35.** $100 - x + y$.
37. (a) $3x + 3.50y + 4z$, (b) $\dfrac{3x + 3.50y + 4z}{x + y + z}$. **39.** $L = 20 + 5.3W$. **41.** (a) $.85p$, (b) $.765p$,
(c) Yes, since $.765p$ is greater than $.75p$.

Exercise 4.3

1. 10. **3.** 371. **5.** 15,000. **7.** 6. **9.** \$200,000. **11.** 300. **13.** 625. **15.** 4%. **17.** 110%.
19. 80. **21.** 5. **23.** 42.5%. **25.** \$8000. **27.** 48 of A, 80 of B. **29.** $5\frac{1}{3}$. **31.** 96.
33. 140,000. **35.** \$72.50. **37.** 3 mi. **39.** 1600.

Exercise 4.4

1. $11\frac{1}{4}$ km. **3.** $\frac{2}{3}$ h or 40 min. **5.** 4 days; 56 in. **7.** 1027 mi. **9.** 20 lb.
11. 420 gal of 20% solution, 280 gal of 30% solution. **13.** $1\frac{2}{3}$ qt. **15.** $66\frac{2}{3}$ gal/min.

Exercise 4.5

1. $x > 2$.

3. $x \leq 3$. **5.** $x \leq -\frac{1}{2}$. **7.** $s < -\frac{2}{5}$.

9. $x < 3$. **11.** $x \geq -\frac{7}{5}$. **13.** $y > 0$. **15.** $x > -\frac{2}{7}$.

17. No solution.

19. $x < 6$.

21. $y \leq -5$.

23. All real numbers.

25. $t > \frac{17}{9}$. **27.** $x \geq -\frac{14}{3}$. **29.** $r > 0$. **31.** $y < 0$.

33. 12,400 mi. **35.** $t > 36.5$. **37.** \$4.50, \$1160.

REVIEW PROBLEMS—CHAPTER 4

1. $x = \frac{1}{2}$. **3.** $y = \frac{3}{2}$. **5.** $x = -\frac{3}{2}$. **7.** $z = -\frac{20}{9}$. **9.** $x = \frac{25}{4}$. **11.** $u = -\frac{11}{14}$. **13.** $x = -32$.
15. $x = \frac{37}{21}$. **17.** $y = \frac{5}{2}$. **19.** $x < \frac{1}{3}$. **21.** $x \leq -2$. **23.** $t > 1$. **25.** $x \geq 40$. **27.** $a = \dfrac{v - v_0}{t}$.
29. .6. **31.** 120. **33.** 12%. **35.** 8250. **37.** 4500.
39. 112 tons of A, 42 tons of B, 14 tons of C.
41. 17,500. **43.** $2\frac{2}{3}$ oz of shellac, $13\frac{1}{3}$ oz of alcohol. **45.** $6\frac{3}{7}$ mi.
47. 8000 kg of industrial strength and 4000 kg of household strength. **49.** 1000.

Exercise 5.1

1. $x^2+8x+16$. **3.** $y^2+20y+100$. **5.** $16x^2+8x+1$. **7.** $x^2+x+\frac{1}{4}$. **9.** $x^2-12x+36$.
11. $9x^2-12x+4$. **13.** $4x^2-4xy+y^2$. **15.** $9x^2+18x+9$. **17.** $4-16y+16y^2$. **19.** x^2-9.
21. x^2-81. **23.** $1-x^2$. **25.** $9x^4-25$. **27.** $4y^2-9x^2$. **29.** $36-x^2$. **31.** $x^2+11x+24$.
33. x^2+5x+4. **35.** x^2-x-2. **37.** $t^2+2t-35$. **39.** x^2-5x+6. **41.** $x^2-9x+20$.
43. y^2+y-6. **45.** $2x^2+2x-12$. **47.** $8x^2-8x+2$. **49.** $6x^2-3x-3$. **51.** $10x^2-13x+4$.
53. $2+9t-35t^2$. **55.** $x^2y^4+2axy^2+a^2$. **57.** x^4-y^4. **59.** x^2-7. **61.** $4x^2-4x+1$.
63. $4x^2+24x+36$. **65.** $2xy^2-18x$. **67.** $a^4b^2-4a^2bm^2n+4m^4n^2$. **69.** x^4-16. **71.** 0.
73. $5x^2-9x+11$. **75.** x^3-2x^2-4x+8. **77.** $16x^4-96x^3+144x^2$. **79.** $6x+1$.
81. $x^3+6x^2+12x+8$. **83.** $8x^3-36x^2+54x-27$. **85.** $2mt^2+2mt+\frac{1}{2}m$. **87.** r^2-R^2.

Exercise 5.2

1. $8(x+1)$. **3.** $5(2x-y+5)$. **5.** $x(5c+9)$. **7.** $3x(2y+z)$. **9.** $x^2(2x-1)$.
11. $x^3y^3(2+x^2y^2)$. **13.** $4mx^3(m-2x)$. **15.** $3a^2y^3(3a^2+y^2-2ayz)$. **17.** $(x+1)(x-1)$.
19. $(x+1)(x+3)$. **21.** $(x-4)(x-5)$. **23.** $(y-4)(y+6)$. **25.** $(y+6)(y-6)$. **27.** $(x+6)^2$.
29. $(x-8)(x+4)$. **31.** $(y-5)^2$. **33.** $(3x+1)(x+2)$. **35.** $(2y-1)(y-3)$. **37.** $(4x+1)^2$.
39. $(3+2xy)(3-2xy)$. **41.** $(4y-1)(y+2)$. **43.** $(2x-5)(3x+2)$. **45.** $2(x+3)(x-1)$.
47. $3x(x+3)^2$. **49.** $4s^2t(2t+1)(2t-1)$. **51.** $2(2y+3)(y-3)$. **53.** $2(x+3)^2(x+1)(x-1)$.
55. $2(x+4)(x+1)$. **57.** $(x^2+4)(x+2)(x-2)$. **59.** $(y^4+1)(y^2+1)(y+1)(y-1)$.
61. $(x^2+2)(x+1)(x-1)$. **63.** $x(x+1)^2(x-1)^2$.

Exercise 5.3

1. $(1+y)(x+4)$. **3.** $(y-2)(x-4)$. **5.** $(2x-3)(x+2)(x-2)$. **7.** $(x+y)(x-y)(a+b)$.
9. $(x+2+a)(x+2-a)$. **11.** $(x+3)(x^2-3x+9)$. **13.** $(2y-3)(4y^2+6y+9)$.
15. $3(2x+y^2)(4x^2-2xy^2+y^4)$. **17.** $(a-b)(a^2+b^2+3a+3b+ab+3)$. **19.** $P=\dfrac{S}{1+rt}$.
21. $x=\dfrac{d-b}{a-c}$. **23.** $m=\dfrac{2c}{2gh+v^2}$. **25.** $s=\dfrac{7+at-bt}{a+b}$. **27.** $2\pi r(r+h)$. **29.** $y=\dfrac{ax}{1+abx}$.

Exercise 5.4

1. $x^3+12x^2+48x+64$. **3.** $y^4-8y^3+24y^2-32y+16$. **5.** $243x^5+405x^4+270x^3+90x^2+15x+1$.
7. $16z^4-32z^3y+24z^2y^2-8zy^3+y^4$. **9.** $x^{100}+100x^{99}+4950x^{98}$.
11. $128x^7-1344x^6y+6048x^5y^2$. **13.** $y^{20}-50y^{18}+1125y^{16}$. **15.** $243z^{10}+405z^8+270z^6$.

REVIEW PROBLEMS—CHAPTER 5

1. $x^2+12x+36$. **3.** $x^2-10x+25$. **5.** $4x^2+16xy+16y^2$. **7.** x^2-64. **9.** $9x^2-4$.
11. $4x^2-16y^2$. **13.** $x^2-2x-24$. **15.** $x^2-13x+42$. **17.** $4x^2-14x+12$. **19.** y^4-16.
21. $2x^4+2x^3-24x^2$. **23.** $8y-18$. **25.** $16x^3+32x^2-9x-18$. **27.** $2xy^4(3x^2+2y^2)$.
29. $(x-5)(x-6)$. **31.** $(4+y)(4-y)$. **33.** $x(x-8)(x+7)$. **35.** $(3x-2)(x+4)$.
37. $2(2x+5)(2x-5)$. **39.** $2(z-4)(z^2+4z+16)$. **41.** $(x^2+2)(x+2)(x-2)$.
43. $2x^3(x-6)(x-3)$. **45.** $(x+4)(2x^2-3)$. **47.** $x=\dfrac{3b-2a}{a-b}$.
49. $x^4-16x^3+96x^2-256x+256$. **51.** $32x^5+80x^4+80x^3+40x^2+10x+1$.

Exercise 6.1

1. $x=-2,-1$. **3.** $x=3,-\frac{1}{2}$. **5.** $t=3,4$. **7.** $z=-3,1$. **9.** $x=6$. **11.** $x=0,8$.
13. $x=0,-5$. **15.** $t=0,\frac{7}{3}$. **17.** $x=\pm 2$. **19.** $x=\pm 5$. **21.** $x=\pm\sqrt{6}$. **23.** $x=\pm 2\sqrt{3}$.
25. $z=\pm 3$. **27.** $x=\frac{1}{3},-\frac{3}{2}$. **29.** $t=\pm\dfrac{\sqrt{15}}{2}$. **31.** $x=\pm\sqrt{7}$. **33.** $x=\frac{1}{2},-4$. **35.** $x=5,-2$.
37. $x=-\frac{1}{2}$. **39.** $x=-1$. **41.** $t=-5,1$. **43.** $x=2,3$. **45.** $x=0,-1$. **47.** $x=4,-2$.
49. $y=-\frac{1}{6},-\frac{1}{4}$. **51.** $x=7,-1$. **53.** $x=-4\pm 2\sqrt{2}$. **55.** $y=\frac{1}{2},-\frac{3}{2}$. **57.** $x=0,1,-2$.
59. $x=2,-1$. **61.** $x=0,2,-3,4$. **63.** $x=0,\pm 1$. **65.** $x=0,\pm 8$. **67.** $y=0,-4,-2$.
69. $x=\pm 1,\pm 3$. **71.** $x=0,\pm 1$. **73.** $x=\pm 3,\pm 2$.

Exercise 6.2

1. $x=-3\pm\sqrt{10}$. **3.** $x=2\pm\sqrt{3}$. **5.** $y=\dfrac{3}{2}\pm\dfrac{\sqrt{13}}{2}$. **7.** $x=-\dfrac{1}{2}\pm\dfrac{\sqrt{17}}{2}$. **9.** $x=\dfrac{7}{2}\pm\dfrac{\sqrt{47}}{2}$.
11. $x=-\dfrac{1}{4}\pm\dfrac{\sqrt{17}}{4}$. **13.** $x=-1\pm\sqrt{5}$.

Exercise 6.3

1. $x=-\dfrac{3}{2}\pm\dfrac{\sqrt{5}}{2}$. **3.** $x=3$. **5.** $y=-\dfrac{3}{4}\pm\dfrac{\sqrt{41}}{4}$. **7.** $x=-\frac{5}{2}$. **9.** $x=-1\pm\dfrac{\sqrt{10}}{5}$.
11. $x=-\dfrac{1}{3}\pm\dfrac{\sqrt{7}}{3}$. **13.** $x=\pm 6$. **15.** $x=1,-7$. **17.** $z=0,2$. **19.** $x=-2\pm\sqrt{2}$. **21.** $y=-8$.
23. $x=\dfrac{2}{3}\pm\dfrac{\sqrt{10}}{3}$. **25.** $x=\frac{1}{5},4$. **27.** $x=-2\pm\sqrt{5}$. **29.** $x=0,-\frac{4}{5}$. **31.** $y=2\pm 2\sqrt{2}$.
33. $y=\frac{1}{3}$. **35.** $x=\dfrac{7}{2}\pm\dfrac{\sqrt{13}}{2}$. **37.** $x=\frac{3}{2},-4$. **39.** $y=1,-5$. **41.** $z=-1\pm\sqrt{5}$. **43.** $x=\frac{1}{2}$.
45. $x=2\pm\dfrac{\sqrt{10}}{2}$. **47.** $x=-\frac{1}{2},6$.

Exercise 6.4

1. $9i$. **3.** $\frac{1}{4}i$ or $\frac{i}{4}$. **5.** $2i\sqrt{3}$. **7.** $4i\sqrt{2}$. **9.** $i\sqrt{2}$. **11.** $-5i$. **13.** $x = 2 \pm i$.

15. $x = -1 \pm i\sqrt{2}$. **17.** $x = 1 \pm i\sqrt{3}$. **19.** $x = \pm 2i$. **21.** $r = -\frac{2}{3} \pm \frac{i\sqrt{2}}{6}$. **23.** $r = \frac{5}{6} \pm \frac{i\sqrt{23}}{6}$.

25. $x = \pm i$. **27.** $x = \frac{1}{2} \pm \frac{i\sqrt{15}}{6}$. **29.** 49, two different real solutions, $x = 2, -\frac{1}{3}$.

31. 0, one real solution, $t = -\frac{2}{3}$. **33.** -20, two different imaginary solutions, $x = \frac{2}{3} \pm \frac{i\sqrt{5}}{3}$.

35. 33, two different real solutions, $x = -\frac{1}{2} \pm \frac{\sqrt{33}}{6}$.

Exercise 6.5

1. $3\frac{1}{2}$ s. **3.** 1.3 or 2.4 s. **5.** $q = 10$. **7.** 50%. **9.** -4 or 3. **11.** 1 m. **13.** 40.

15. \$310, or \$290. **17.** 80 ft by 140 ft. **19.** 9 cm long, 4 cm wide. **21.** 25 ft. **23.** 5%.

25. 700.

REVIEW PROBLEMS—CHAPTER 6

1. $x = 5$. **3.** $x = 6, -4$. **5.** $x = \frac{1}{6}, \frac{3}{2}$. **7.** $x = \pm 2\sqrt{3}$. **9.** $x = 0, \frac{1}{2}$. **11.** $x = 5 \pm 2\sqrt{6}$.

13. $x = -\frac{3}{2} \pm \frac{\sqrt{11}}{2}$. **15.** $x = 3 \pm \sqrt{2}$. **17.** $x = -\frac{1}{2}$. **19.** $x = \frac{1}{2} \pm \frac{3i}{2}$. **21.** $x = \pm \frac{3}{4}$.

23. $y = 4, -6$. **25.** $z = 2, -10$. **27.** $t = \frac{3}{4} \pm \frac{i\sqrt{15}}{4}$. **29.** $x = -\frac{3}{2} \pm \frac{\sqrt{17}}{2}$. **31.** $x = -\frac{5}{4}$.

33. 4 in. **35.** 50. **37.** Either \$440 or \$460.

Exercise 7.1

1. $\frac{x}{y}$. **3.** $\frac{a}{bc}$. **5.** $-\frac{x}{y+z}$. **7.** $\frac{x+2}{3}$. **9.** $\frac{2(2x+1)}{x-3}$. **11.** $\frac{y^2+1}{y^3(y-4)}$. **13.** $\frac{1}{x+6}$. **15.** $\frac{x+2}{x+1}$.

17. $\frac{z+1}{2z+1}$. **19.** $-\frac{5+x}{x+3} = -\frac{x+5}{x+3}$. **21.** $\frac{6x^2}{(x-1)(x+2)}$. **23.** $-\frac{1}{x}$. **25.** $\frac{x}{2(x+3)}$. **27.** $\frac{(x-3)^2}{4}$.

29. $x^3(x+3)$. **31.** $\frac{x+2}{(x-3)(x+1)}$. **33.** $\frac{x^2+4}{x(x-2)}$. **35.** $\frac{x+1}{x+2}$. **37.** $\frac{x}{2}$. **39.** $\frac{1}{2n}$. **41.** $\frac{2}{3}$. **43.** 1.

45. $-x$. **47.** $\frac{x-2}{x-1}$. **49.** $\frac{2x+3}{2x}$. **51.** $-\frac{(x+2)(2+3x)}{9}$. **53.** $\frac{8(x+2)}{15(x-2)}$. **55.** $x = 10$.

57. $y = -4$. **59.** $x = -2$. **61.** $x = \frac{3}{14}$. **63.** $y = \frac{10}{3}$. **65.** $x = \frac{69}{5}$. **67.** $\frac{Wa}{2g}$. **69.** 7.

Exercise 7.2

1. $(x-4)^5$. **3.** x^2y^3. **5.** $(x+3)^2(x-3)$. **7.** $2x(x+1)$. **9.** $\frac{x+5}{x-3}$. **11.** 2. **13.** $\frac{2x-1}{x-1}$.
15. $\frac{2y+3x}{xy}$. **17.** $\frac{x-8}{18}$. **19.** $\frac{3y-4}{2xy}$. **21.** $\frac{8x-21}{(x-2)(x-3)}$. **23.** $\frac{5y^2-2x+3x^2}{x^2y}$. **25.** $\frac{5x-1}{x-1}$.
27. $\frac{x^2+2xy-y^2}{(x-y)(x+y)}$. **29.** $\frac{x^2+18x+9}{2(x-3)(x+3)}$. **31.** $\frac{x+3}{x+1}$. **33.** $\frac{-1}{(x+1)(x-1)}=-\frac{1}{(x+1)(x-1)}$.
35. $\frac{x^2-2x+9}{(x+3)^2(x-3)}$. **37.** $\frac{-x^2-2x+1}{(x+5)(x+2)(x+1)}$. **39.** $\frac{x^2+4x-4}{(x-2)(x+2)}$.
41. $\frac{2x^2+x-1}{x-1}=\frac{(2x-1)(x+1)}{x-1}$. **43.** $\frac{-x^3+3x^2-x+3}{x^2(x+1)(x+2)}$.
45. $\frac{2y^2-y-6}{(2y+1)^2(y+3)}=\frac{(2y+3)(y-2)}{(2y+1)^2(y+3)}$. **47.** $\frac{x^2+y^2+4xy}{(x+y)^2(x-y)}$. **49.** $\frac{x^2y^2+2xy+1}{y^2}$. **51.** $\frac{xy}{xy+1}$.
53. $\frac{k_2+k_1}{k_1k_2}$. **55.** $\frac{4k^2+k-2}{4k(k-1)}$.

Exercise 7.3

1. $\frac{1}{x}$. **3.** $\frac{2x-3}{x}$. **5.** $\frac{14}{6x-1}$. **7.** $\frac{x(3y-1)}{y(2x+1)}$. **9.** $\frac{x+3}{2x-3}$. **11.** $\frac{3(x+2)}{x+3}$. **13.** $\frac{3x-8}{3x-10}$.
15. $\frac{2x+1}{x^2}$. **17.** $\frac{y^2-2xy+x^2}{x^2y^2}$. **19.** $\frac{2av}{2at_rv-v^2+2al}$.

Exercise 7.4

1. $x=\frac{1}{4}$. **3.** $x=\frac{3}{2}$. **5.** $r=\frac{2}{3}$. **7.** $x=3, -6$. **9.** No solution. **11.** $y=-\frac{1}{2}$. **13.** $x=-\frac{1}{4}, -\frac{1}{2}$.
15. $x=0$. **17.** $x=6, -2$. **19.** $x=5, -2$. **21.** $x=-2$. **23.** $x=\frac{1}{8}$. **25.** $y=3$.
27. $x=-1, \frac{13}{10}$. **29.** $x=1, -\frac{18}{5}$. **31.** No solution. **33.** $x=2, -\frac{4}{5}$. **35.** 24 cm. **37.** 10.
39. 12 Ω. **41.** $\frac{12}{13}$ h. **43.** 15 mi/h. **45.** 60. **47.** 1 year and 10 years.

REVIEW PROBLEMS—CHAPTER 7

1. $\frac{6(x-2)}{x(x-6)}$. **3.** $\frac{x-8}{2x}$. **5.** $\frac{3}{2}$. **7.** $\frac{-x^2+3x-5}{(x-2)(x-3)}$. **9.** $-3x(x+1)$. **11.** $\frac{3x^2-1}{(x-1)(x+1)}$. **13.** 1.
15. -2. **17.** $\frac{x-2}{x+2}$. **19.** $\frac{2x}{(x-1)^2}$. **21.** $-\frac{3+x}{(x+4)(x+2)}$. **23.** $\frac{x(x+1)}{x+2}$. **25.** $-\frac{x+4}{x+3}$.
27. $\frac{x}{1+x^2}$. **29.** $\frac{1-2x}{x}$. **31.** $x=-15$. **33.** $x=3$. **35.** $x=4$. **37.** No solution.
39. $a=\frac{bE-P}{EP}$. **41.** 12 min.

Exercise 8.1

1. 5. **3.** $\frac{1}{9}$. **5.** 9. **7.** $\frac{1}{9}$. **9.** 16. **11.** 8. **13.** -2. **15.** 16. **17.** $x^{1/2}$. **19.** $x^{2/3}$. **21.** $x^{3/4}y^{5/4}$. **23.** $x^{5/6}y^2$. **25.** $x^{9/4}$. **27.** $\frac{3}{x^{1/2}}$. **29.** $(x^2-5x)^{2/3}$. **31.** x^2. **33.** x. **35.** x^4. **37.** $x^{3/4}$. **39.** $3xy$. **41.** $x^{3/2}$. **43.** y^2. **45.** $\frac{2}{x^4}$. **47.** $\frac{8y}{x^6}$. **49.** $a^2b^{21/4}c^{9/4}$. **51.** $x+3x^{1/3}$. **53.** $\frac{1}{3x^5}$. **55.** $-\frac{2}{x^2}$. **57.** $\frac{x^{1/3}}{y}$. **59.** $\frac{x^2}{y^2}$. **61.** $\frac{x^4}{y^2}$. **63.** $\frac{1}{x^{8/3}}$. **65.** $4x^2$. **67.** $-3x^{1/2}$. **69.** $\frac{2}{x^{1/3}}$. **71.** $x^{1/3}+x-x^3$. **73.** $x^{1/4}-3x^{1/12}$. **75.** $4y^3$. **77.** .20.

Exercise 8.2

1. x^2. **3.** $2x^4$. **5.** $3x^8y^9$. **7.** xy^2z^3. **9.** $\frac{x^3}{y^4}$. **11.** x^2. **13.** x. **15.** $2\sqrt{3}$. **17.** $4\sqrt{2}$. **19.** $2\sqrt[3]{2}$. **21.** $x^3\sqrt{x}$. **23.** $2x^2\sqrt[3]{3}$. **25.** $x^2\sqrt[4]{xy^2}$. **27.** $x^2z\sqrt[3]{yz}$. **29.** $2ay\sqrt[3]{y^2}$. **31.** $x^4y^2z\sqrt[5]{x^3z}$. **33.** $9yzw^2\sqrt{xz}$. **35.** $\frac{\sqrt{2}}{2}$. **37.** $\frac{\sqrt[3]{50}}{5}$. **39.** $\frac{\sqrt[3]{x^2}}{y}$. **41.** $\frac{\sqrt{x}}{y^2}$. **43.** $\frac{\sqrt{2xy}}{y}$. **45.** $\frac{\sqrt[3]{2x^2y}}{xy}$. **47.** $\frac{\sqrt[4]{24xy^3z^2}}{2x^2yz}$. **49.** $\sqrt[3]{x}$. **51.** $\sqrt{3}$. **53.** $2x\sqrt{y}$. **55.** $\frac{\sqrt{xy}}{y}$. **57.** $\sqrt[6]{xyz^5}$. **59.** $x^2\sqrt{xy}$. **61.** $x^7\sqrt[3]{x^2}$. **63.** $x^2w^4\sqrt[6]{y^5w}$. **65.** $\frac{2x\sqrt[4]{x}}{y^2}$. **67.** $\frac{\sqrt{2}}{2}$. **69.** $\frac{\sqrt{xy}}{y^2}$. **71.** $\sqrt{x}$. **73.** 50%.

Exercise 8.3

1. $3\sqrt[3]{3}$. **3.** $2x^2\sqrt{2x}$. **5.** $11\sqrt{3}$. **7.** $20\sqrt{2}$. **9.** $-y\sqrt{x}$. **11.** $-8\sqrt[3]{2}$. **13.** 0. **15.** $\sqrt[3]{4}$. **17.** $-2x\sqrt{2}$. **19.** 4. **21.** $2\sqrt{3}$. **23.** $18\sqrt{2}$. **25.** $3\sqrt[3]{12}$. **27.** $x\sqrt{6x}$. **29.** 3. **31.** $16x\sqrt[3]{x}$. **33.** $2\sqrt{5}$. **35.** $6\sqrt{2}-12$. **37.** $-10-2\sqrt{3}$. **39.** -3. **41.** $9+4\sqrt{5}$. **43.** $2x+3\sqrt{x}-5$. **45.** $3xy^3\sqrt{2x}$. **47.** $31-10\sqrt{6}$. **49.** $46\sqrt{6}$.

Exercise 8.4

1. $\frac{3\sqrt{7}}{7}$. **3.** $\frac{2\sqrt{2x}}{x}$. **5.** $\frac{\sqrt[3]{4}}{2}$. **7.** $\frac{\sqrt[3]{9x^2}}{3x}$. **9.** 4. **11.** $a\sqrt{2}$. **13.** $\frac{2\sqrt[4]{x^3}}{x}$. **15.** $\frac{\sqrt[3]{12x^2}}{2x}$. **17.** $\frac{\sqrt[3]{50y}}{5y}$. **19.** $2-\sqrt{3}$. **21.** $-\frac{\sqrt{6}+2\sqrt{3}}{3}$. **23.** $-4-2\sqrt{6}$. **25.** $\frac{x-\sqrt{5}}{x^2-5}$. **27.** $3-2\sqrt{2}$. **29.** (a) $440\sqrt[6]{2}$ vibrations/s, (b) $220\sqrt{2}$, 311 vibrations/s.

Exercise 8.5

1. $x=27$. **3.** $y=\frac{41}{2}$. **5.** $x=4$. **7.** $x=4$. **9.** No solution. **11.** $x=7$. **13.** $z=4, 8$. **15.** $x=2$. **17.** $x=\pm 5$. **19.** 81 m^2. **21.** 67 ft.

Exercise 8.6

1. $-3-i$. **3.** $-2+3i$. **5.** $-3+5i$. **7.** $1+i$. **9.** $5+5i$. **11.** $26+7i$. **13.** i. **15.** 18. **17.** $10+8i$. **19.** $5+12i$. **21.** 58. **23.** $2+\frac{3}{2}i$. **25.** $\frac{11}{17}-\frac{7}{17}i$. **27.** $\frac{7}{10}-\frac{1}{10}i$. **29.** $\frac{2}{17}+\frac{8}{17}i$. **31.** $-\frac{3}{2}i$. **33.** $2+4i$. **35.** $6\sqrt{3}+(2+4\sqrt{2})i$. **37.** $-2\sqrt{10}$. **39.** $-40-19i$. **41.** $-16-30i$. **43.** $\frac{12}{29}+\frac{30}{29}i$. **45.** (a) $Z=3+2i$, (b) $Z=\frac{9}{13}+\frac{7}{13}i$.

REVIEW PROBLEMS—CHAPTER 8

1. 10. **3.** 8. **5.** $\frac{1}{4}$. **7.** $\frac{1}{32}$. **9.** $3z^3$. **11.** $\frac{9t^2}{4}$. **13.** $a^{3/4}b^{3/2}c^{9/4}$. **15.** $-27xy^2\sqrt{x}$. **17.** $\frac{2\sqrt[6]{x^5}}{y^3}$. **19.** $x-y$. **21.** $4\sqrt{2}$. **23.** $x\sqrt[3]{2}$. **25.** $4x^2$. **27.** $14\sqrt{2}$. **29.** $\sqrt[3]{t^2}$. **31.** $\frac{\sqrt[4]{24x^2y}}{2xy}$. **33.** $2\sqrt{2}$. **35.** $\sqrt{2}-2\sqrt{3}$. **37.** 4. **39.** $\frac{2\sqrt{7}}{7}$. **41.** $\frac{3\sqrt[4]{x^3}}{x}$. **43.** $3x\sqrt{x}$. **45.** x^2. **47.** $x\sqrt{3}$. **49.** $\frac{x\sqrt{y}}{y^2}$. **51.** $\frac{3\sqrt[3]{x^2y}}{xy}$. **53.** $\frac{\sqrt[3]{12x}}{2}$. **55.** $\frac{\sqrt{6}+2}{2}$. **57.** $x=10$. **59.** $x=5$. **61.** No solution. **63.** $x=10$. **65.** $20+5i$. **67.** $-2+12i$. **69.** $-5+10i$. **71.** 5. **73.** $\frac{7}{5}+\frac{1}{5}i$. **75.** $2+i\sqrt{3}$. **77.** $-5+5i$. **79.** $v=\frac{\sqrt{5}c}{3}$.

Exercise 9.1

1. (1, 1). **3.** (−4, −1). **5.** (0, 3). **7., 9., 11., 13., 15., 17.** See diagram.

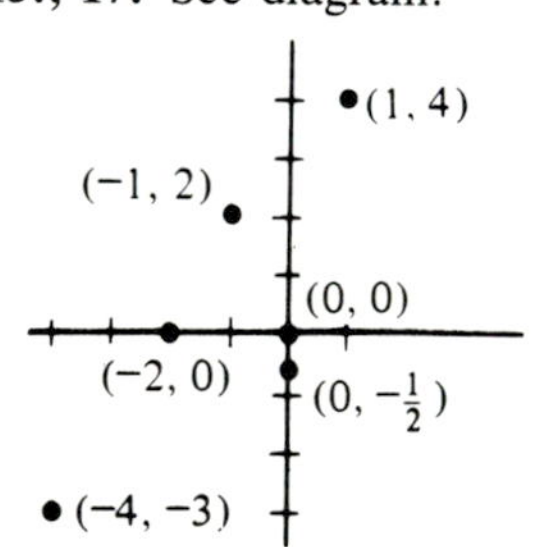

19. I. **21.** IV. **23.** II. **25.** III. **27.** II. **29.** I.

Exercise 9.2

1.

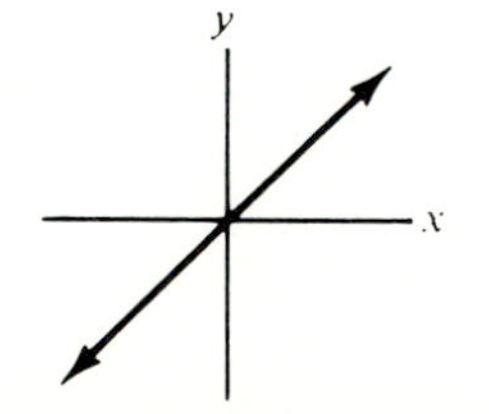

3.

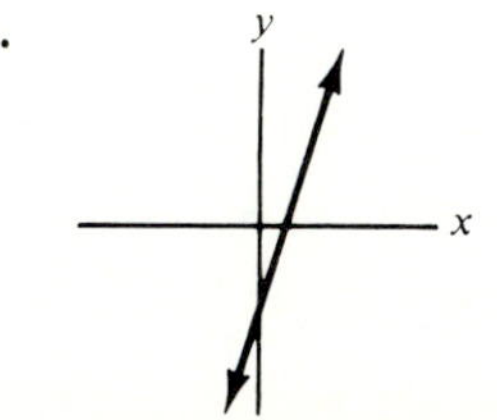

5.

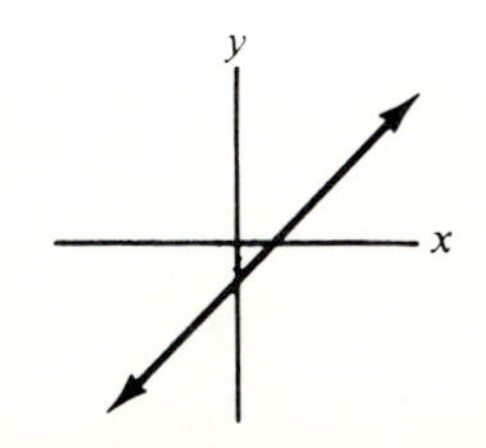

7.

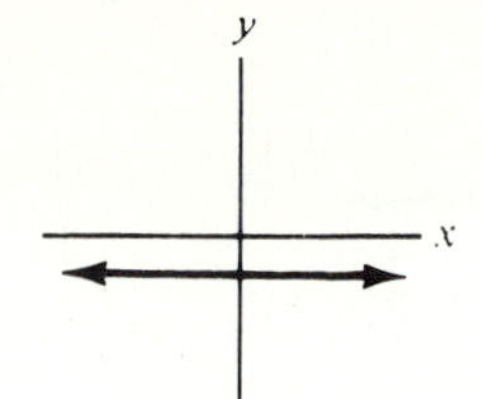

9.

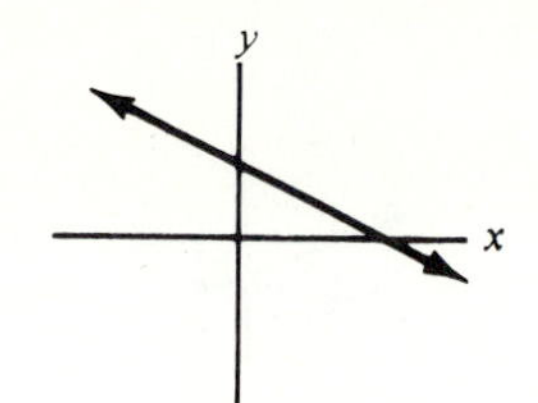

11.

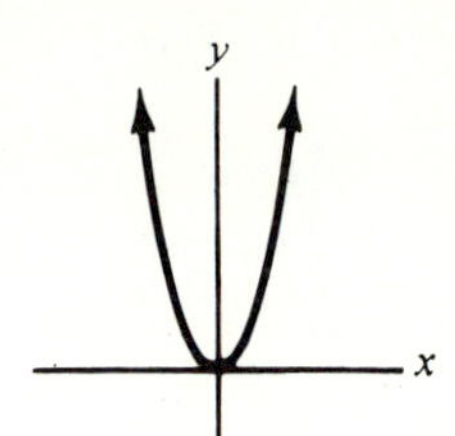

13.

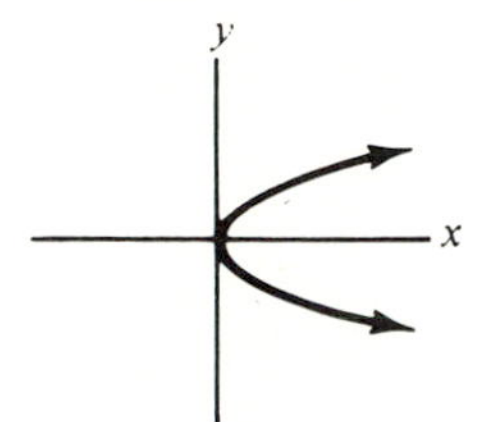

15.

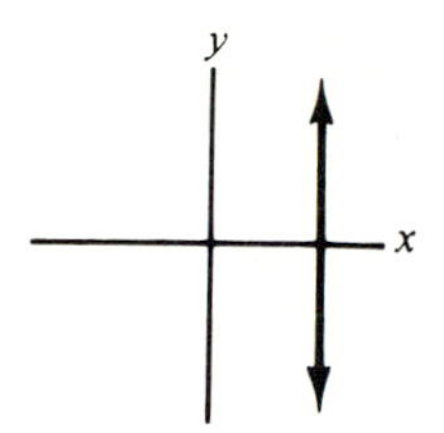

17.

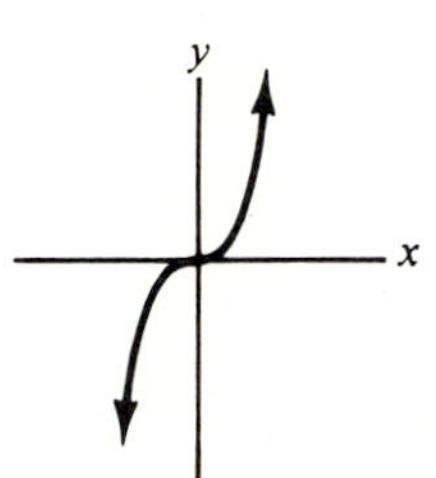

19.

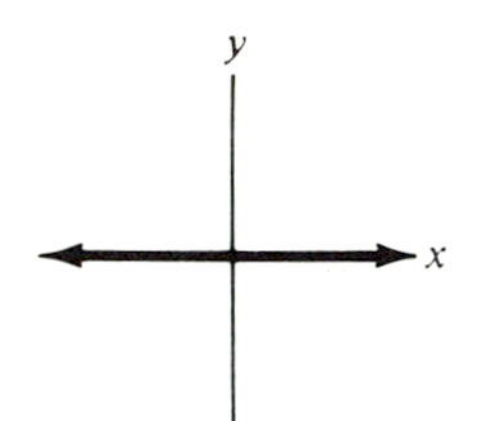

21.

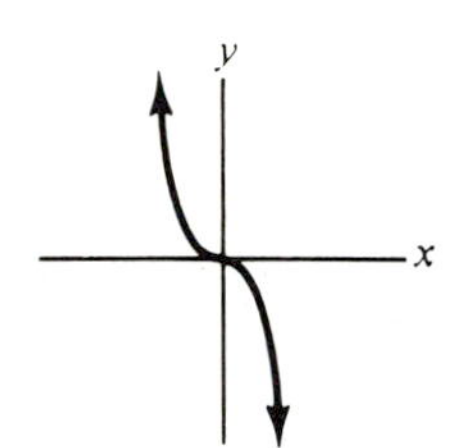

23.

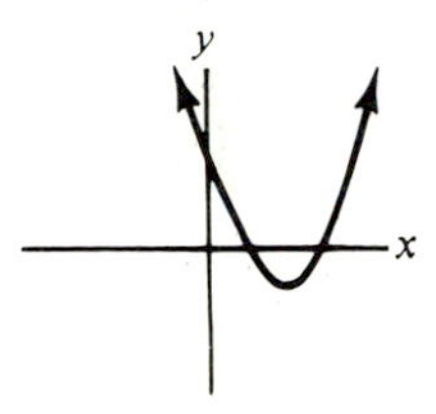

25.

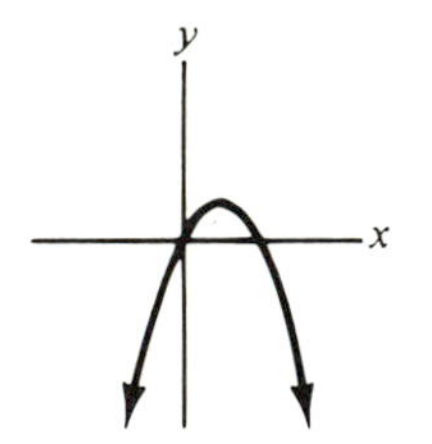

27.

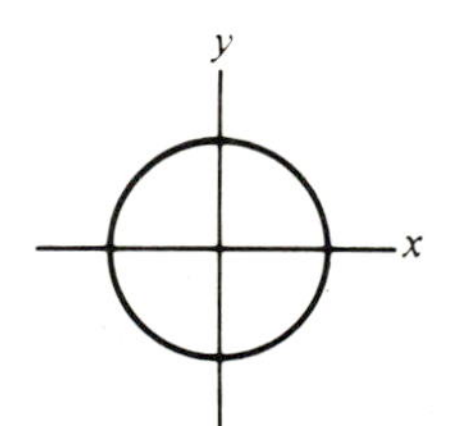

29.

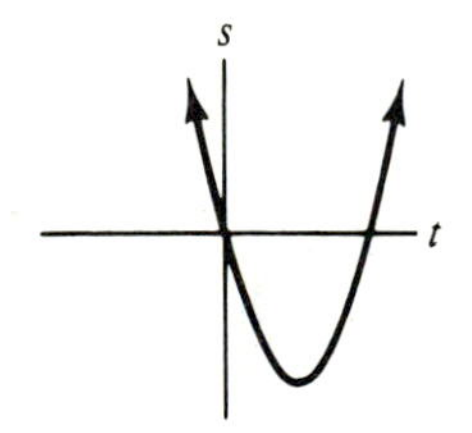

Exercise 9.3

1. $\frac{3}{2}$. **3.** $-\frac{4}{5}$. **5.** Undefined. **7.** 0. **9.** $y = x$. **11.** $y = -\frac{1}{4}x + \frac{9}{2}$. **13.** $y = \frac{1}{3}x - 7$. **15.** $y = -5x$. **17.** $y = \frac{1}{2}x + 3$. **19.** $y = -x$. **21.** $y = 8x - 25$. **23.** $y = -7$. **25.** $(5, -4)$.

Exercise 9.4

1. $y = 2x + 4$.

3. $y = -\frac{1}{2}x - 3$.

5. $y = -2$.

7. $x = 2$.

9. $y = 2x - 1$; 2; -1.

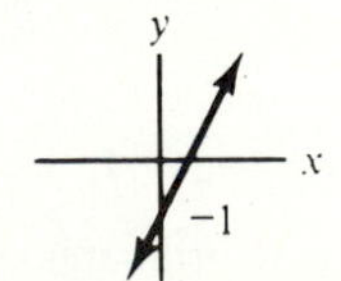

11. $y = -3x + 2$; -3; 2.

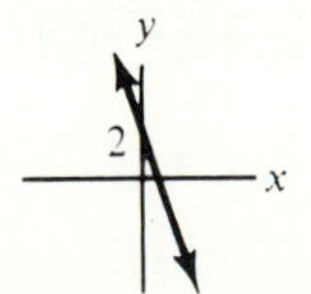

13. $y = 4x$; 4; 0. **15.** $y = -\frac{1}{3}x - 1$; $-\frac{1}{3}$; -1. **17.** $y = 1$; 0; 1.

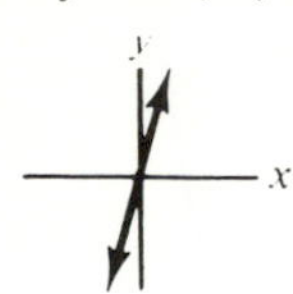

19. $x - y - 6 = 0$. **21.** $x + 2y - 2 = 0$. **23.** Yes. **25.** $s = -\frac{5}{2}t + 70$; 10. **27.** $c = 3q + 10$; \$115. **29.** $x + 10y = 100$. **31.** (a) $r = -\frac{35}{2}l + 195$, (b) 177.5. **33.** (a) $p = .059t + .025$, (b) .556. **35.** $w = .66d + 20$, (b) 53 kg.

Exercise 9.5

1. Parallel. **3.** Parallel. **5.** Perpendicular. **7.** Parallel. **9.** Neither. **11.** $y = 4x + 7$. **13.** $y = 1$. **15.** $y = -\frac{1}{3}x + 5$. **17.** $x = 7$. **19.** $y = -\frac{2}{3}x - \frac{29}{3}$. **21.** (a) $y = -\frac{1}{3}x + \frac{1}{6}$, (b) $y = 3x - \frac{3}{2}$.

REVIEW PROBLEMS—CHAPTER 9

1.

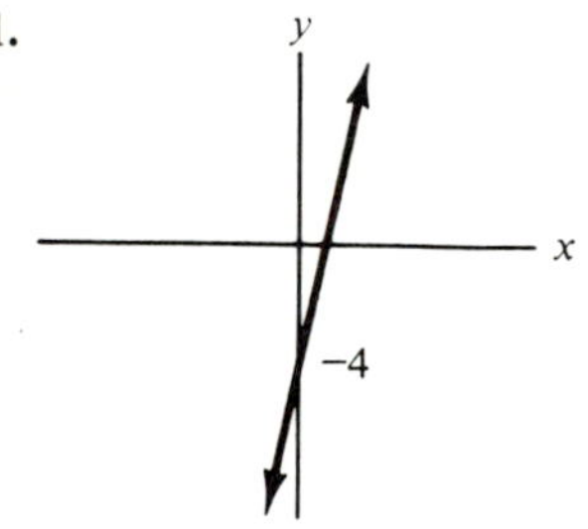

3.

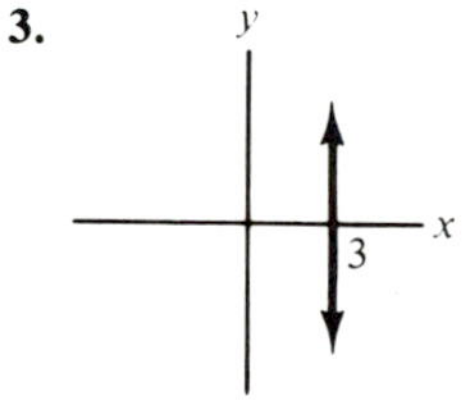

5.

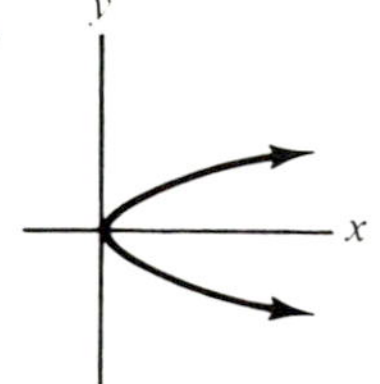

7.

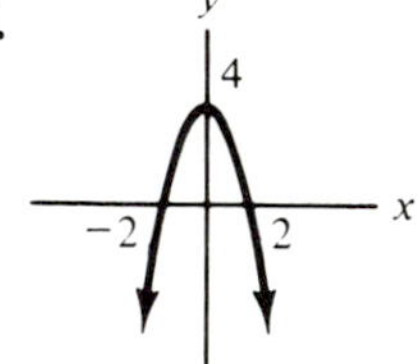

9. $-\frac{1}{3}$. **11.** 0. **13.** $y = -2x + 1$. **15.** $y = \frac{1}{3}x + \frac{11}{3}$. **17.** $y = 2$. **19.** $x = 0$. **21.** $y = 3x - 4$. **23.** $y = -\frac{3}{5}x + \frac{13}{5}$. **25.** $y = -5x + 7$. **27.** Perpendicular. **29.** Neither. **31.** Parallel. **33.** $y = \frac{3}{2}x - 2$; $3x - 2y - 4 = 0$; $\frac{3}{2}$. **35.** $a + b - 3 = 0$; 0.

Exercise 10.1

1. $x = -1, y = 1$. **3.** $x = 3, y = 2$. **5.** $x = 2, y = -2$. **7.** $x = \frac{1}{2}, y = \frac{3}{2}$. **9.** $x = 4, y = -5$. **11.** $x = 3, y = -1$. **13.** $p = 0, q = 18$. **15.** The coordinates of any point on the line $2x + 6y = 3$.

17. No solution. **19.** $x = 12$, $y = -12$. **21.** 420 gal of 20% solution, 280 gal of 30% solution. **23.** 60 units of Argon I, 40 units of Argon II. **25.** 275 mi/h (speed of airplane in still air), 25 mi/h (speed of wind). **27.** 240 units (early American), 200 units (contemporary). **29.** $T = \$100$ million, $L = \$80$ million. **31.** 800 at Exton, 700 at Whyton. **33.** $q = 200$, $p = \$12$. **35.** 1800.

Exercise 10.2

1. $x = 4$, $y = 2$, $z = 0$. **3.** $x = 13$, $y = 22$, $z = -1$. **5.** $x = \frac{1}{2}$, $y = \frac{1}{2}$, $z = \frac{1}{4}$. **7.** $x = 0$, $y = \frac{4}{3}$, $z = 2$. **9.** 100 chairs, 100 rockers, 200 chaise lounges. **11.** 20, 40, and 40 lb of \$2.20, \$2.30, and \$2.60 per lb coffee, respectively. **13.** $a = 1$, $b = -2$, $c = 0$. **15.** $i_1 = -2$, $i_2 = 1$, $i_3 = -1$.

Exercise 10.3

1. 1. **3.** 36. **5.** $-\frac{25}{12}$. **7.** $a - 3b$. **9.** -3. **11.** -6. **13.** -12. **15.** 6. **17.** -16. **19.** -89. **21.** 6. **23.** -8. **25.** 1. **27.** 2.

Exercise 10.4

1. $x = 1$, $y = 1$. **3.** $x = \frac{9}{5}$, $y = -\frac{2}{5}$. **5.** $x = \frac{7}{16}$, $y = \frac{13}{8}$. **7.** $x = -\frac{1}{3}$, $y = -1$. **9.** $x = \frac{6}{5}$, $z = \frac{16}{5}$. **11.** $x = 4$, $y = 2$, $z = 0$. **13.** $x = 0$, $y = 2$, $z = -4$. **15.** $x = \frac{2}{3}$, $y = -\frac{28}{15}$, $z = -\frac{26}{15}$. **17.** $r = 1$, $s = 3$, $t = 5$.

19. Since $\begin{vmatrix} 1 & 1 \\ 1 & 1 \end{vmatrix} = 0$, Cramer's rule does not apply. But the equations in $\begin{cases} x + y = 2 \\ x + y = -3 \end{cases}$ represent different parallel lines. Thus there is no solution.

Exercise 10.5

1. $x = 0$, $y = 1$; $x = -1$; $y = 0$. **3.** $x = -5$, $y = -3$; $x = 0$, $y = 2$. **5.** $x = 6$, $y = -8$. **7.** $q = 4$, $p = 8$; $q = -1$, $p = 3$. **9.** No solution. **11.** $x = 2$, $y = 4$. **13.** $w = 3\sqrt{2}$, $z = 2$; $w = -3\sqrt{2}$, $z = 2$; $w = \sqrt{15}$, $z = -1$; $w = -\sqrt{15}$, $z = -1$. **15.** $x = -2$, $y = -\frac{1}{3}$.

REVIEW PROBLEMS—CHAPTER 10

1. $x = 2$, $y = -1$. **3.** No solution. **5.** $x = \frac{1}{2}$, $y = 1$. **7.** The coordinates of any point on the line $6x = 3 - 9y$. **9.** $x = 3$, $y = 1$, $z = -2$. **11.** $r = 1$, $s = 3$, $t = \frac{1}{2}$. **13.** 18. **15.** -10. **17.** 3.

19. -12. **21.** $x = 1, y = 2$. **23.** $x = 1, y = -2, z = 1$. **25.** $x = -3, y = -4; x = 2, y = 1$.
27. $x = -1, y = 4$. **29.** 2.5 kg of 20% copper alloy, 12.5 kg of 50% copper alloy.
31. 40 semiskilled workers, 20 skilled workers, 10 shipping clerks.

Exercise 11.1

1. $1 < x < 5$. **3.** $x < -1$ or $x > 3$. **5.** $-1 < x < 1$. **7.** $x < -2$ or $x > 3$. **9.** $s \le 0$ or $s \ge 5$.
11. $-3 < x < -2$. **13.** $x \le -5$ or $x \ge 4$. **15.** $-\frac{3}{2} < z < 4$. **17.** $x < -1$ or $x > -1$.
19. $-2 < x < 1$ or $x > 4$. **21.** $x < -5$. **23.** $y \le -1$ or $0 \le y \le 1$. **25.** $x = 0$ or $x \ge 2$.
27. $x < -1$ or $x > 1$. **29.** $x < -8$ or $x > 4$. **31.** $t \le 0$ or $t > 3$. **33.** $x < -3$ or $-1 < x < 1$.
35. $x < -4$ or $-1 < x \le 1$ or $x \ge 4$. **37.** $-2 < x < 3$. **39.** $-\frac{11}{2} < x < -4$.
41. $-1 \le x < 1$ or $x \ge 3$.

Exercise 11.2

1. $-3 < x < 3$. **3.** $x > 6$ or $x < -6$. **5.** $-1 \le x \le 1$. **7.** $x \ge 3$ or $x \le -3$. **9.** $-12 < x < 20$.
11. $y \ge 5$ or $y \le -7$. **13.** $\frac{4}{3} \le x \le 2$. **15.** $-\frac{1}{3} < x < 1$. **17.** $t < 0$ or $t > 1$. **19.** $x \le 0$ or $x \ge \frac{16}{3}$.
21. The absolute value of any quantity is never negative. **23.** (a) $|x - 7| < 3$; (b) $|x - 2| < 3$;
(c) $|x + 7| \le 5$; (d) $|7 - x| = 4$; (e) $|x + 4| < 2$; (f) $|x| < 3$; (g) $|x| > 6$; (h) $|x - 6| \ge 4$; (i) $|x - 105| < 3$;
(j) $|x - 1000| < 100$. **27.** $|x - .01| \le .005$.

Exercise 11.3

1.

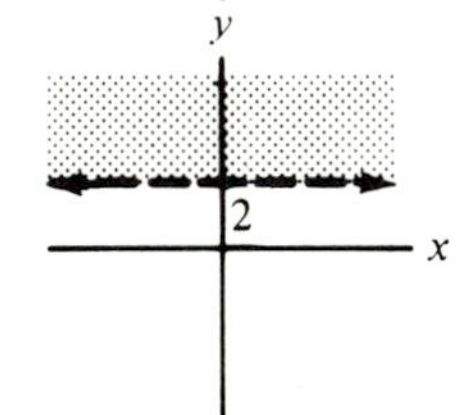

3.

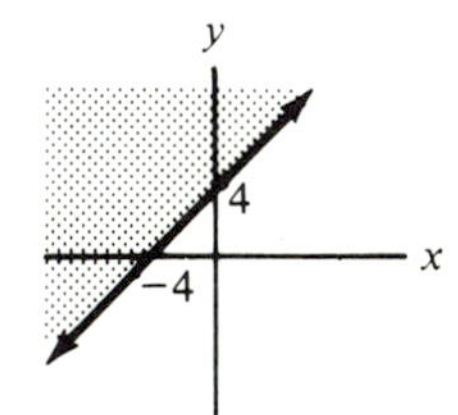

5.

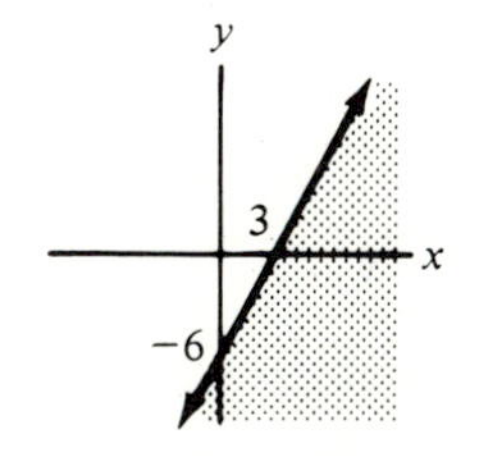

7.

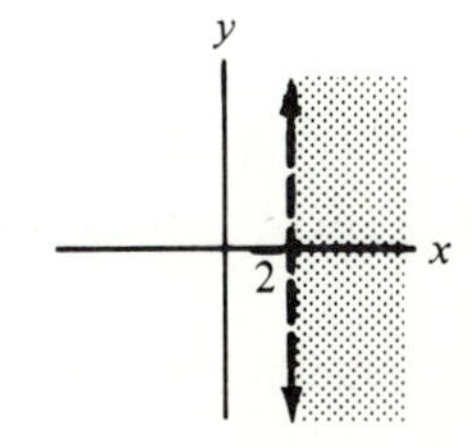

9.

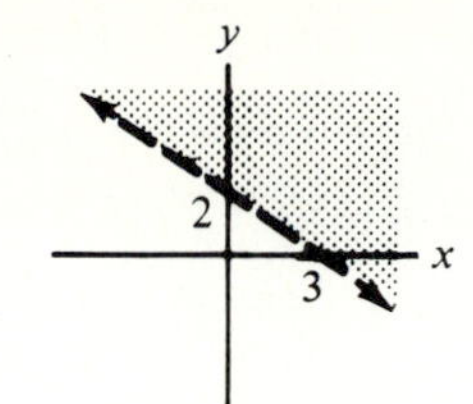

11.

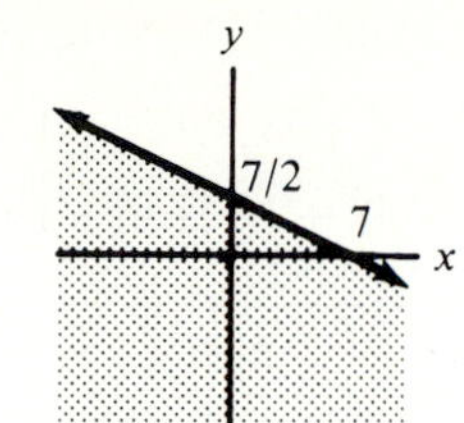

13.

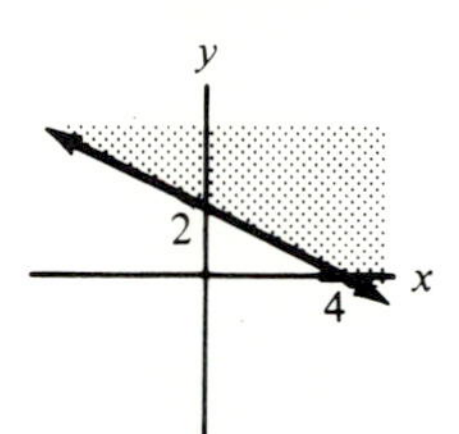

15.

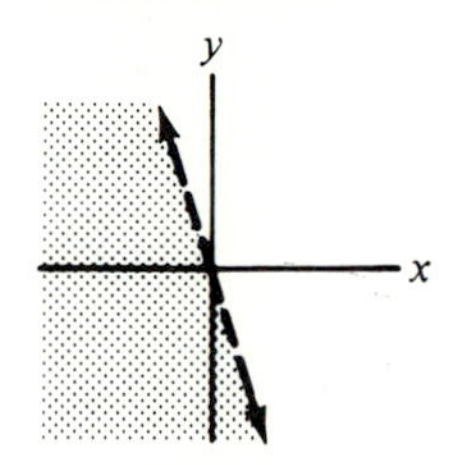

17.

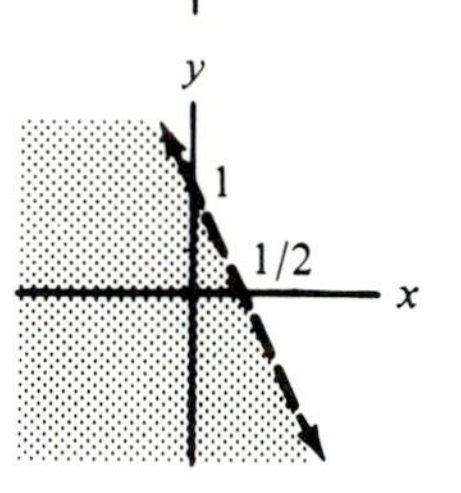

19.

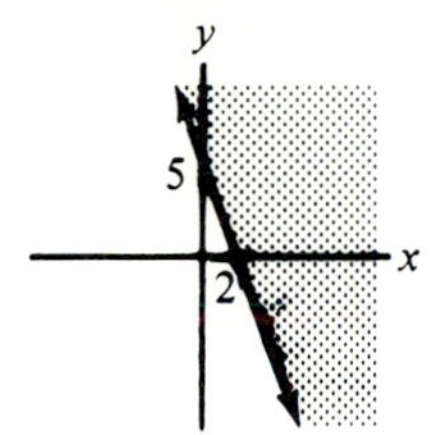

REVIEW PROBLEMS—CHAPTER 11

1. $-6<x<2$. **3.** $y<0$ or $y>6$. **5.** $x \le -5$ or $x \ge -4$. **7.** $-1<x<\frac{1}{3}$.

9. $-4<x<5$ or $x>9$. **11.** $p \le 0$ or $0 \le p \le 8$; more simply $p \le 8$. **13.** $x \le -9$ or $x > -2$.

15. $x < -5$ or $-2<x<3$ or $x>3$. **17.** $x<-1$ or $x>2$. **19.** $x<-4$ or $-2 \le x \le 4$.

21. $x>2$ or $x<-2$. **23.** $0<x<\frac{1}{2}$. **25.** $x \le 0$ or $x \ge 4$. **27.** $-2 \le x \le 1$.

29.

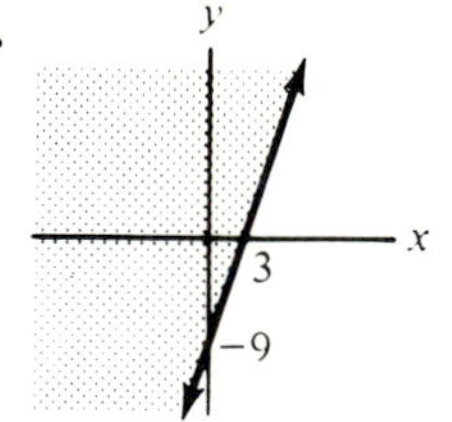

31.

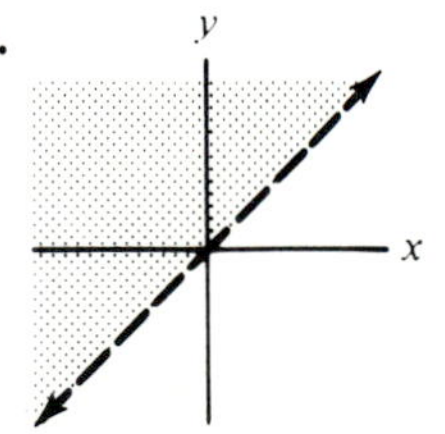

Exercise 12.1

1. All real numbers except 0. **3.** All real numbers. **5.** All nonnegative numbers.

7. All real numbers. **9.** All real numbers. **11.** All real numbers except -2.

13. All real numbers except $-\frac{5}{2}$. **15.** All real numbers greater than or equal to $-\frac{3}{4}$.

17. All real numbers except 0 and 1. **19.** All real numbers except -1 and -5.

21. All real numbers except 4 and $-\frac{1}{2}$. **23.** 0, 15, −20. **25.** 1, 6, $\frac{5}{2}$.

27. $7s$, $7(t+1)=7t+7$, $7(x+3)=7x+21$. **29.** 2, $(2v)^2+2v=4v^2+2v$, $(x^2)^2+x^2=x^4+x^2$.

31. 12, 12, 12. **33.** 4, 0, $(x+h)^2+2(x+h)+1=x^2+2xh+h^2+2x+2h+1$.

35. 0, $\dfrac{3x-5}{(3x)^2+4}=\dfrac{3x-5}{9x^2+4}$, $\dfrac{(x+h)-5}{(x+h)^2+4}=\dfrac{x+h-5}{x^2+2xh+h^2+4}$. **37.** 1, −1, 0, −1. **39.** 8, 3, 3, 1.

41. (a) $3x+3h-4$; (b) 3. **43.** (a) 2; (b) 0. **45.** y is a function of x; x is a function of y.

47. y is a function of x; x is not a function of y. **49.** 2, $\frac{1}{2}$, 3.

51. 20; the price per unit at which 30 units are demanded; 30, 60, 78. **53.** Yes.

55. 32 ft, 32 ft, no, yes. **57.** $x \le -6$, or $x \ge 2$. **59.** $t < -5$ or $t \ge 1$.

Exercise 12.2

1. Domain: all real numbers.
Range: all real numbers.

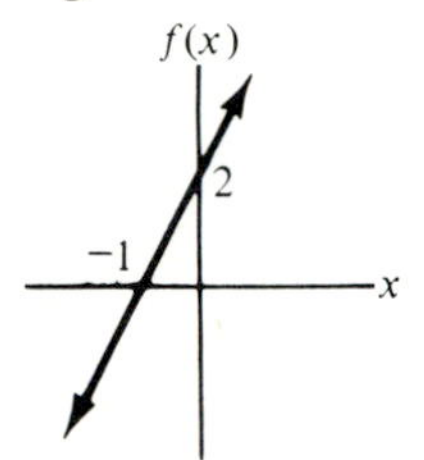

3. Domain: all real numbers.
Range: all nonnegative numbers.

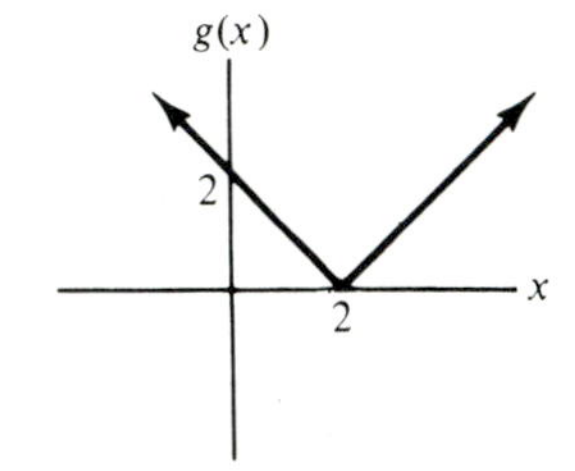

5. Domain: all nonnegative numbers.
Range: all nonnegative numbers.

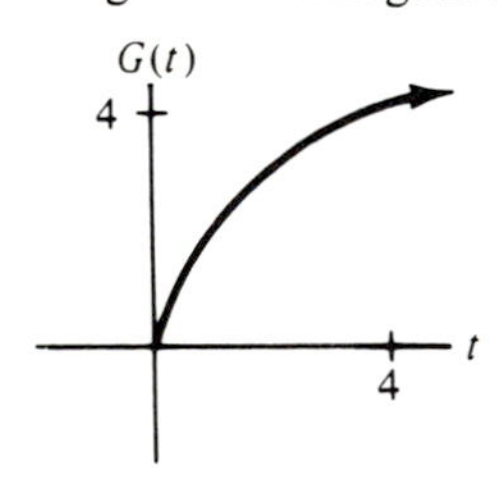

7. Domain: all real numbers.
Range: 4.

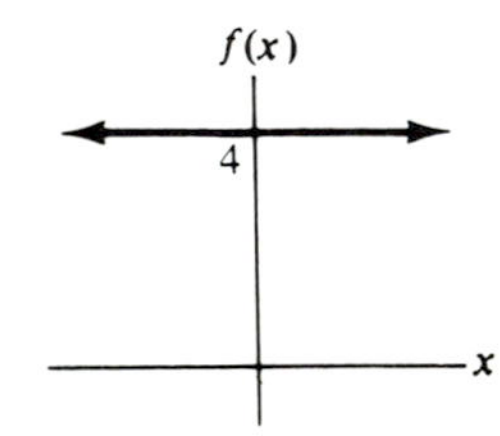

9. Domain: all real numbers.
Range: all numbers less than or equal to 4.

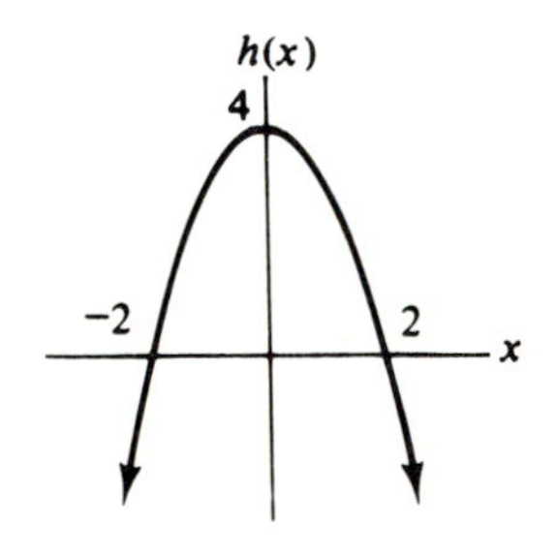

11. Domain: all real numbers except 0.
Range: all real numbers except 0.

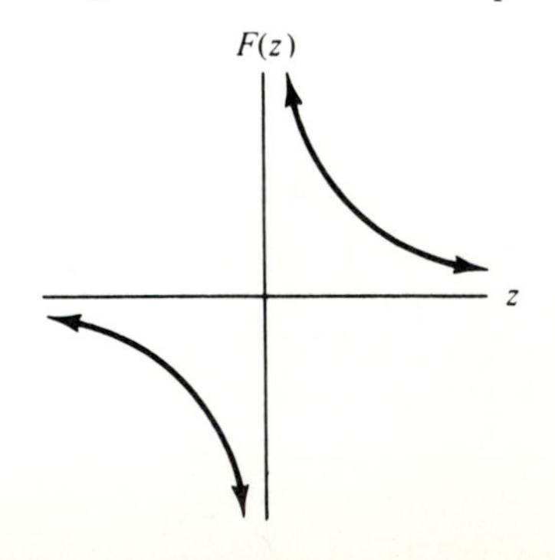

13. Domain: all real numbers except 4.
Range: all real numbers except 0.

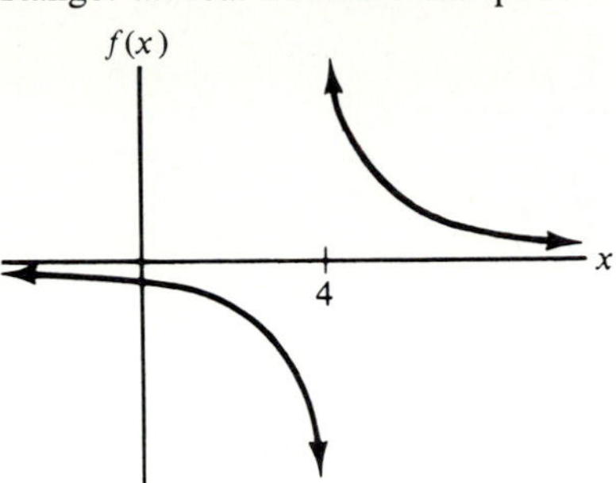

15. Domain: all positive real numbers.
Range: 4, -1, and all real numbers y such that $0 < y \leq 3$.

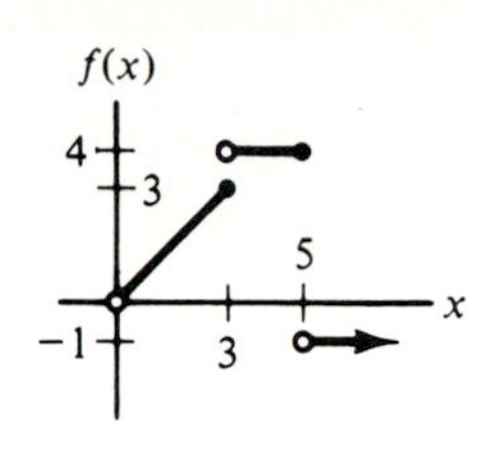

17. (a) 1, 2, 3, 0; (b) all real numbers; (c) all real numbers; (d) -2.

19. (a) 0, -1, -1; (b) all real numbers; (c) all $y \leq 0$; (d) 0.

21. (a) 20, 10, 5, 4;
(b)

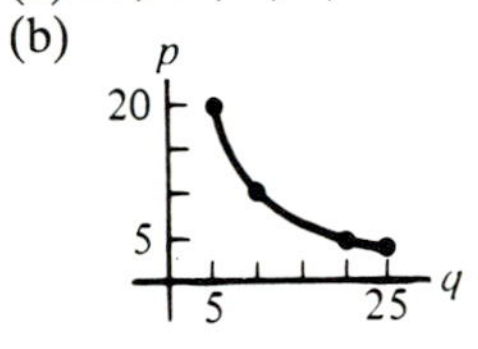

23. a, b, d.

25.

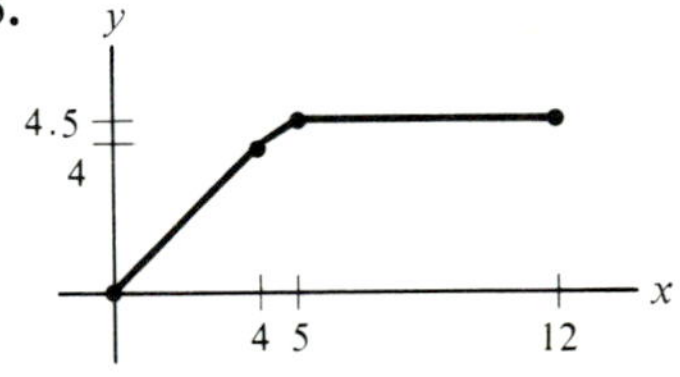

Exercise 12.3

1. Not quadratic. **3.** Quadratic. **5.** Not quadratic. **7.** Quadratic. **9.** (a) (1, 11); (b) Highest.

11. (a) -8; (b) -4, 2; (c) $(-1, -9)$.

13. Vertex: $(3, -4)$.

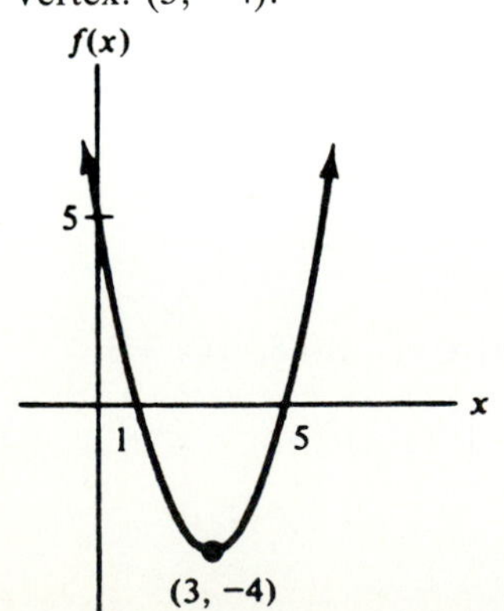

15. Vertex: $(-\frac{3}{2}, \frac{9}{2})$.

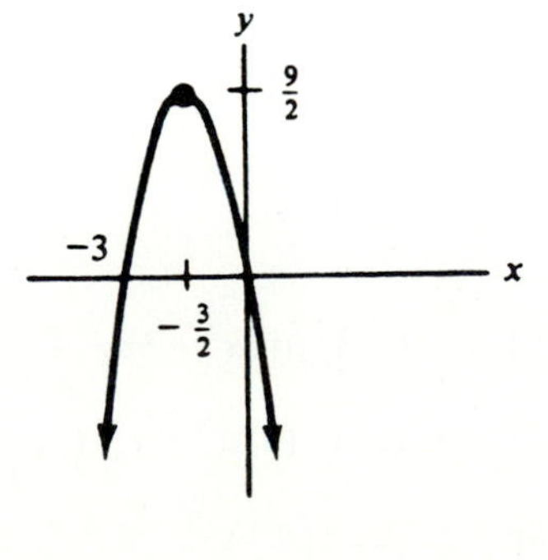

17. Vertex: $(-1, 0)$.

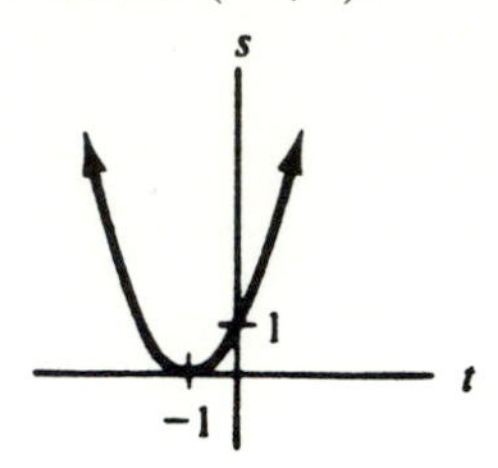

19. Vertex: $(2, -1)$.

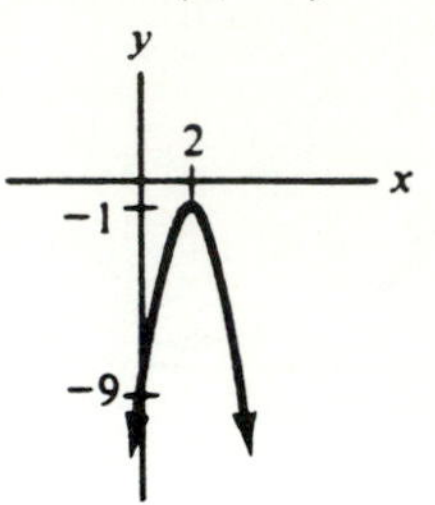

21. Vertex: $(4, -3)$.

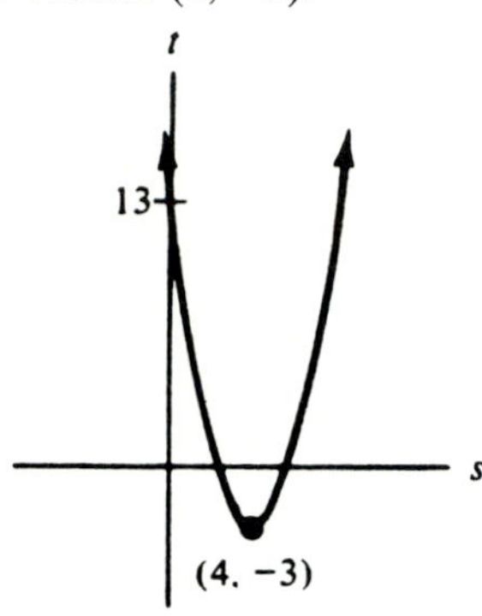

23. Vertex: $(\frac{1}{2}, \frac{25}{4})$; y-intercept: 6; x-intercepts: 3, -2.

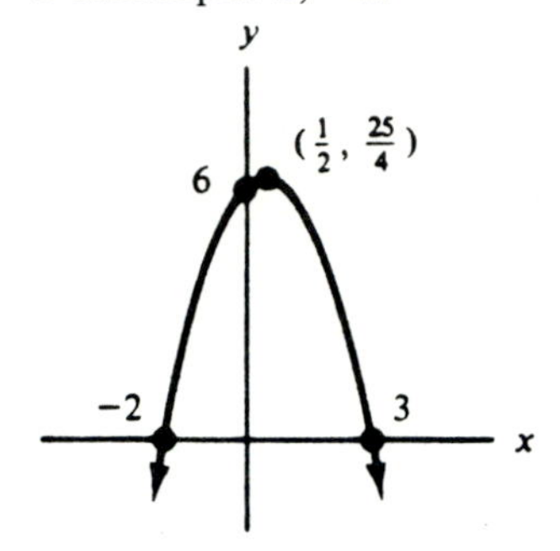

25. Vertex: $(3, -2)$; y-intercept: 7; x-intercepts: $3 + \sqrt{2}$, $3 - \sqrt{2}$.

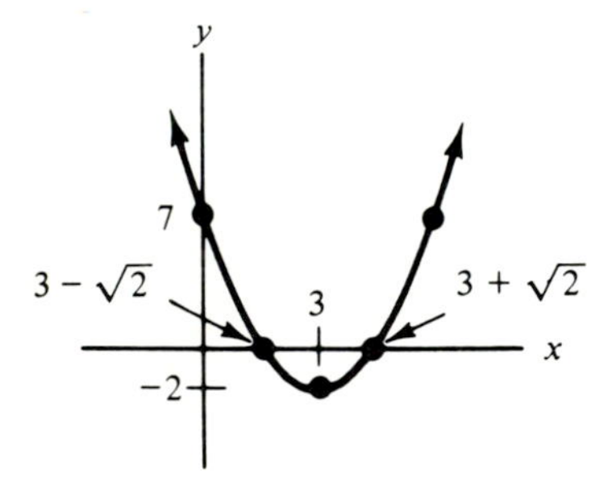

27. Minimum; 24. **29.** Maximum; -10.

31. 6 s; 176.4 m. **33.** 250 units; \$125,000. **35.** 50 yd by 100 yd. **37.** \$350.

39. 300 ft by 250 ft.

Exercise 12.4

1. (a) $2x + 5$, (b) 5, (c) -3, (d) $x^2 + 5x + 4$, (e) -2, (f) $\dfrac{x+1}{x+4}$, (g) $x + 5$, (h) 8, (i) $x + 5$.

3. (a) $2x^2 + x$, (b) $-x$, (c) $\frac{1}{2}$, (d) $x^4 + x^3$, (e) $\dfrac{x^2}{x^2 + x} = \dfrac{x}{x+1}$, (f) -1, (g) $(x^2 + x)^2 = x^4 + 2x^3 + x^2$, (h) $x^4 + x^2$, (i) 90. **5.** $\sqrt{x-1}$, $\sqrt{x} - 1$. **7.** $\dfrac{1}{x^2 - 1}$, $\left(\dfrac{1}{x}\right)^2 - 1 = \dfrac{1 - x^2}{x^2}$.

9. $f(x) = x^2$, $g(x) = x^2 + 2$. **11.** $f(x) = x^2 + 2x$, $g(x) = x^2 - 1$. **13.** $f(x) = \sqrt[5]{x}$, $g(x) = \frac{x+1}{3}$.
15. $400m - 10m^2$; the total revenue received when the total output of m employees is sold.

REVIEW PROBLEMS—CHAPTER 12

1. All real numbers except 1 and 2. **3.** All real numbers. **5.** All nonnegative numbers except 1.
7. 7, 46, 62, $3t^2 - 4t + 7$. **9.** 0, 3, $\sqrt{t}$, $\sqrt{x^2 - 1}$. **11.** $\frac{3}{5}$, 0, $\frac{\sqrt{x+4}}{x}$, $\frac{\sqrt{u}}{u-4}$. **13.** -8, 4, 4, -92.
15. (a) $3 - 7x - 7h$; (b) -7.

17.

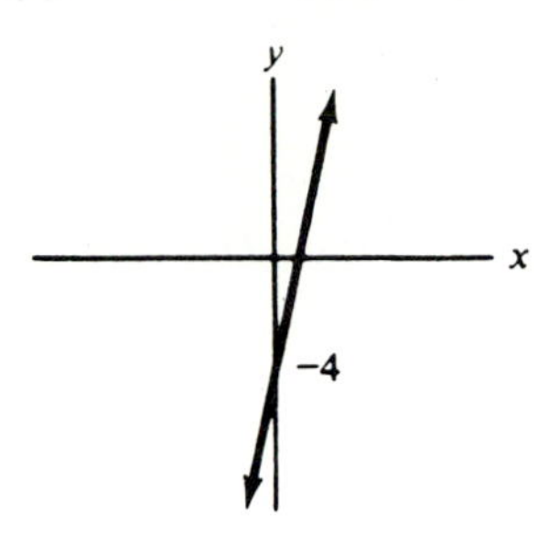

19. Vertex: (0, 9).

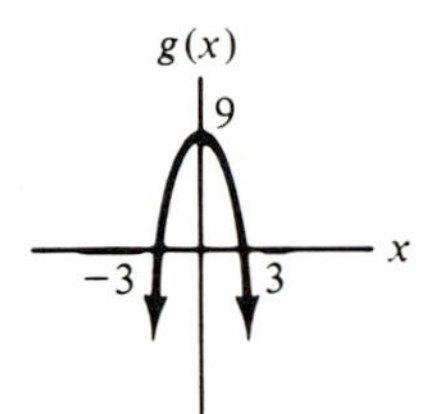

21.

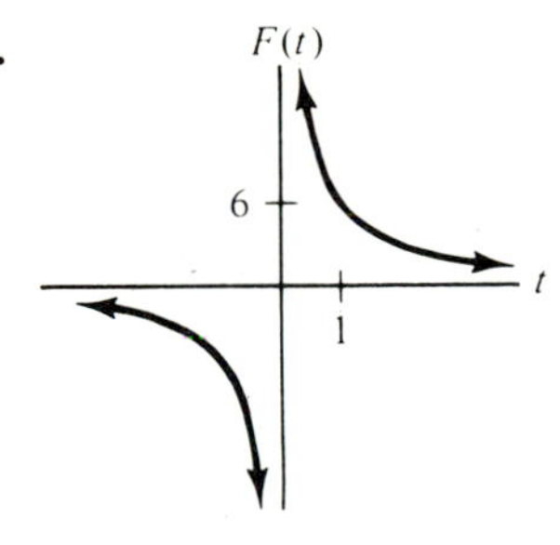

23.

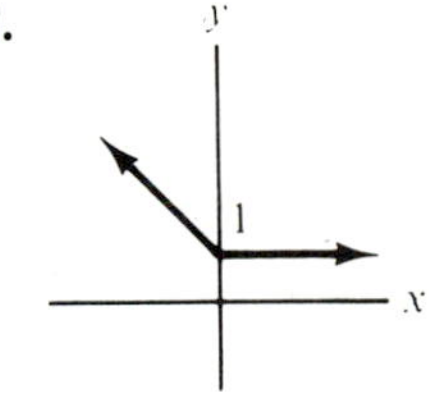

25. Vertex: (2, -9).

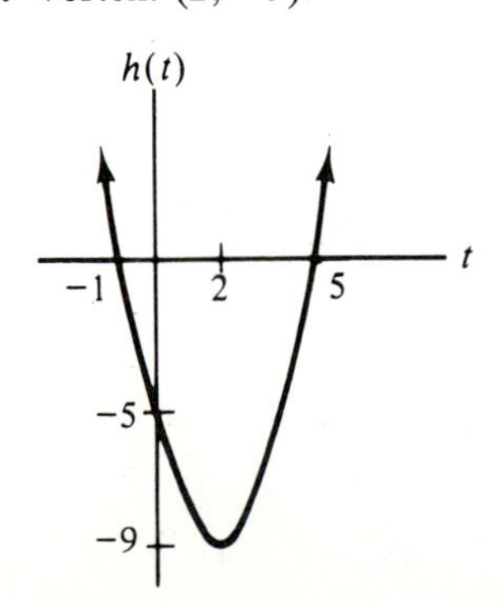

27. Vertex: (-1, -2).

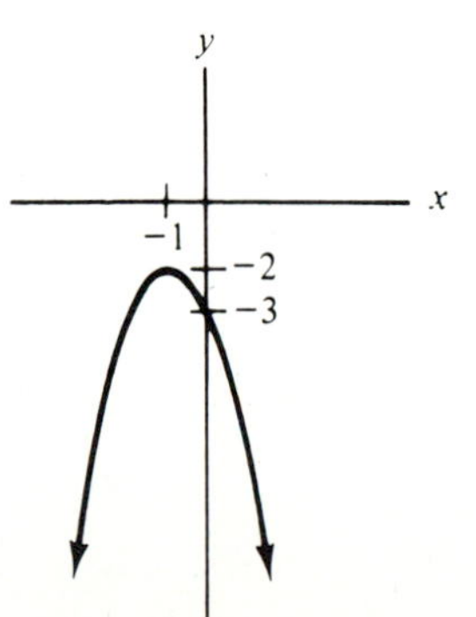

29. (a) $5x + 2$, (b) 22, (c) $x - 4$, (d) $6x^2 + 7x - 3$, (e) 10, (f) $\frac{3x-1}{2x+3}$, (g) $3(2x + 3) - 1 = 6x + 8$, (h) 38, (i) $2(3x - 1) + 3 = 6x + 1$. **31.** $\frac{1}{x-1}$, $\frac{1}{x} - 1 = \frac{1-x}{x}$. **33.** $x^3 + 2$, $(x + 2)^3$.

35. 15; the price per unit at which 10 units are supplied; 20, 60, 260. **37.** (a) 2, 2, 2, 2; (b) all real numbers; (c) 2. **39.** -1. **41.** 40,000. **43.** \$4.50; 2250.

Exercise 13.1

1.

3.

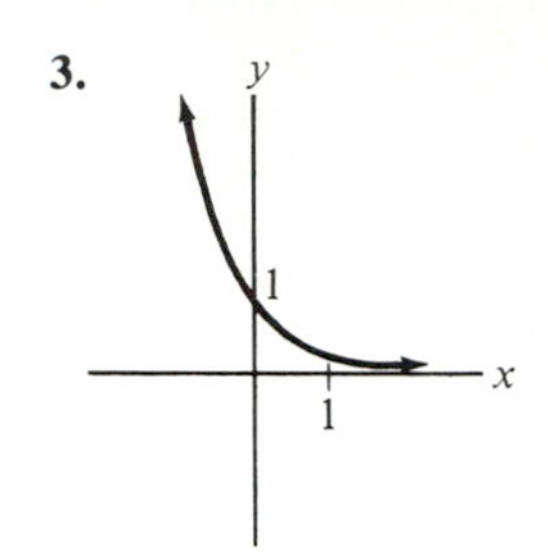

5. 20.086. **7.** 0.2331. **9.** 81, $\frac{1}{81}$, 3. **11.** 12, 6, $\frac{3}{2}$. **13.** 3, $\frac{33}{32}$, $\frac{3}{2}$. **15.** 140,000. **17.** .2241. **19.** 50 mg. **21.** \$1491.80. **23.** (a) 91 units, (b) 432 units. **25.** 17.3837.

Exercise 13.2

1. $\log_4 64 = 3$. **3.** $\log 100{,}000 = 5$. **5.** $2^6 = 64$. **7.** $8^{1/3} = 2$. **9.** $\log_6 1 = 0$. **11.** $2^{14} = x$. **13.** $\ln 7.3891 = 2$. **15.** $e^{1.0986} = 3$. **17.**

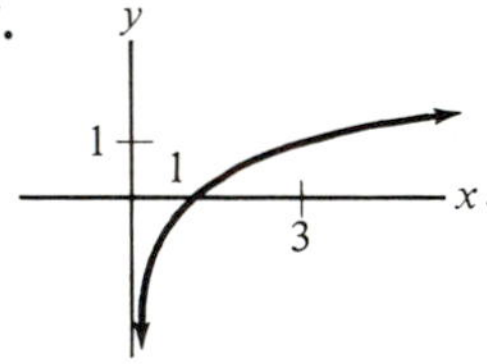

19. 1.60944. **21.** 2.00013. **23.** 2. **25.** 3. **27.** $\frac{1}{2}$. **29.** 1. **31.** -2. **33.** 0. **35.** -3. **37.** 16. **39.** 125. **41.** $\frac{1}{10}$. **43.** e^2. **45.** 2. **47.** 6. **49.** 2. **51.** 4. **53.** $\frac{1}{2}$. **55.** $\frac{1}{81}$. **57.** 9. **59.** 3. **61.** 11. **63.** 41.50. **65.** -4.1045.

Exercise 13.3

Since there are different ways to find the values in Problems 1–33, and since the entries in Table 13-1 are only approximations, your answers may differ *slightly* from those given below.

1. .6990. **3.** 1.3222. **5.** 1.5441. **7.** 1.3980. **9.** .3521. **11.** $-.3010$. **13.** 4.5155. **15.** -2.7960. **17.** .1505. **19.** .5188. **21.** .4471. **23.** $-.2711$. **25.** 4. **27.** -2. **29.** 2.4771. **31.** 1.0458. **33.** -2.2677. **35.** 48. **37.** $\frac{3}{4}$. **39.** 4. **41.** -4. **43.** $\frac{9}{2}$. **45.** $-\frac{1}{3}$. **47.** 3. **49.** $\log 28$. **51.** $\log_2 \frac{x+2}{x+1}$. **53.** $\ln \frac{25}{64}$. **55.** $\log_4(27\sqrt{2})$. **57.** $\log_3 \frac{xy}{z}$. **59.** $\log_2 \frac{45}{7}$. **61.** $\ln \frac{x^2y^3}{z^4w^2}$. **63.** $\log x + 2\log(x+1)$. **65.** $2\log x - 3\log(x+1)$. **67.** $3[\log x - \log(x+1)]$.

69. $\ln x - \ln(x+1) - \ln(x+2)$. **71.** $\frac{1}{2}\ln x - 2\ln(x+1) - 3\ln(x+2)$.

73. $\frac{2}{5}\ln x - \frac{1}{5}\ln(x+1) - \ln(x+2)$. **75.** 3.5229. **77.** 10. **79.** 10 db. **81.** 5.52146.

83. -9.16155.

Exercise 13.4

1. $t = .8047$. **3.** $t = .5694$. **5.** $x = 1.771$. **7.** $x = 3.183$. **9.** $t = 4.322$. **11.** $x = -4.055$.

13. $x = 1003$. **15.** $x = \frac{20}{3}$. **17.** $x = \frac{299}{97}$. **19.** $x = \frac{1}{2}$. **21.** $t = 20.48$. **23.** (a) 6, (b) 28.

25. 7. **27.** 1210 cm.

Exercise 13.5

1. .6826. **3.** 1.585. **5.** 1.107. **7.** .3010. **9.** 1.5057. **11.** 1.0913.

REVIEW PROBLEMS—CHAPTER 13

1. 5, 13, $\frac{13}{3}$. **3.** 100, $100e^4 = 5459.8$, $100e^{-3/2} = 22.31$.

5.

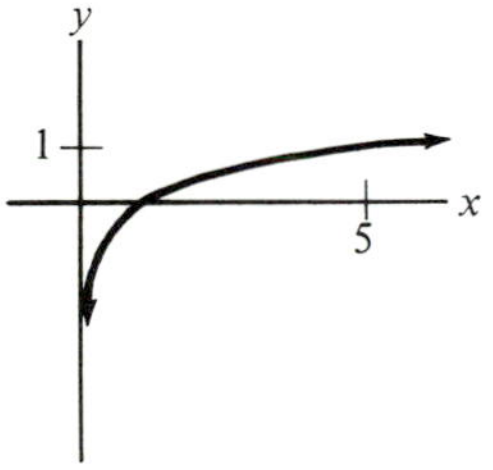

7. $\log_3 243 = 5$. **9.** $16^{1/4} = 2$.

11. $\ln 54.598 = 4$. **13.** 3. **15.** -4. **17.** -2. **19.** -2. **21.** $\frac{1}{100}$. **23.** 9. **25.** 3.

Since there are different ways to find the values in Problems 27–37, and since the entries in Table 13-1 are only approximate, your answer may differ *slightly* from those given below.

27. 1.2042. **29.** .4260. **31.** .2258. **33.** 2.6990. **35.** $-.1639$. **37.** .1990. **39.** 0. **41.** $\frac{1}{3}$.

43. $\frac{5}{2}$. **45.** $\log \frac{25}{27}$. **47.** $\ln \frac{x^2 y}{z^3}$. **49.** $\log_2 \frac{x^{9/2}}{(x+1)^3(x+2)^4}$. **51.** $2\ln x + \ln y - 3\ln z$.

53. $\frac{1}{3}(\ln x + \ln y + \ln z)$. **55.** $\frac{1}{2}(\ln y - \ln z) - \ln x$. **57.** (a) \$134,060; (b) \$109,760.

59. 1.465. **61.** $x = .2311$. **63.** $x = 10$. **65.** 180.

Exercise 14.1

1. 3, 6, 9, 12. **3.** 2, 6, 12, 20. **5.** 0, $\frac{1}{2}$, $\frac{8}{11}$, $\frac{5}{6}$. **7.** x^2, x^4, x^6, x^8. **9.** 1, -4, 9, -16. **11.** n.

13. $3n$. **15.** 3^{n-1}. **17.** $\frac{(-1)^{n+1}}{n+1}$. **19.** 35. **21.** 14. **23.** 9. **25.** $-\frac{3}{16}$. **27.** $\frac{7}{4}$.

Exercise 14.2

1. 22. **3.** −71. **5.** Not arithmetic. **7.** $\frac{11}{3}$. **9.** 4. **11.** 110. **13.** −360. **15.** −108. **17.** 620. **19.** 135. **21.** (a) 44, (b) 142. **23.** $\frac{32}{15}$. **25.** $a_1 = 2$, $d = 3$. **27.** 5050. **29.** $a_{16} = 48$, $S_{16} = 528$. **31.** $n = 23$, $a_n = 36$. **33.** 7 cm. **35.** \$4.65. **37.** 110.

Exercise 14.3

1. 48. **3.** $-\frac{3}{16}$. **5.** Not geometric. **7.** $-\frac{1}{64}$. **9.** $\frac{243}{32}$. **11.** 1092. **13.** −126. **15.** .3333333. **17.** $\frac{255}{128}$. **19.** $\frac{1}{4}$. **21.** 6561°C. **23.** $a_1 = 64$, $a_6 = 2$. **25.** −2. **27.** $a_1 = 81$, $a_5 = 1$. **29.** 59%. **31.** $250(1.065)^n$ dollars. **33.** 6. **35.** $\frac{1000}{11}$. **37.** 12 m.

REVIEW PROBLEMS—CHAPTER 14

1. 48. **3.** $-\frac{1}{128}$. **5.** 25. **7.** $\frac{127}{64}$. **9.** 11. **11.** $-\frac{4}{11}$. **13.** $\frac{4}{3}$. **15.** $\frac{243}{128}$ m. **17.** 156.

Exercise 15.1

1. 5. **3.** $\sqrt{53}$. **5.** $2\sqrt{2}$. **7.** 7. **9.** $\frac{\sqrt{5}}{2}$. **11.** 13.

Exercise 15.2

1. $(x-2)^2 + (y-3)^2 = 36$; $x^2 + y^2 - 4x - 6y - 23 = 0$.

3. $(x+1)^2 + (y-6)^2 = 16$; $x^2 + y^2 + 2x - 12y + 21 = 0$. **5.** $x^2 + y^2 = \frac{1}{4}$; $x^2 + y^2 - \frac{1}{4} = 0$.

7. $C = (0, 0)$, $r = 3$.

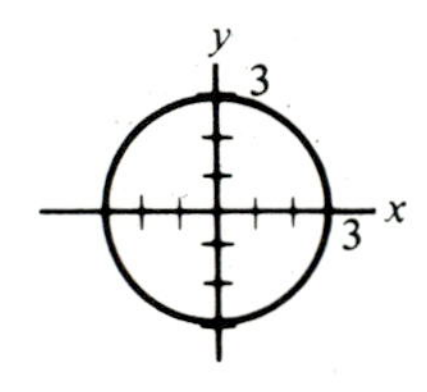

9. $C = (3, -4)$, $r = 2$.

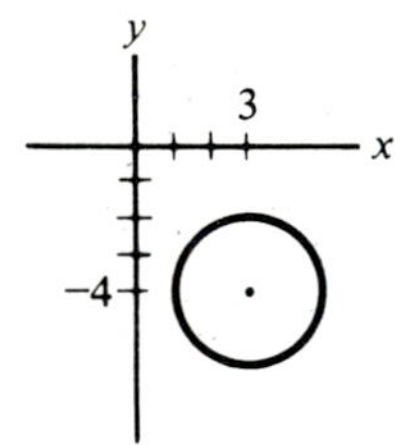

11. $C = (-2, 0)$, $r = 1$.

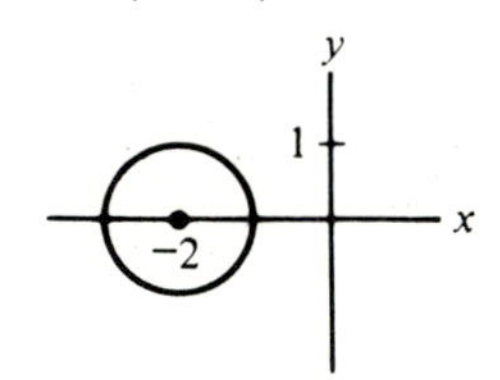

13. Circle; $C = (1, 2)$, $r = 3$. **15.** Circle; $C = (0, -3)$, $r = 2$. **17.** No graph.
19. Circle; $C = (7, -2)$, $r = 4$. **21.** Point $(\frac{3}{2}, -\frac{1}{2})$. **23.** Circle; $C = (1, -\frac{7}{4})$, $r = \frac{7}{4}$.

Exercise 15.3

1. Parabola.

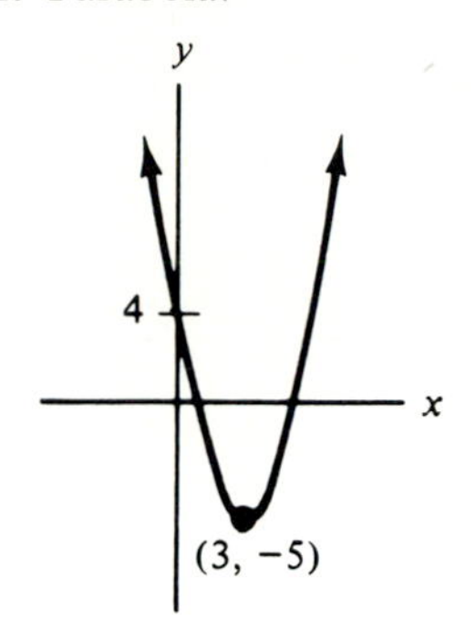

3. Ellipse.

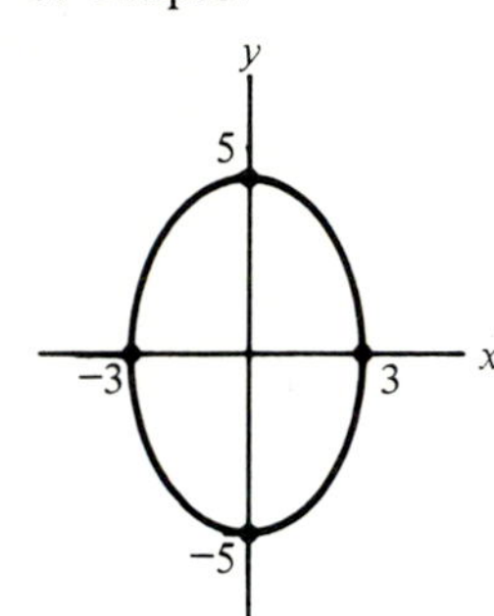

5. Parabola.

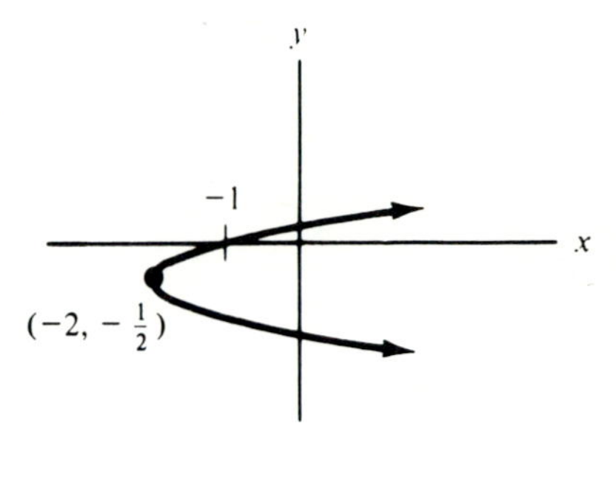

7. Parabola.

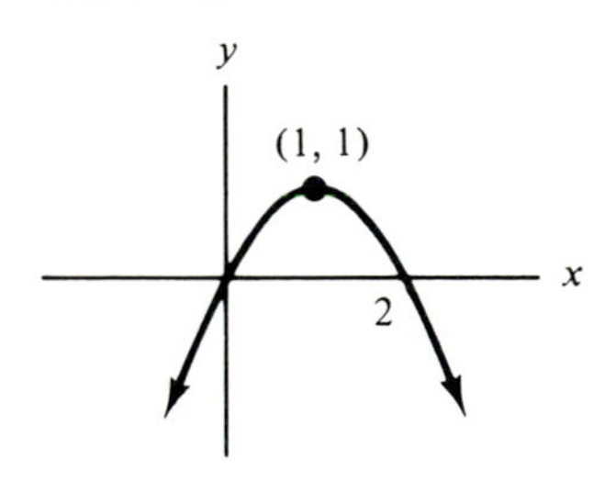

9. Ellipse.

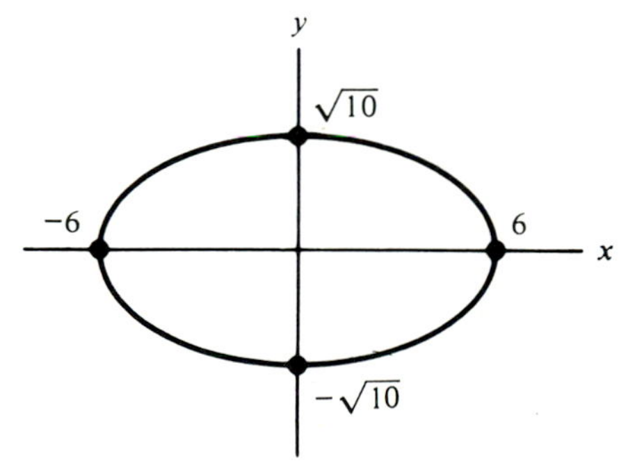

11. Ellipse.

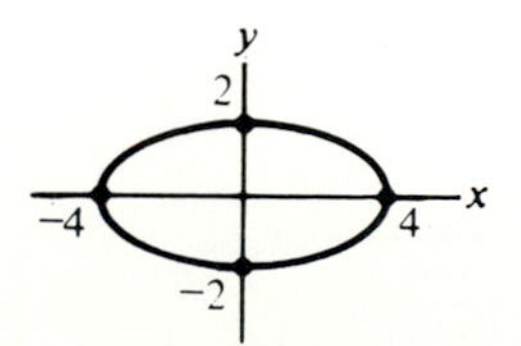

13. Parabola.

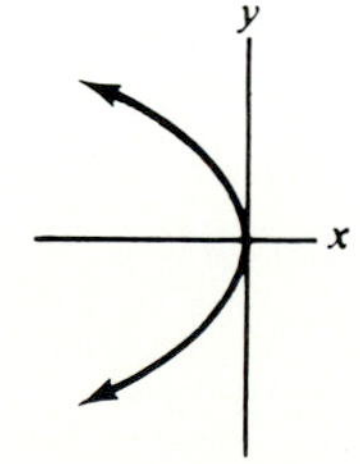

15. Ellipse.

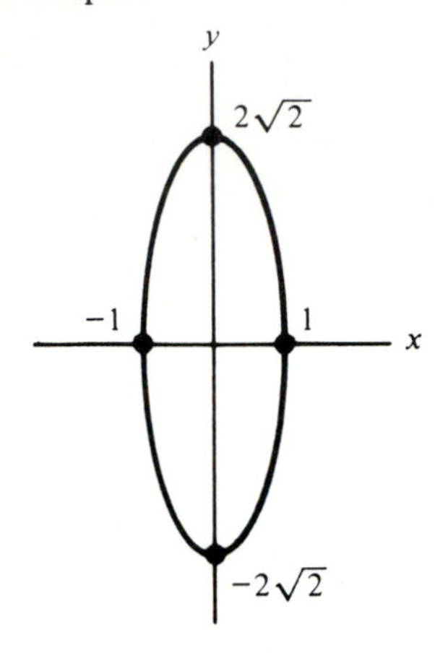

17. Parabola.

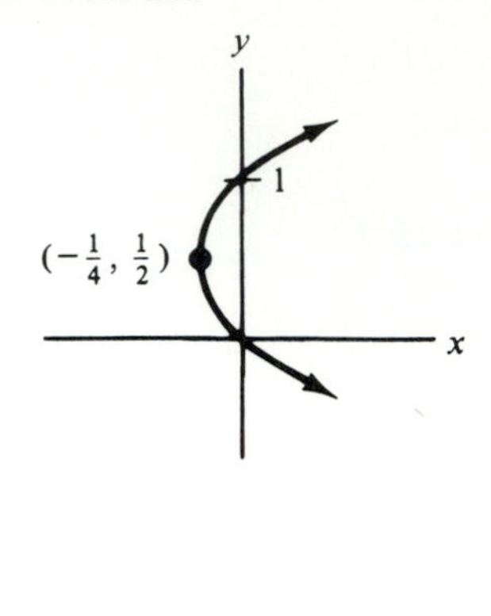

Exercise 15.4

1. Hyperbola.

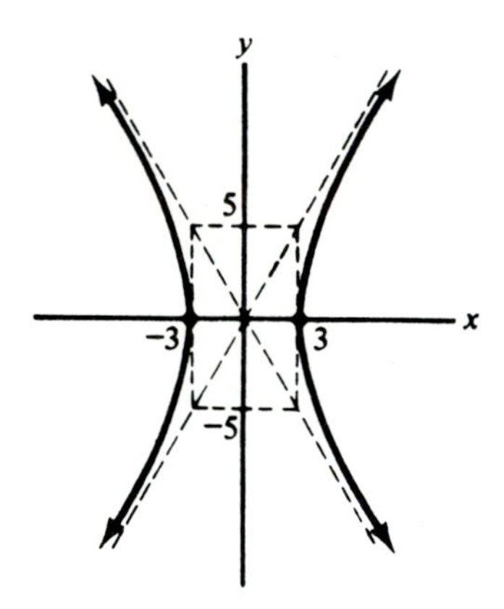

3. Hyperbola.

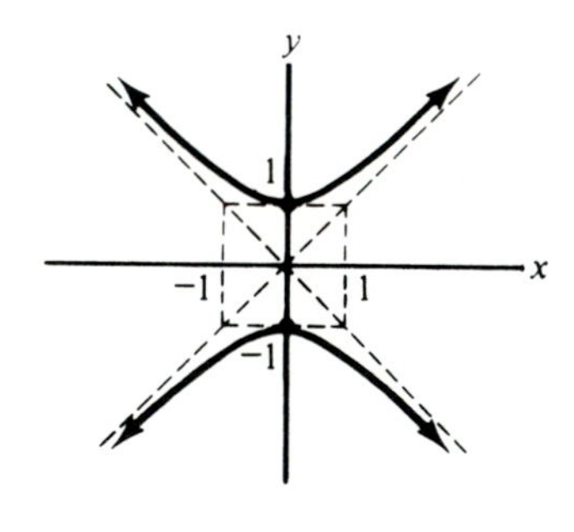

5. Circle.

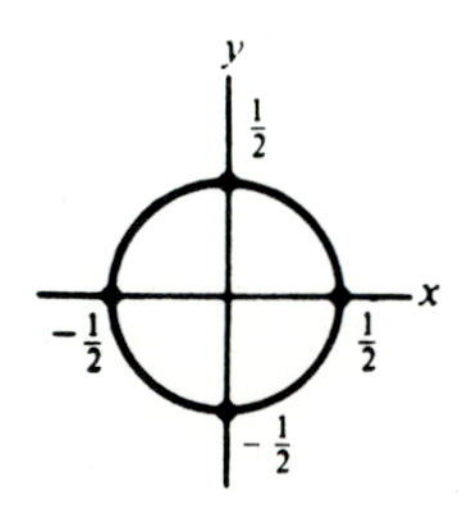

7. Parabola.

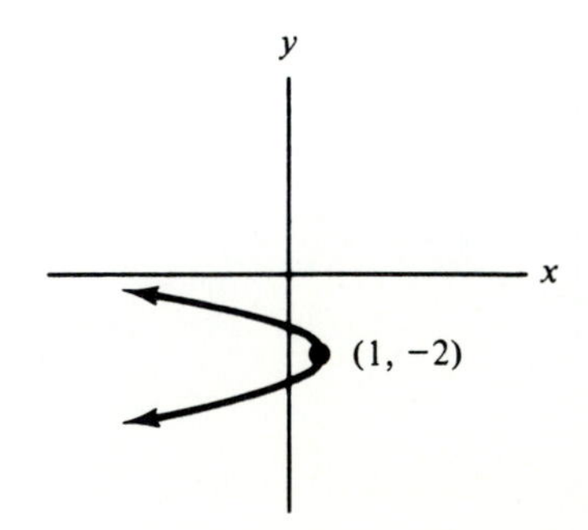

9. Ellipse.

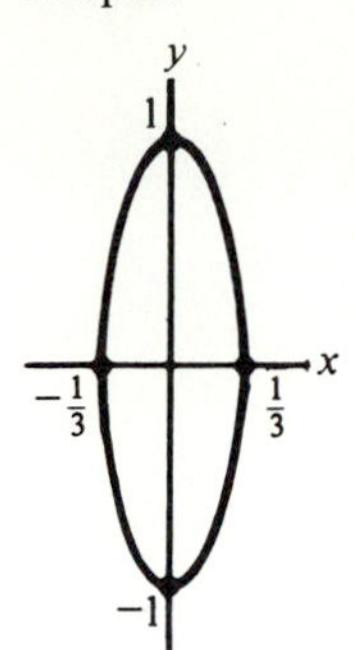

11. Hyperbola.

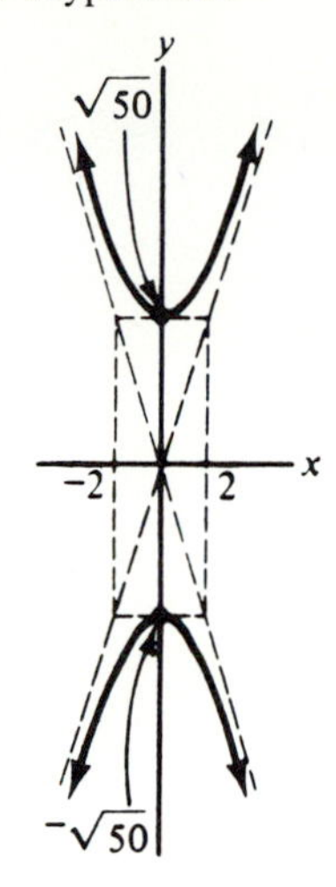

13. Parabola.

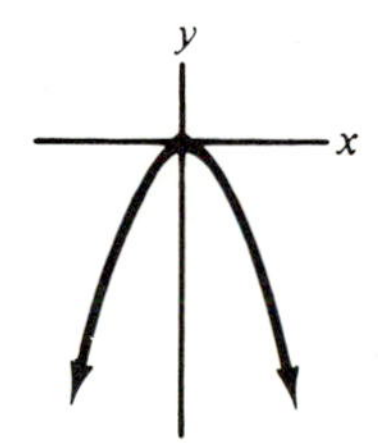

15. Ellipse.

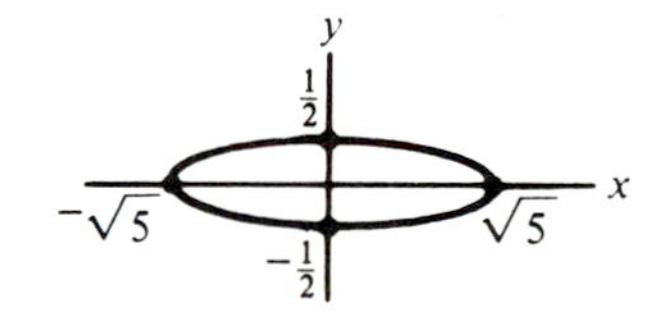

REVIEW PROBLEMS—CHAPTER 15

1. $2\sqrt{5}$. **3.** 10. **5.** $x^2 + y^2 = 25$; $x^2 + y^2 - 25 = 0$. **7.** $(x - 1)^2 + (y + 1)^2 = \frac{1}{4}$; $x^2 + y^2 - 2x + 2y + \frac{7}{4} = 0$. **9.** Circle; $C = (0, 0)$, $r = 4$. **11.** Circle; $C = (-2, 1)$, $r = \sqrt{7}$. **13.** Circle; $C = (-\frac{5}{2}, 3)$, $r = \dfrac{\sqrt{53}}{2}$. **15.** Point $(1, \frac{1}{3})$.

17. Hyperbola.

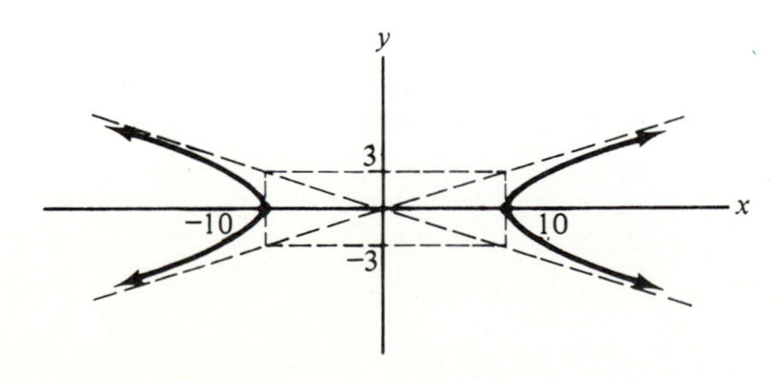

19. Ellipse.

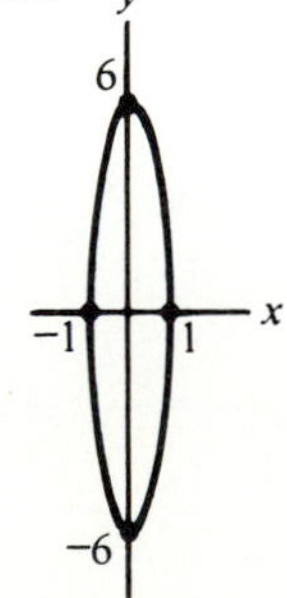

21. Parabola.

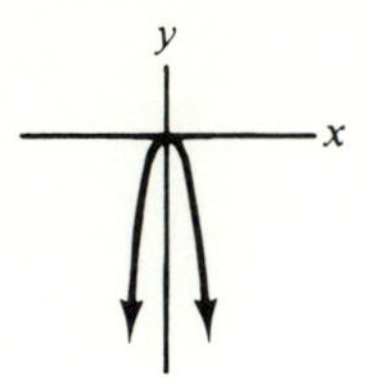

23. Circle.

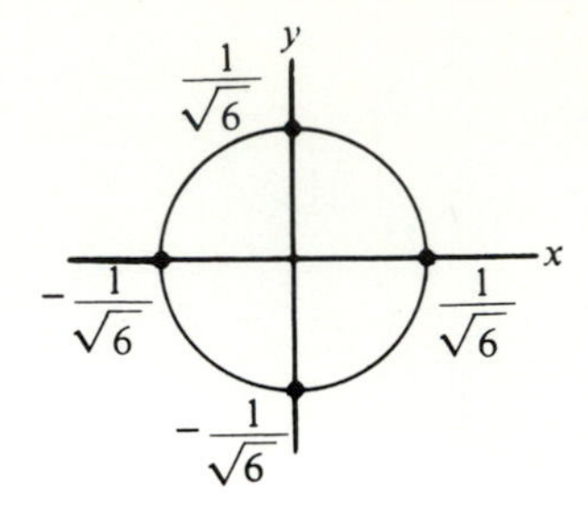

25. Parabola.

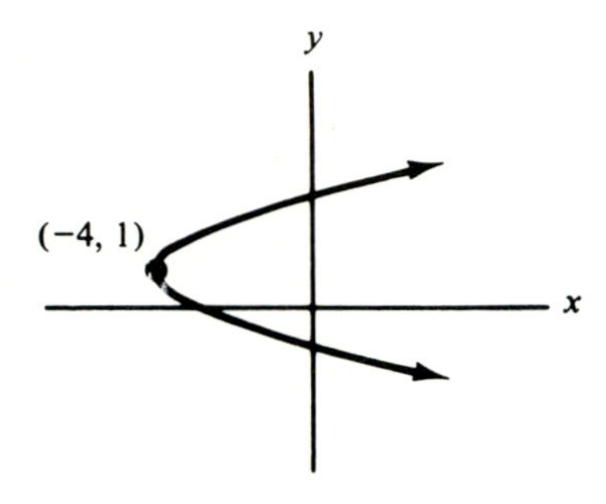

Index